T0402143

Mechanics of Natural Solids

Dimitrios Kolymbas and Gioacchino Viggiani (Eds.)

Mechanics of Natural Solids

Prof. Dr. Dimitrios Kolymbas
Universität Innsbruck
Fak. Bauingenieurwesen
und Architektur
Inst. f. Geotechnik und Tunnelbau
Technikerstr. 13 6020 Innsbruck
Austria
E-mail: dimitrios.kolymbas@uibk.ac.at

Prof. Dr. Gioacchino Viggiani
Université Joseph Fourier
Laboratoire 3S-R
Grenoble, France
E-mail: cino.viggiani@hmg.inpg.fr

ISBN 978-3-642-03577-7 e-ISBN 978-3-642-03578-4

DOI 10.1007/978-3-642-03578-4

Library of Congress Control Number: Applied for

Typesetting: Data supplied by the authors

Production: Scientific Publishing Services Pvt. Ltd., Chennai, India

Cover Design: eStudio Calamar, Steinen-Broo

Printed in acid-free paper

9 8 7 6 5 4 3 2 1

springer.com

Preface

The field of geosciences is full of scientific fascination and questions that are crucial for humanity. Our fluid environment (the atmosphere, oceans, rivers, etc.) is responsible for climate, hurricanes, floods and other phenomena characterised by rapid changes. These have to be contrasted with the permanence of our solid underground made of soil, rock, ice and snow. However, this permanence is only apparent as shown for example by earthquakes and landslides, but also by a number of other processes of deformation. Such processes are nowadays of high importance whenever we look to the future (think for example of disposal of radioactive waste or carbon dioxide) or to the past (prospection of oil, gas and ore). But also short range extrapolations are important if we look e.g. at the prediction and mitigation of landslides or the foundation of oil and gas production plants offshore or on glaciers.

Geosciences are pronouncedly multidisciplinary as they comprise perhaps the most widespread collection of disciplines, such as geology, geophysics, physics, geochemistry, geography, geotechnical and geoenvironmental engineering, petroleum engineering, soil mechanics and rock mechanics. Unfortunately, often these disciplines operate rather independent of each other and the increasing quest for transdisciplinary exchange is inhibited by the different languages and views prevailing in the various disciplines.

It appears thus that mechanics could offer a substantial link across disciplines, at least with respect to geotechnical engineering and geology. As a first step toward this direction, It seems interesting to create a link between the mechanics of the main groups of natural solids of the earth: soil, snow, ice and rock, the common mechanical properties of which are astonishingly many. To a first exchange in the field of mechanics of natural solids was devoted a symposium in the Greek village of Horto which took place from 7th to 9th of September 2009, and these Proceedings contain most of the papers presented in this symposium.

The editors wish to thank the Greek company OTM guided by Panos Vettas and the Foundation G. Angelinis - Pia Hadjinikos for sponsoring this event, Dr. Ansgar Kirsch for his thorough processing of the manuscripts as well as Ms. Christine Neuwirt, Ms. Sarah-Jane Loretz-Theodorine and Daniel Renk for the organization and financial administration of the symposium.

Horton,
September 2009

Dimitrios Kolymbas
Cino Viggiani

Contents

Sand as an archetypical natural solid

Dimitrios Kolymbas

Dedicated to the Choir of our University and the wonderful music we enjoyed together in the time I was writing this paper.

Abstract Sand, a product of erosion, cannot be broken into parts because it is already a broken ('clastic') material. Being a solid, it can sustain shear stresses at rest but it can also undergo large plastic deformations without considerable change of its properties, behaving thus like a fluid. The pronounced deformability of sand gave rise not only to a large diversity of experimental investigations in Soil Mechanics but rendered also sand a model material for physical simulations of deformation processes of the earth crust: Sand box models serve to simulate not only folding and faulting processes of the earth crust but also processes of deformation of the earth mantle. There are also similarities between magma volcanism and the so-called sand boils or sand volcanoes that appear subsequent to liquefaction of water-saturated loose sand. The complex behaviour of sand is a permanent object of study not only by Soil Mechanics but —in recent time— also by Physics. In this paper, the ability of sand to model the behaviour of other geomaterials is elucidated and a new theoretical frame is presented to describe mathematically the behaviour of sand based on its asymptotic properties.

1 The magic of sand

Specialists in Soil Mechanics having spent a large part of their lives trying to decipher the mysteries of sand do quite well understand Antoine de Saint-Exupéry, who wrote in his book "The Wisdom of the Sands (Citadelle)"

> Nous nous sommes nourris de la magie des sables, d'autres peut-être y creuseront leurs puits de pétrole et s'enrichiront de leurs marchandises. Mais ils seront venus trop tard. Car

Dimitrios Kolymbas
University of Innsbruck, Institute of Infrastructure, Division of Geotechnical and Tunnel Engineering, Technikerstr. 13, A-6020 Innsbruck, Austria, e-mail: dimitrios.kolymbas@uibk.ac.at

les palmeraies interdites ou la poudre vierge des coquillages nous ont livré leur part la plus précieuse : elles n'offraient qu'une vaine ferveur, et c'est nous qui l'avons vécue.

In fact, David Muir Wood named his remarkable Bjerrum lecture of 2005 "The magic of sand" [21]. In 2008 the German journal DER SPIEGEL published an article entitled "The magic of intelligent sand" [7]. There, a special sand is envisioned whose grains are equipped with mini-computers such that, say, an initial cube of sand can attain upon demand the form of a trumpet, see Fig. 1. Judging from the difficulties to decipher the mechanical behaviour of sand we may state that sand is indeed an intelligent material, but his intelligence resides not upon the individual grains but upon the interaction and the resulting mean behaviour of large assemblies of grains.

The concept of matter consisting of innumerable small particles has been introduced by Democrit and obtained an important role in modern physics, which succeeded in explaining the behaviour of e.g. gases as resulting from the motion of molecules. The related branch of physics, statistical physics, grew to an admirable refinement but it left aside sand, whose behaviour is completely different from ideal gas. Sand became the realm of Soil Mechanics, a branch of Civil Engineering, and followed a completely different trajectory from ideal and real gases. Although its foundations were laid down by Coulomb, the father of electrostatics, sand was disregarded by physicists until recently, when they detected that its behaviour is fascinating. Prof. M. Liu (Theoretical Physics, University of Tübingen) stated that sand is more difficult than fluid helium or liquid crystals. Another statement is that *"Many properties of sand and other granular materials are as puzzling to scientists as the big bang is"* [20]. In the meanwhile, Soil Mechanics has carried out an amazingly large collection of laboratory experiments and obtained a large collection of theoretical results.

2 Sand as a model of other geomaterials

2.1 Sand as a model for rock

Sand covers large parts of the earth and is, thus, in itself an important geomaterial. Another important soil type, clay, consists also of particles and its behaviour is in many aspects identical to sand. More precisely, dense sand is similar to so-called overconsolidated clay, and loose sand is similar to normally consolidated clay. The majority of rocks consist also of grains (e.g. granite), as shown in the thin cut of sandstone shown in Fig. 2. In difference to soils, the grains of rock are densely packed and are also 'glued' (cemented) to each other. However, density and cementation do not hinder large deformation of rock associated with grain re-arrangement, provided that the deformation is sufficiently slow.

Another similarity of sand and rock can be found in the fact that both are frictional materials, i.e. their strength increases with normal stress. Byerlee [5] ob-

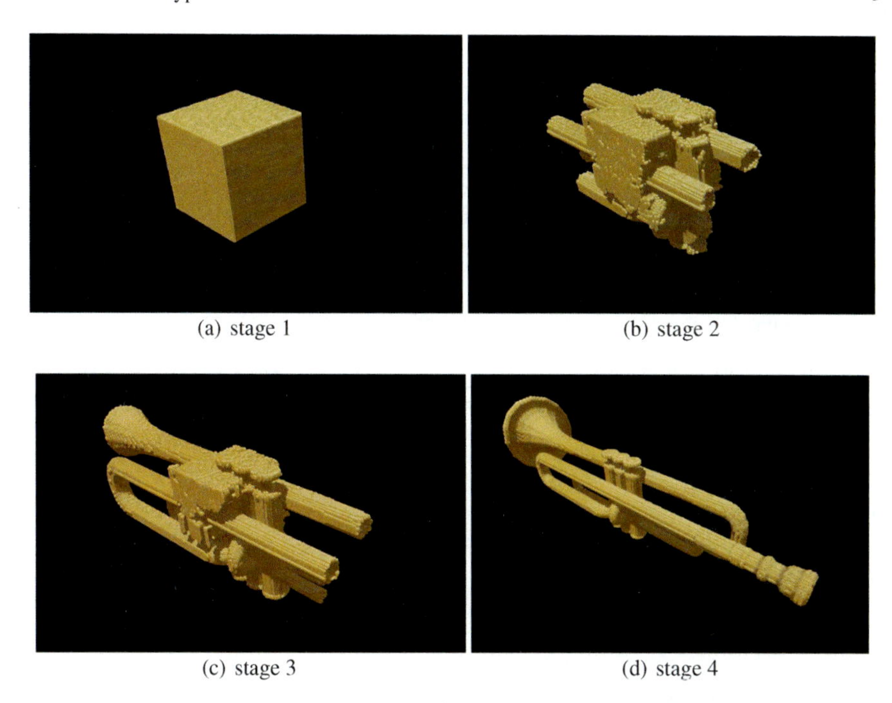

(a) stage 1 (b) stage 2

(c) stage 3 (d) stage 4

Fig. 1 Several fictitious transformation stages of a cube made of 'intelligent sand' [7]

Fig. 2 Thin cut of sandstone. Courtesy of Prof. Chr. Lempp, Halle.

served that the coefficient of friction of almost every type of crustal rock falls within the surprisingly narrow range of 0.6 - 1.0, a range that is commonly referred to as Byerlee's Law;

> "The experimental results show that at the low stresses encountered in most civil engineering problems the friction of rock can vary between very wide limits and the variation is mainly because at these low stresses friction is strongly dependent on surface roughness. At intermediate pressure such as encountered in mining engineering problems and at high stresses involved during sliding on faults in the deep crust the initial surface roughness has little or no effect on friction. At normal stresses up to 2 kbar ($= 2 \cdot 10^5$ kN/m^2) the shear stress required to cause sliding is given approximately by the equation $\tau = 0.85\sigma_n$[1]. At normal stresses above 2 kbar the friction is given approximately by $\tau = 0.5 + 0.6\sigma_n$. These equations are valid for initially finely ground surfaces, initially totally interlocked surfaces or on irregular faults produced in initially intact rocks. Rock types have little or no effect on friction"[5].

The capability of sand to undergo large plastic deformations renders it an ideal material to scale down (in geometry and time) tectonic deformation processes. This fact has been exploited in the so-called sandbox models (Figs. 3, 4 and 5). Also in numerical simulations sand proves to be an appropriate material to model tectonic processes (Figs. 6 and 7).

Fig. 3 Sandbox model for the simulation of tectonic processes [1]

[1] i.e. $\tau = 1.7 \cdot 10^5$ kN/m^2. Compare with the cohesion of granite of the order of magnitude of $2 \cdot 10^3$ kN/m^2.

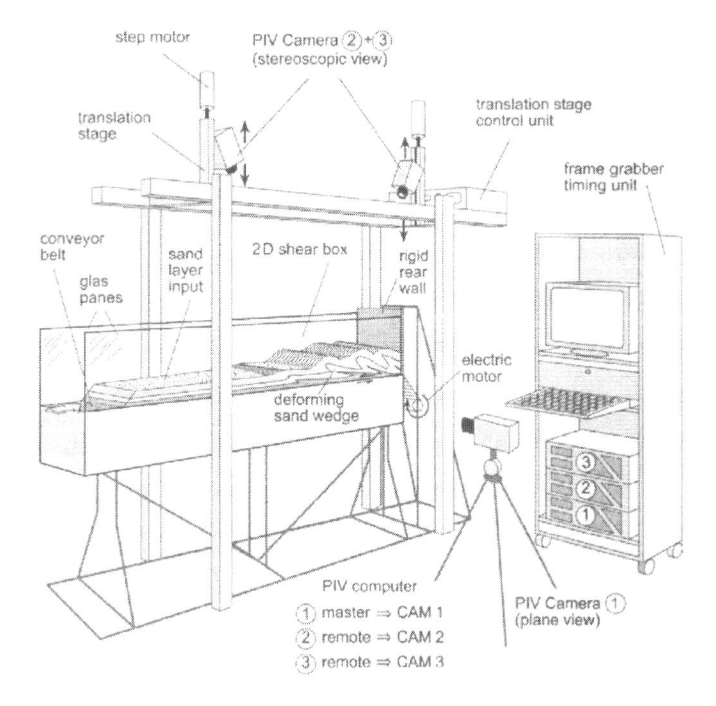

Fig. 4 Sandbox apparatus [2]

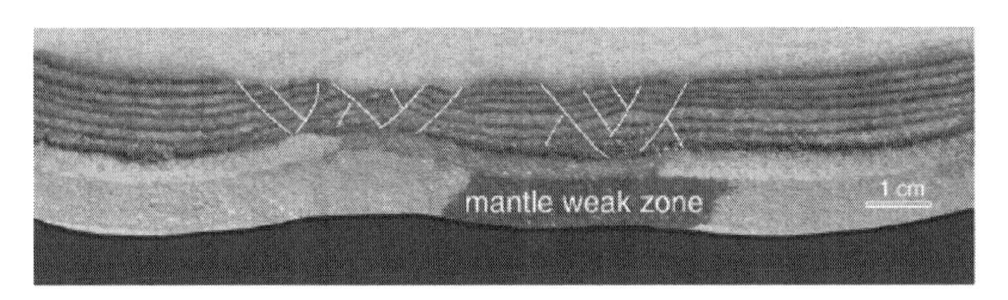

Fig. 5 Modelling of extension of lithosphere [27]

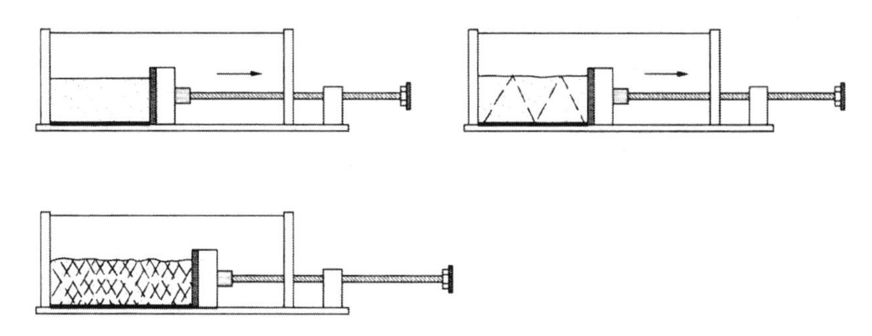

Fig. 6 Stretching the base of a granular layer [22]. Compare the obtained pattern with the numerical simulation shown in Fig. 7.

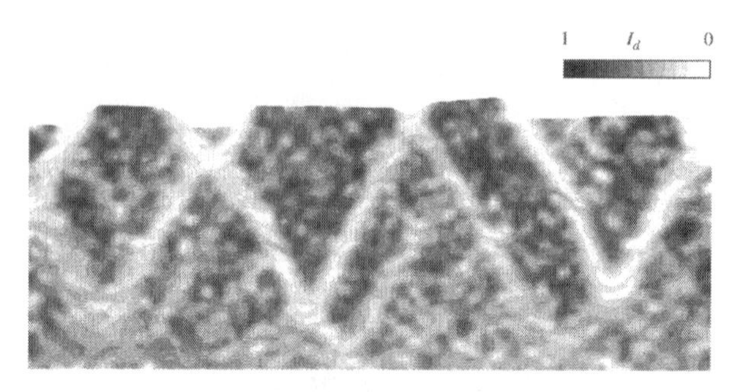

Fig. 7 Calculated pressure-adjusted density index I_d after base deformation $\Delta l/l = 20\%$ [22]

Also clay is a good material to model deformations of rock strata. The first to conduct simulations of tectonic processes with clay cakes was the geologist HANS CLOOS, see Fig. 8.

The simulation of tectonic processes in sandbox models shows that sand is, in fact, a good model for rock strata but does not explain why this is so. BYERLEE's law explains that in large geometric scale the strength of rock is controlled by friction, which is a genuine property of sand, whereas cohesion is of minor importance. Nevertheless, it is still difficult to grasp how rock can undergo large deformation without exhibiting brittle failure, as one would expect based on the experience with small geometry and time scales. Even if we recognise both materials, sand and rock (e.g. sandstone), as granular ones we have to admit that in rock the grains are (i) in a much denser state, and (ii) cemented to each other, so that grain-rearrangement is not possible without breakage of the bonds. It turns however, that ductile grain re-arrangement is not possible *in short time scales* but still possible for very slow deformation. Non-brittle, i.e. ductile deformation, of rock is considered, mainly in Structural Geology [30]. In macroscopic terms, ductile deformation is often manifested as creep. For stationary creep, two cases are distinguished:

1. Power-law creep (for low and moderate stress): $|\mathbf{D}| \propto \left(\sqrt{\mathrm{tr}\mathbf{T}^{\star 2}}\right)^n \cdot \exp\frac{-E}{RT}$
2. Exponential creep law (for high stress): $|\mathbf{D}| \propto \exp(\beta\sqrt{\mathrm{tr}\mathbf{T}^{\star}}) \cdot \exp\frac{-E}{RT}$.

Herein, $\mathbf{D}$ indicates the so-called stretching tensor, i.e. the symmetric part of the velocity gradient grad $\mathbf{v} = v_{i,j}$, and $\mathbf{T}^\star$ is the deviatoric part of the CAUCHY stress $\mathbf{T}$. The term $\exp\frac{-E}{RT}$ indicates that creep is a thermally activated process. In microscopic terms, several types of creep have been identified, such as diffusion and solution creep, including the so-called Nabarro-Herring creep, coble creep and others. It must be remarked, however, that these types of creep are not (yet) integrated into a global model, so that their consideration appears rather confusing from a macroscopic point of view.

Note that Structural Geology uses some terms different from Soil Mechanics:

Granular flow: Deformation based on grain re-arrangement. This is the main type of deformation considered in Soil Mechanics. In large depths, dilatancy and grain

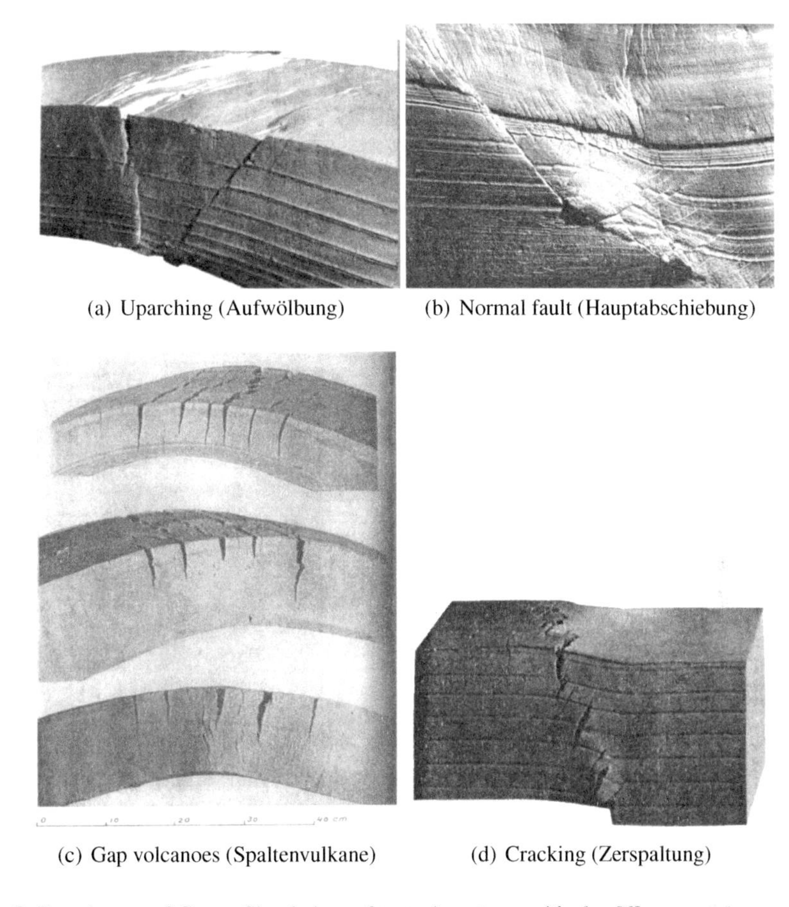

(a) Uparching (Aufwölbung) (b) Normal fault (Hauptabschiebung)

(c) Gap volcanoes (Spaltenvulkane) (d) Cracking (Zerspaltung)

Fig. 8 Experiments of Cloos: Simulations of tectonic patterns with clay [6]

re-arrangement may be suppressed by the large prevailing pressure. However, large pore fluid pressures may reduce the effective stresses so that granular flow can freely develop.

Cataclastic flow: Granular flow with grain crushing. Mainly assumed to occur within faults.

Large ductile deformation can be explained by the fact that rock is capable to heal manifestations of brittle failure when deformed slowly enough. This can be demonstrated by the cobble shown in Fig. 9.

Fig. 9 Cobble collected by the author at the Greek island of Kythera. The slip surface does not affect the integrity of the cobble.

2.2 Sand as model of the earth mantle

The mechanical properties of the earth mantle can only be indirectly assessed by interpretation of seismograms. It is generally accepted that the mantle is in perpetual motion, which is the cause of continental drift and other tectonic deformations of the earth crust. What is the reason of this perpetual motion? — Besides thermal convection, tidal deformation has been assumed (e.g. by WEGENER) to sustain a monotonic motion of the mantle. In fact, the daily rotation of the earth in the gravitational fields of the sun and the moon causes a cyclic loading of the earth. If the earth behaved elastically, cyclic loading would cause a (certainly minute) cyclic deformation but no monotonic deformation of the earth. If, however, the earth (mainly, the earth mantle) behaves anelastically, then cyclic loading may result in a perpetual monotonic motion of the mantle. The daily anelastic deformation, how small it may be, accumulates over centuries and attains considerable values. REVUZHENKO has carried out a series of spectacular tests with sand, as a modelling material of the earth mantle [26]. Unable to simulate tidal forces (i.e. mass forces with rotating direction) in the laboratory, he invented a 2D test where a flexible containment was filled with dry sand and subjected to cyclic deformation by means of an imposed rotating mold with elliptical cross section (Figs. 10 a-c). Thus, the mass forces exerted by the attraction of moon and sun were replaced by surface tractions that produced more or less the same deformation. The result was a monotonic rotational deformation of the sand, as shown with embedded coloured sand in Figs. 11 and 12. Similar results were obtained with 3D tests with an analogous deformation of a sphere (Fig. 10 d).

2.3 Lava volcanoes - sand volcanoes

There is a striking similarity of magma and sand volcanoes. Soil is a granular material consisting of grains and pores, the later being usually filled with water. The analogy with melting rock has been pointed out by McKENZIE [19]: Partially molten

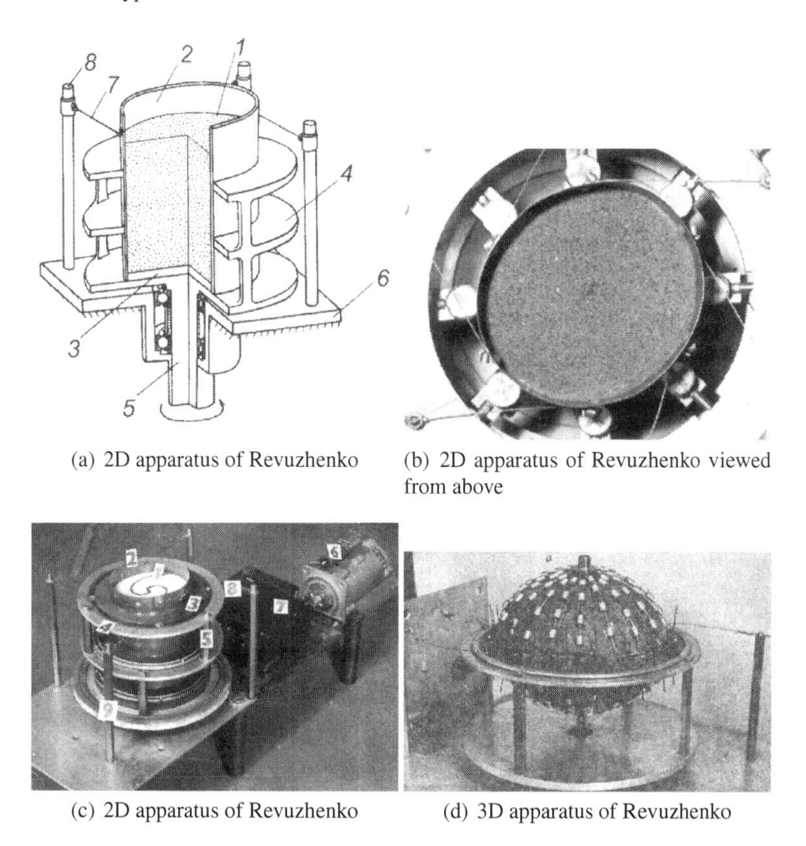

(a) 2D apparatus of Revuzhenko

(b) 2D apparatus of Revuzhenko viewed from above

(c) 2D apparatus of Revuzhenko

(d) 3D apparatus of Revuzhenko

Fig. 10 Apparatuses of Revuzhenko [26] intended to simulate the tidal deformation of the earth by the application of surface tractions to dry sand bodies.

rock consists of a deformable matrix (corresponding to the grain-skeleton of soils) and pore fluid, which in this case is molten rock. Thus, both materials, i.e. partially molten rock and water-saturated soil, are two-phase materials consisting of a solid matrix and interstitial fluid.

Whereas the detrimental action of magma volcanoes is widespread known, sand volcanoes (also called 'sand boils') are less known. They appear in liquefied soil. Soil liquefaction is a peculiar phenomenon that occurs when loose water-saturated sand undergoes rapid deformation, as this is the case due to e.g. earthquakes or (underground) explosions. Sand volcanoes accompany liquefaction and the resulting craters are (besides the devastation of buildings) the only remnant of liquefaction in green fields. Japanese authors [18] describe sand volcanoes as follows:

> "The most violent water spouting during the Earthquake No. 4 was observed at the site ... near the Shonai river where water with sand ejected high above 2 m from wells and sands deposited over the roofs of nearby houses."

Similar reports are available from India (Bihar earthquake, 1934, cited in [10]):

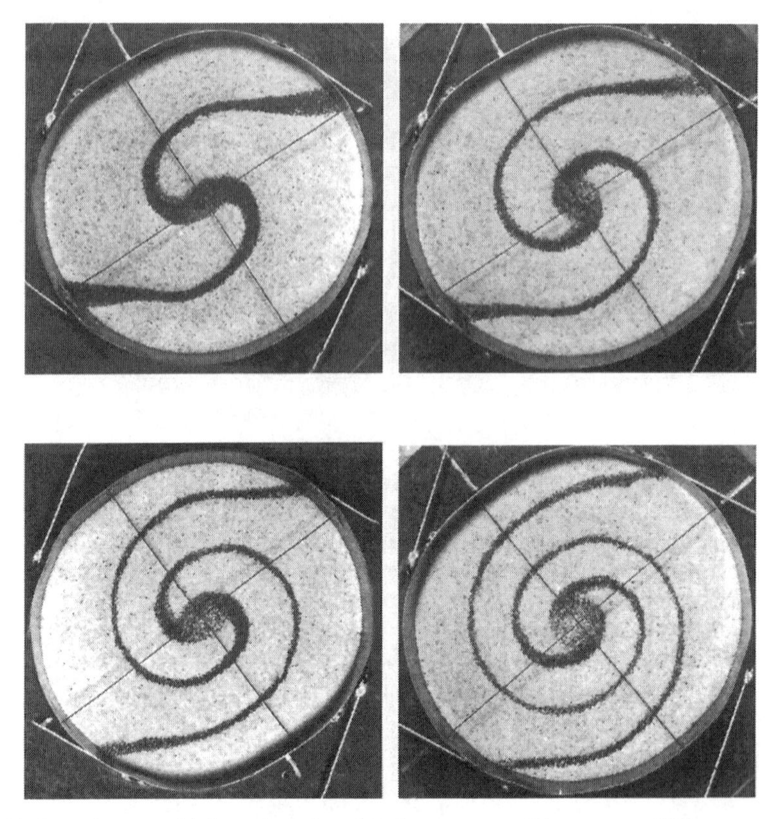

Fig. 11 Several stages of deformation of sand in the apparatus of Revuzhenko [26]

Fig. 12 Initial and deformed sand bodies in the 2D apparatus of Revuzhenko [26]

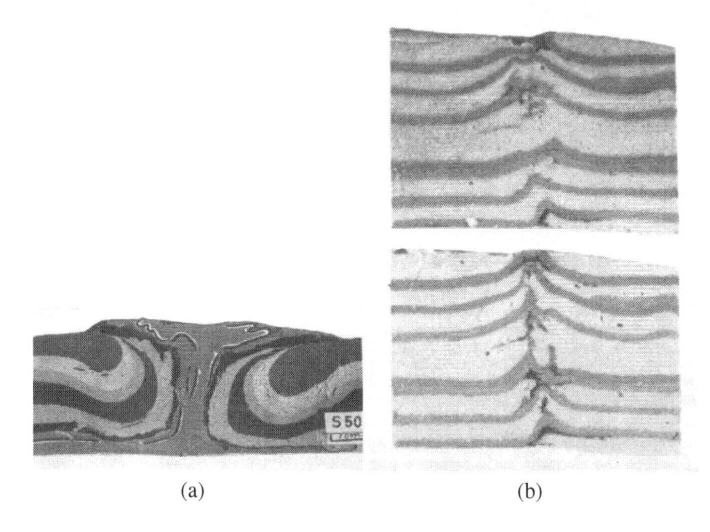

(a) (b)

Fig. 13 Simulations with clay-oil mixtures by RAMBERG [25]

> "...then water spouts, hundreds of them throwing up water and sand, were to be observed on the whole face of the country, the sand forming miniature volcanoes, while the water spouted out of the craters; some of the spouts were were quite six feet high. In a few minutes, on both sides of the road as far as the eye could see, was vast expanse of sand and water ..."

The places of appearance of sand volcanoes are unpredictable and, therefore, photos can be rarely taken (Fig. 14). Craters remain after the eruption to witness liquefaction (Fig. 15).

3 The mechanical behaviour of sand

3.1 The grain skeleton and its physics

It is tempting to conceive a grain skeleton as a framework. In case of elastic grains, a linear relation between the forces $\Delta\mathbf{F}$ and displacements $\Delta\mathbf{x}$ would then prevail: $\Delta\mathbf{F} = \mathbf{K}\,\Delta\mathbf{x}$, i.e. the stiffness $\mathbf{K}$ depends on the deformation $\Delta\mathbf{x}$. However, the individual grains are not fixed to each other (except for the case of cementation) and, thus, the grain skeleton is not a framework but rather a sort of bad-quality ball-bearing. The individual grains are displaced and rotated during deformation and the points of contact permanently change, i.e. deformation is *always accompanied with grain re-arrangement* i.e. change of geometry. This means that the force-displacement relation is non-linear: $\Delta\mathbf{F} = \mathbf{K}(\Delta\mathbf{x})\Delta\mathbf{x}$. From a microscopic point of view, this non-linearity is geometric, but from a macroscopic point of view, which disregards grain

Fig. 14 Sand boil after liquefaction. Photograph taken by the author. The liquefaction was released by underground explosion.

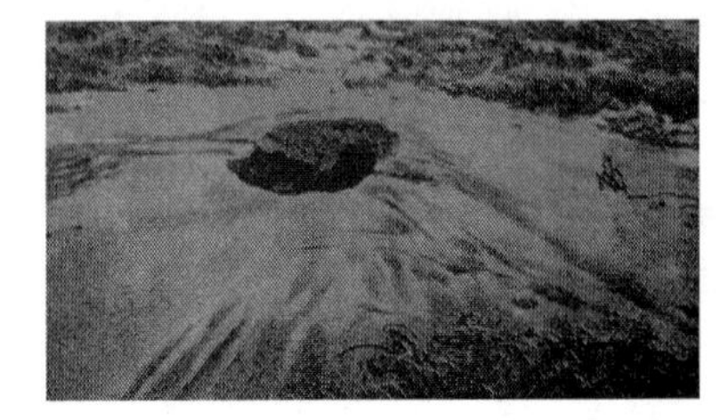

Fig. 15 Crater from sand boil

re-arrangement, the non-linearity is inherent (material) and 'incremental', i.e. it cannot be linearised 'in the small'.

If we idealise a granular material as a continuum, then the considered displacements $\mathbf{u}$ are a mean field that is always accompanied with a scatter $\Delta\mathbf{u}$ due to grain re-arrangement. The mean square of this scatter can be interpreted as 'granular temperature' T_g and this notion can be used to extend thermodynamic concepts to the mechanics of sand. Since any macroscopic deformation is connected with granular temperature, it is natural to set $T_g \propto \sqrt{\mathrm{tr}\mathbf{D}^2}$.

The inner friction of granulates has little or nothing to do with the friction between two adjacent grains. If the contacts were controlled by friction and if there was no grain re-arrangement, then a proportional loading of a skeleton consisting of stiff grains would not cause any considerable deformation. In reality, however, every loading (including proportional ones) causes grain re-arrangement and macroscopic deformation.

Internal friction (and, hence, shear strength) of sand is not a manifestation of intergranular friction but rather a result of force diffusion within the irregular grain skeleton.

4 Strain localisation and pattern formation

At first glance, sand being an amorphous material, seems incapable to exhibit any pronounced structure. All the same, sand was the first medium to store information (remember of Archimedes drawing his sketches on sand), thanks to its irreversible or hysteretic mechanical behaviour. Besides of imposed patterns, sand is capable to spontaneously produce fascinating patterns. These are only visible if the sand has been marked before (say by colour) or if the pattern appears at the surface, as is the case in Figs. 16 and 17.

Fig. 16 Pattern formation on the horizontal surface of sheared sand (REVUZHENKO [26])

The majority of the patterns are related to strain localisation into thin shear bands. Their spontaneous formation, especially in the course of tests with initially homogeneous deformation of the samples, has been extensively studied. The sample deformation is imposed to the sample via prescribed displacements of the sample boundaries, say within a HAMBLY-type biaxial apparatus (Fig. 18).

It is remarkable that the imposed information propagates from the boundaries to the interior of the sample and manages that the deformation is equally distributed all over the sample. This propagation occurs with wave velocity, even if the deformation is quasistatic and no waves are perceivable. However, if a wave velocity happens to vanish, homogeneous deformation is no more possible. It then happens something peculiar: the sample gets subdivided into several blocks, each of which obeys the displacement imposed upon its own external boundary. The several blocks then glide relatively to each other along shear bands.

Fig. 17 Localisation pattern in nature (Photograph: Intern. Soc. Rock. Mech.). The pattern formation in this photograph is similar to the one in Fig. 16 that was obtained in the laboratory.

Fig. 18 Biaxial apparatus of the HAMBLY type [17, 28]

4.1 Experimental observations—Proportional loading

From true triaxial tests carried out with sand GOLDSCHEIDER inferred two rules.[2] They refer to proportional paths (PP)[3] [9]:

[2] GOLDSCHEIDER, impressed by the complexity of the stress and strain paths obtained with a true triaxial apparatus, expressed (in the seventies of the past century) to the author the opinion that it is, presumably, impossible to find a constitutive relation of sand. It is, though, remarkable to see that the constitutive relation for sand is more or less included in the aforementioned rules, as will be shown in this paper.

[3] Proportional stress and strain paths are characterised by constant ratios of the principal values $\sigma_1 : \sigma_2 : \sigma_3$ and $\varepsilon_1 : \varepsilon_2 : \varepsilon_3$, respectively.

Rule 1: Proportional strain paths (PεP, also called 'consolidations', as they are connected with densification) starting from the stress[4] $\mathbf{T} = \mathbf{0}$ are associated to proportional stress paths (PσP).
Rule 2: Proportional strain paths starting from $\mathbf{T} \neq \mathbf{0}$ lead asymptotically to the corresponding proportional stress paths obtained when starting at $\mathbf{T} = \mathbf{0}$ (which act, so to say, as attractors).

Rule 1 also applies to much simpler materials, e.g. linear elastic ones, for which straight stress paths are obtained with PεP starting from *any* stress state. The specialty of cohesionless granular materials arises from the fact that the feasible stress states and paths are limited within a fan with apex at $\mathbf{T} = \mathbf{0}$. Therefore, PεP starting from $\mathbf{T} \neq \mathbf{0}$ correspond to stress paths which, being limited within this fan, either approach asymptotically PσP's or end at $\mathbf{T} = \mathbf{0}$ (Fig. 19). The latter ones are characterised by volumetric dilation.

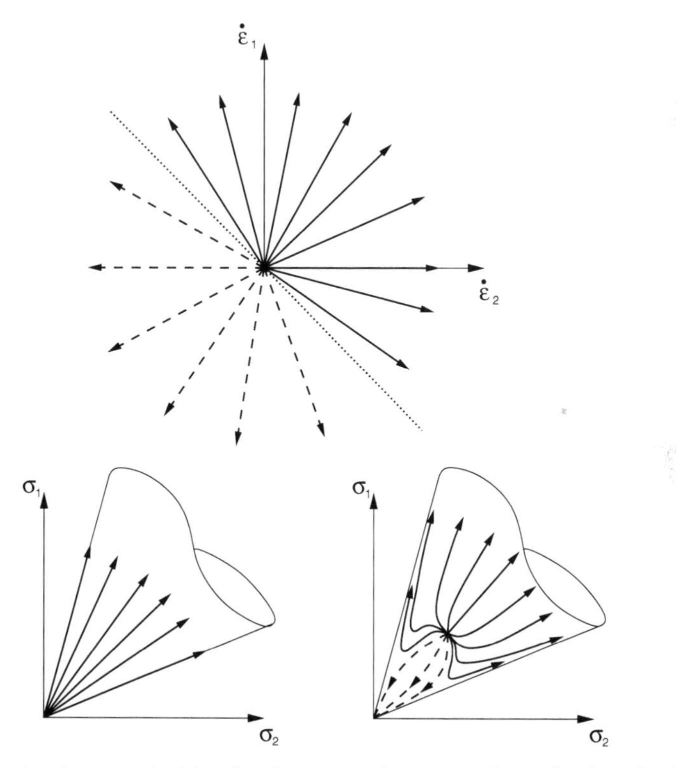

Fig. 19 Proportional strain paths (above) and corresponding stress paths starting from $\mathbf{T} = \mathbf{0}$ (below left) and $\mathbf{T} \neq \mathbf{0}$ (below right)

The second rule is, more or less, a consequence of the limitation of feasible stress states within the aforementioned fan. In this paper it is shown how a constitutive

[4] In compliance with [29], the CAUCHY stress is denoted with $\mathbf{T}$ instead of σ.

equation for granular materials can be inferred from these two rules. A constitutive relation that holds for all possible PεP's starting from *arbitrary* stress states can be expected to describe the entire mechanical behaviour of granular materials.

4.2 Other experimental evidence

In Figures 20 to 23 are shown test results that demonstrate some principal features of the mechanical behaviour of sand. Figure 20 shows the behaviour of sand at oedometric compression, i.e. at compression with no lateral strain. There are shown the stress path and the stress-strain curve for two load cycles of loading and unloading.

Figure 21 shows the behaviour of sand at drained triaxial compression, i.e. compression in axial direction with lateral expansion. Note that a constant-stress condition prevails at the mantle of the cylindrical sand sample. The upper curve shows a series of four loadings and unloadings whereas the lower curve shows the corresponding volumetric strain ε_{vol} $(= \Delta V/V)$ vs. the axial strain ε_1.

Figure 22 shows the behaviour of sand at undrained (i.e. constant volume) triaxial deformation of sand at various stress levels. Figure 23 shows experimental results that reveal the rate sensitivity of sand, and Fig. 24 shows sand at cyclic undrained triaxial deformation.

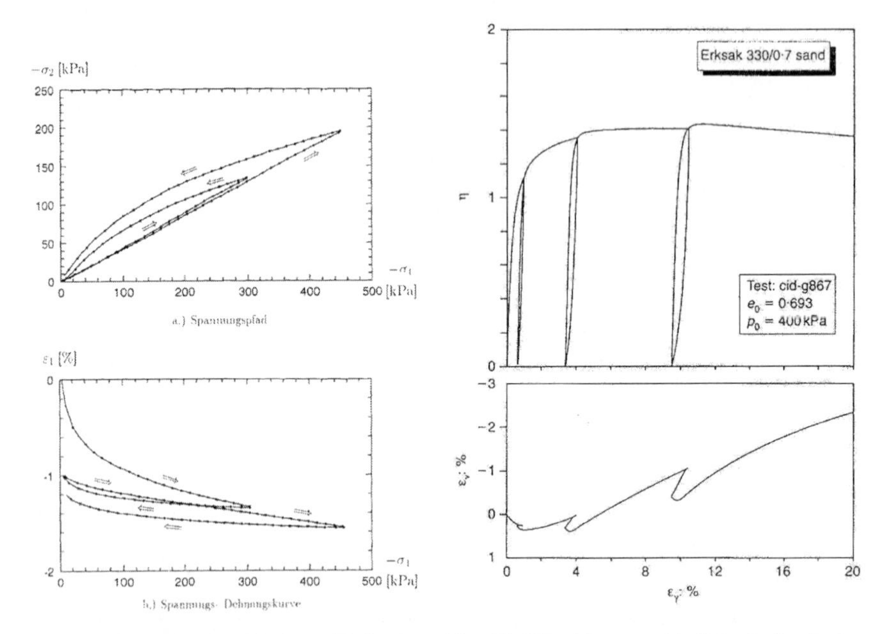

Fig. 20 Oedometric test with dense Karlsruher Sand (initial void ratio e_0 = 0.55). BAUER, 1992 [4]

Fig. 21 Triaxial test with Erksak Sand, JEFFERIES, 1997 [11]. $\eta := \frac{\sigma_1 - \sigma_2}{(\sigma_1+\sigma_2+\sigma_3)/3}$

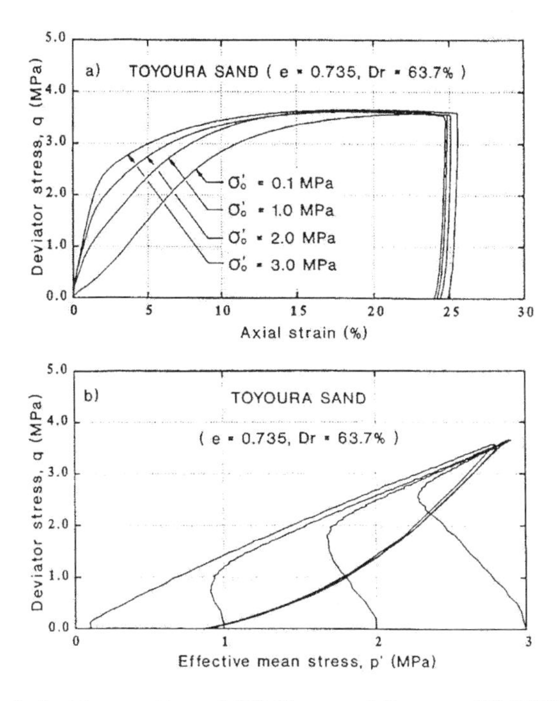

Fig. 22 Undrained triaxial tests with $e = 0.833$, VERDUGO & ISHIHARA, 1996 [31]

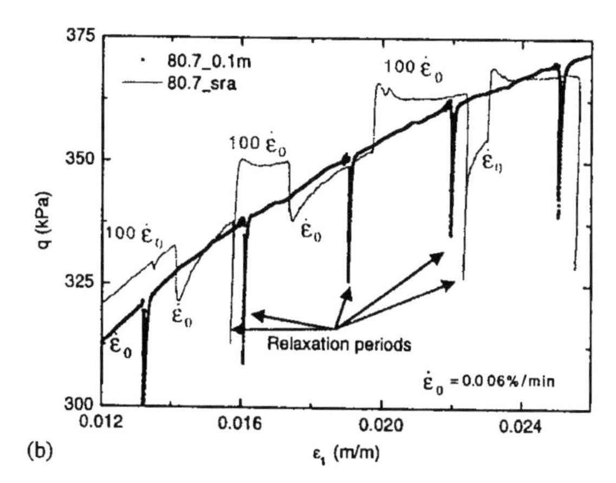

Fig. 23 CD triaxial tests on dry Hostun sand with jumps of deformation rate, PHAM VAN BANG et al., 2007 [3]

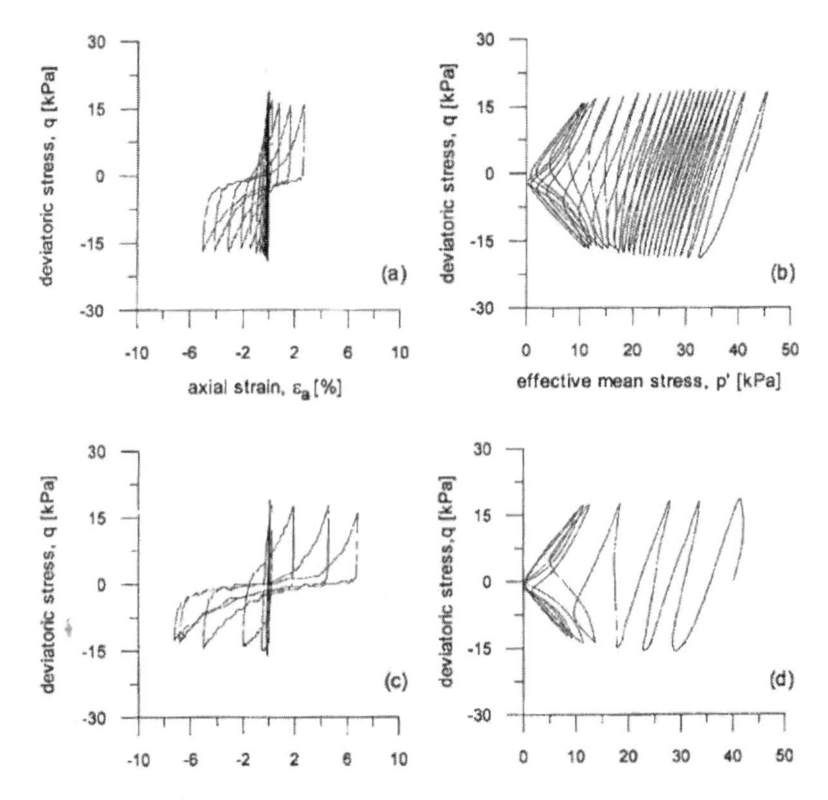

Fig. 24 Gravel and sand, reconstituted (water sedimentation (top row), air pluviation (bottom row)), GHIONNA & PORCINO, 2006 [8]

5 Mathematical models

There is a large number of mathematical models to describe the mechanical behaviour of sand. Most of them are formulated in the frame of thc theory of elastoplasticity, i.e. elastic behaviour is assumed to prevail as long as the stress is within the so-called yield surface. Plastic strains occur only when the stress lies upon the yield surface. The pertinent definitions of loading, unloading and reloading are built upon considerations of the yield surface and its transformations in the stress space. Contrary to elastoplastic theories, hypoplasticity [13, 14, 15, 16] concentrates on the representation of the constitutive relation as a single evolution equation for stress, built along the principles of Rational Mechanics [29]. In this paper is introduced 'barodesy', a new branch of hypoplasticiy.

5.1 *Barodesy*

5.1.1 Mathematical model of Goldscheider's first rule

As known, the stretching tensor $\mathbf{D}$ is defined as the symmetric part of the velocity gradient grad$\mathbf{v}$. For rectilinear extensions, $\mathbf{D}$ can be conceived as the time rate of the logarithmic strain, i.e. $\mathbf{D} = \dot{\varepsilon}_{ij}$.[5] A proportional strain path (PεP) is obtained by the application of a constant stretching $\mathbf{D}$. The equations of barodesy are here derived on the basis of rectilinear extensions, motions with rotation of the principal axes are, however, automatically included in the obtained equations, which are written for fully occupied 3×3 tensors. All proportional stress paths (PεP), i.e. all consolidations, form a fan $\mathcal{F}_\varepsilon$ in the strain space, and the corresponding PσP's form a fan $\mathcal{F}_\sigma$ in the stress space.

GOLDSCHEIDER's first rule relates $\mathbf{D}$ with a tensor $\mathbf{R}$ which points in the direction of the associated stress path. This relation can be expressed as:

$$\mathbf{R}(\mathbf{D}) := (\mathrm{tr}\mathbf{D}^0)\, \mathbf{1} + \alpha \exp(\beta\, \mathbf{D}^0) \quad , \tag{1}$$

where α and β are material constants. Equation 1 has been obtained by trial and error. Herein, the exponent 0 denotes normalisation of a tensor $\mathbf{X}$, i.e. $\mathbf{X}^0 := \mathbf{X}/|\mathbf{X}|$, with $|\mathbf{X}| := \sqrt{\mathrm{tr}\mathbf{X}^2}$. Obviously, $\mathbf{R}(\mathbf{D})$ is homogeneous of the zeroth degree with respect to $\mathbf{D}$.

5.1.2 Adaptation to the second rule

Using equ. 1, a constitutive law of the rate type that is valid only for PP starting from $\mathbf{T} = \mathbf{0}$ can be formulated:

$$\dot{\mathbf{T}} = \mathbf{R} \cdot |\mathbf{T}| \cdot |\mathbf{D}| \; . \tag{2}$$

Obviously, the constitutive relation 2 is homogeneous of the first degree with respect to $\mathbf{T}$ and $\mathbf{D}$.[6] Now, adding to the right part of equ. 2 a term which is proportional to stress $\mathbf{T}$ will incorporate also rule 2 without breaking rule 1. Rule 2 states that if we start from a stress state $\mathbf{T} \neq \mathbf{0}$ and apply a PεP, i.e. a constant stretching $\mathbf{D}$, the stress will asymptotically approach the line $\mathbf{T} = \mu\, \mathbf{R}$. In other words, the obtained stress rate $\dot{\mathbf{T}}$ will fulfil the equation $\mathbf{T} + \dot{\mathbf{T}}\, \Delta t = \mu \mathbf{R}$.[7] This equation can also be written as

$$\dot{\mathbf{T}} = a_1 \mathbf{R} + a\, \mathbf{T} \; . \tag{3}$$

---

[5] In the general case, $\mathbf{D}$ is *not* the time derivative of any strain measure.

[6] As known, homogeneity of the degree 1 with respect to $\mathbf{D}$ implies rate-independence.

[7] Note that for rectilinear extensions, as considered in this section, $\overset{\circ}{\mathbf{T}} \equiv \dot{\mathbf{T}}$ holds, with $\overset{\circ}{\mathbf{T}}$ being a co-rotational stress rate.

To preserve homogeneity of the first degree with respect to $\mathbf{D}$, the scalar factor a must be homogeneous of the first degree with respect to $\mathbf{D}$. It proves that $a = a_2\,|\mathbf{D}|$ is the best choice:

$$\dot{\mathbf{T}} = |\mathbf{T}| \cdot (a_1 \mathbf{R} + a_2 \mathbf{T}^0) \cdot |\mathbf{D}| \ . \tag{4}$$

5.1.3 Homogeneity with respect to stress

The subsequent modification of equation 4 is to change the degree of homogeneity in $\mathbf{T}$ from 1 to γ:

$$\dot{\mathbf{T}} = |\mathbf{T}|^\gamma \cdot (a_1 \mathbf{R} + a_2 \mathbf{T}^0) \cdot |\mathbf{D}| \ . \tag{5}$$

In this way, equation 5 complies with OHDE's relation $d\sigma/d\varepsilon \propto \sigma^\gamma$ [12, 24]. For *proportional paths* (i.e. $\mathbf{D}$ =const), integration of equ. 5 or 2 can be obtained as follows: Setting $\sigma := |\mathbf{T}|$, $\dot{\epsilon} := |\mathbf{D}|$ or $\epsilon := |\int_0^t \mathbf{D} dt'|$, $\zeta_{\mathbf{D}} := |a_1 \mathbf{R} + a_2 \mathbf{T}^0| \cdot |\mathbf{D}|/\mathrm{tr}\mathbf{D}$ and $\eta_{\mathbf{D}} := |a_1 \mathbf{R} + a_2 \mathbf{T}^0|$ and using the known relation[8] $\dot{e} = (1+e)\,\mathrm{tr}\mathbf{D}$ leads to the differential equations

$$\dot{\sigma} = \sigma^\gamma \zeta_{\mathbf{D}} \frac{\dot{e}}{1+e} \tag{6}$$

and

$$\dot{\sigma} = \sigma^\gamma \eta_{\mathbf{D}} \dot{\epsilon} \tag{7}$$

with the general solutions

$$\sigma^{1-\gamma} = \sigma_0^{1-\gamma} + (1-\gamma)\zeta_{\mathbf{D}} \log \frac{1+e}{1+e_0} \tag{8}$$

and

$$\sigma^{1-\gamma} = \sigma_0^{1-\gamma} + (1-\gamma)\eta_{\mathbf{D}}(\epsilon - \epsilon_0) \ . \tag{9}$$

Taking into account that for proportional stress paths the initial stress and strain vanish, $\sigma_0 = 0, \epsilon_0 = 0$, yields finally:

$$\sigma^{1-\gamma} = (1-\gamma)\zeta_{\mathbf{D}} \log \frac{1+e}{1+e_0} \tag{10}$$

and

$$\sigma^{1-\gamma} = (1-\gamma)\eta_{\mathbf{D}} \epsilon \ . \tag{11}$$

e_0 is the maximum void ratio of the considered sand.

[8] This relation holds for incompressible grains.

5.1.4 Virgin strain

The 'virgin' strain $\hat{\mathbf{E}} = \hat{\varepsilon}_{ij} = \hat{\varepsilon}_{ij}(\mathbf{T})$ can be defined as the strain that *uniquely* corresponds to a particular stress $\mathbf{T}$ obtained with a PP starting at $\mathbf{T} = \mathbf{0}$ and $e = e_0$.[9] To determine $\hat{\varepsilon}_{ij}(\mathbf{T})$ we need to know the $\mathbf{D}$-tensor pertinent to $\mathbf{T}$ according to equ. 1. This tensor we call $\hat{\mathbf{D}}$. It can be obtained by numerical inversion of equ. 1.[10]

With $\hat{\mathbf{D}}$ we can easily obtain $\hat{\mathbf{E}}$:

$$\hat{\mathbf{E}} = \epsilon\hat{\mathbf{D}} \tag{12}$$

with

$$\epsilon = \frac{\sigma^{1-\gamma}}{(1-\gamma)\eta_{\hat{\mathbf{D}}}} \ . \tag{13}$$

In a similar way we can obtain the virgin void ratio $\hat{e}$ from equation 10:

$$\hat{e} = \exp\left(\frac{\sigma^{1-\gamma}}{(1-\gamma)\zeta_{\hat{\mathbf{D}}}}\right)(1+e_0) - 1 \tag{14}$$

with $e_0 = e_{max}$.[11] Equations 13 and 14 are only seemingly explicit in ϵ and $\hat{e}$, because $\eta_{\mathbf{D}}$ and $\zeta_{\mathbf{D}}$ depend on a_2 and hence on c, which (via equation 17) depends on $\hat{e}$. Thus, the determination of $\hat{e}$ on the basis of equations 13 has to be achieved numerically.

5.1.5 Barotropy and pyknotropy

The terms barotropy and pyknotropy denote the dependence of the mechanical behaviour (stiffness) of soil on stress level and on density, respectively.[12] This dependence marks the more intricate aspects of soil mechanics. A crucial notion in this context is the so-called critical void ratio e_c, which denotes the void ratio at a state where the stress remains constant under isochoric (undrained) deformation.

Let us consider the fan $\mathcal{F}_\sigma$. Its boundary consists of PσP's that correspond to isochoric (or undrained) consolidations[13]. However, 'undrained consolidation' is a contradiction in itself, because in standard soil mechanics 'consolidation' is associated with volume decrease. How to resolve this contradiction? – In reality, undrained

[9] Such a path can be considered as 'virgin consolidation'.

[10] According to the theorem of Cayley-Hamilton, the quasi-elastic relation $\hat{\mathbf{E}}(\mathbf{T})$ can be represented as $\hat{\mathbf{E}} = \xi_1(\mathbf{T})\mathbf{1} + \xi_2(\mathbf{T})\mathbf{T} + \xi_3(\mathbf{T})\mathbf{T}^2$. However, the analytical expression of this relation on the basis of equ.s 1 and 4 is extremely difficult, if not impossible.

[11] The experimental determination of e_{max} is based on convention, thus the so obtained values of e_{max} should be used with caution.

[12] It is remarkable that barotropy and pyknotropy are interrelated and, therefore, cannot be considered separately.

[13] The term 'consolidation' refers to the fact that along such PσP's the mean stress $p := -(T_1 + T_2 + T_3)/3$ increases.

consolidations can be traced if the actual void ratio is smaller than the critical one. To model this fact, it is necessary that the void ratio e appears in the constitutive equation.

We re-write equ. 5 in the form

$$\dot{\mathbf{T}} = a\,|\mathbf{T}|^{\gamma} \cdot (b\,\mathbf{R} + c\,\mathbf{T}^0) \cdot |\mathbf{D}| \tag{15}$$

with $ab = a_1$ and $ac = a_2$, and require that $c = 1$ for critical states, i.e. for $e = e_c$ and isochoric motions (i.e. $\mathrm{tr}\mathbf{D} = 0$). To obtain vanishing stiffness at virgin states with isochoric deformation we then only need to set

$$b = -\frac{1}{|\mathbf{R}(\mathbf{D}^{0\star})|} \quad , \tag{16}$$

with

$$\mathbf{D}^{0\star} := \mathbf{D}^0 - \frac{1}{3}\mathrm{tr}\mathbf{D}^0\,\mathbf{1} \quad .$$

As mentioned, the scalar quantity c must depend on void ratio e and on stress $\mathbf{T}$ in such a way that $c = 1$ for virgin states. No explicit relation for the critical void ratio e_c will be used, as it proves that the virgin void ratio is a more general concept. To meet the requirements for c we set:

$$c = \exp[c_1(\hat{e} - e)] \quad . \tag{17}$$

With the consitutive relation of barodesy (i.e. with equ.s 17, 18 and 19) the simulations of element tests shown in Figs. 25, 26 and 27 are obtained.

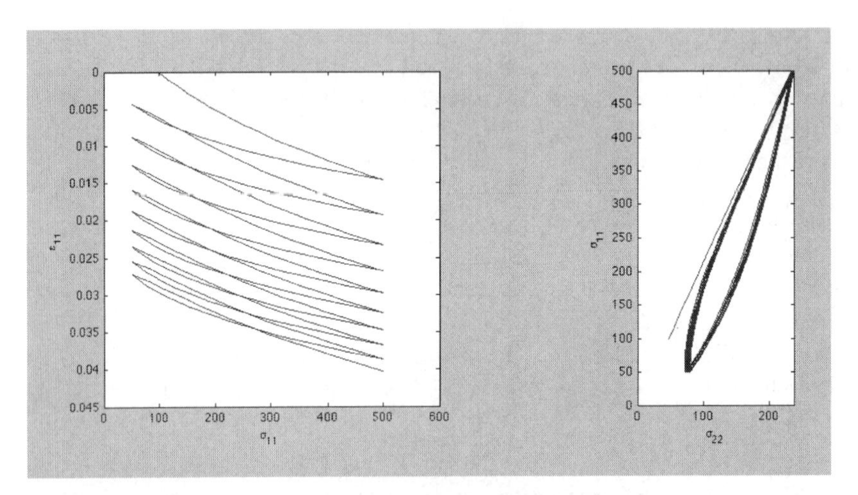

Fig. 25 Stress-strain curve and stress path for cyclic oedometric test on loose sand.

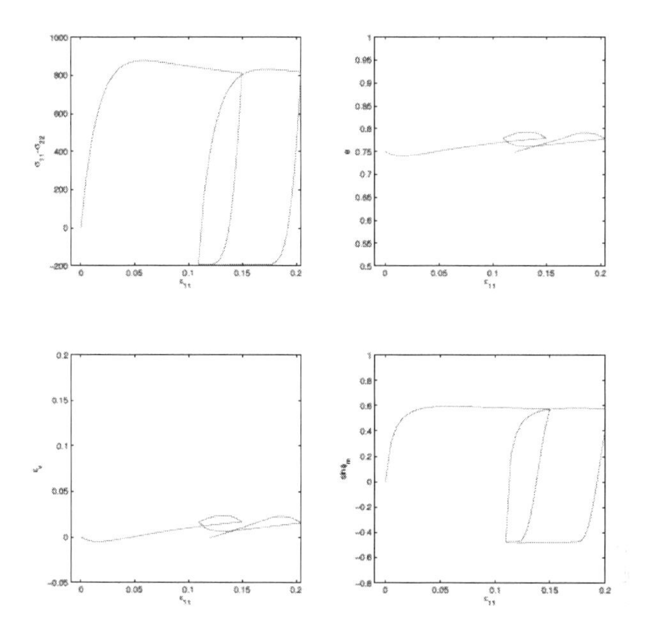

Fig. 26 Simulation of a conventional triaxial test.

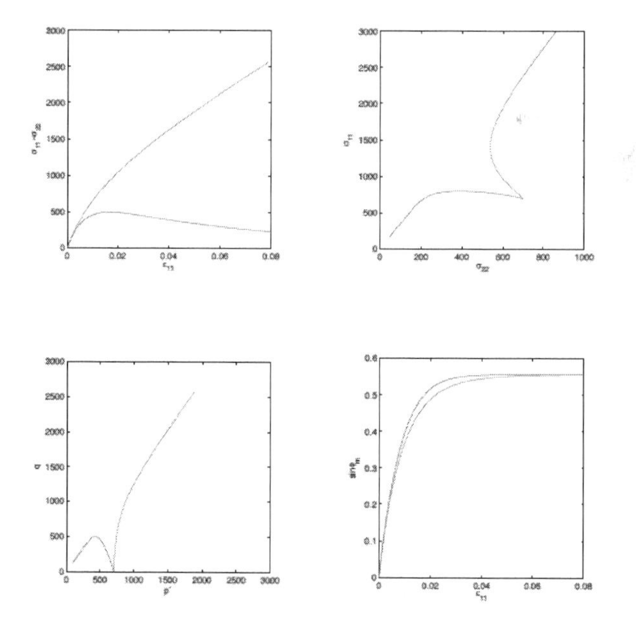

Fig. 27 Undrained triaxial compressions of dense and loose sand.

5.2 *Plasticity theory without yield surfaces*

In 1973 Palmer and Pearce published a paper titled "Plasticity theory without yield surfaces" [23]. Some sentences of this paper deserve being quoted here:

> It was quite natural that the idea of a yield surface should assume such importance in a theory built on experience with metals , since in most metals yield occurs at a fairly well-defined stress level. ...
>
> In soil mechanics the status of the yield surface concept is quite different, both in theory and experiment...
>
> ...strain measurements in clay depend on direct observation of boundary displacements, so that only quite large strain increments are reliably measurable, creep and pore-pressure diffusion confuse results...
>
> ...yield surface motions during strain-hardening are often too complex for the results to be helpful in constructing usable stress-strain relations.
>
> Might it be possible to resolve this (dilemma) by constructing a different kind of plasticity model, in which the yield surface concept had been dropped or relegated to a minor role?
>
> ...it might be useful to idealise clay as a material in which the yield surface has shrunk to a point, so that all deformations are plastic and *any* changes of stress from the current state will produce plastic strain increments.

Palmer and Pearce present in their paper a concept for a plasticity theory without yield surfaces. This concept is based on two postulates by Ilyushin which are, in a sense, precursors of Goldscheider's theorems:

Isotropy postulate: If the strain path is rotated in strain space, then the corresponding stress path is rotated by the same amount. This postulate has actually nothing to do with isotropy, since it considers rotations in the strain and stress spaces, not in the natural space. It is controversial and certainly not valid in the full stress and strain spaces. It is only approximately valid in the deviatoric subspaces: This postulate implies that the deviatoric directions of proportional strain and stress paths coincide. This is, however, not true, according to experimental results by Goldscheider [9].

Delay postulate: The stress at some instant in a loading history does not depend on the whole previous history, but only on the last part of it. This is a postulate of fading memory and is similar to Goldscheider's second theorem.

Based on Ilyushin's postulates, Palmer and Pearce present the following concept:

> The deviatoric stress has two components. The magnitude of the first component is a function of the octahedral shear strain, and its direction coincides with the principal strain vector (referring strain to an isotropically-consolidated initial state). The magnitude of the second component is constant, and its direction coincides with the current strain rate ...
>
> Reversal of the strain path would reverse the second component but not the first ...

The very last sentence appears prophetic, as it strongly resembles to a basic concept of hypoplasticity, to which certainly the authors would have concluded, had they used rate equations instead of finite ones.

References

1. Die visuelle Geschichte der Erde und des Lebens, Gerstenbergs visuelle Enzyklopädie. Gerstenberg Verlag, Hildesheim (1999)
2. Adam, J., Urai, J., Wieneke, B., Oncken, O., Pfeiffer, K., Kukowski, N., Lohrmann, J., Hoth, S., van der Zee, W., Schmatz, J.: Shear localisation and strain distribution during tectonic faulting–new insights from granular-flow experiments and high-resolution optical image correlation techniques. Journal of Structural Geology **27**(2), 283–301 (2005)
3. Bang, D.P.V., Benedetto, H.D., Duttine, A., Ezaoui, A.: Viscous behaviour of dry sand. International Journal for Numerical and Analytical Methods in Geomechanics **31**(15), 1631–1658 (2007)
4. Bauer, E.: Zum mechanischen Verhalten granularer Stoffe unter vorwiegend ödometrischer Beanspruchung. No. 130 in Veröffentlichungen des Institutes für Bodenmechanik und Felsmechanik der Universität Karlsruhe (1992)
5. Byerlee, J.: Friction of rocks. Pure and Applied Geophysics **116**, 615–626 (1978)
6. Cloos, H.: Hebung – Spaltung – Vulkanismus. Geologische Rundschau - XXX - Zwischenheft 4A (1939)
7. Dworschak, M.: Magie des schlauen Sandes. Der SPIEGEL **6** (02.02.2009), 126 (2009)
8. Ghionna, V.N., Porcino, D.: Liquefaction resistance of undisturbed and reconstituted samples of a natural coarse sand from undrained cyclic triaxial tests. Journal of Geotechnical and Geoenvironmental Engineering **132**(2), 194–202 (2006)
9. Goldscheider, M.: Grenzbedingung und Fließregel von Sand. Mech. Res. Comm. **3**, 463–468 (1976)
10. Housner, G.: The mechanisms of sandblows. Bulletin of the Seismological Society of America **48**, 155–161 (1958)
11. Jefferies, M.: Plastic work and isotropic softening in unloading. Géotechnique **47**, 1037–1042 (1997)
12. Jefferies, M., Been, K.: Implications for critical state theory from isotropic compression of sand. Géotechnique **50**(4), 419–429 (2000)
13. Kolymbas, D.: A rate-dependent constitutive equation for soils. Mech. Res. Comm. **4**, 367–372 (1977)
14. Kolymbas, D.: Computer-aided design of constitutive laws. Int. J. Numer. Anal. Methods Geomech. **15**, 593–604 (1991)
15. Kolymbas, D.: An outline of hypoplasticity. Archive of Applied Mechanics **61**, 143–151 (1991)
16. Kolymbas, D.: Introduction to hypoplasticity. No. 1 in Advances in Geotechnical Engineering and Tunnelling. Balkema, Rotterdam (2000)
17. Kuntsche, K.: Materialverhalten von wassergesättigten Tonen bei ebenen und zyklischen Verformungen. No. 91 in Veröffentlichungen des Institutes für Bodenmechanik und Felsmechanik der Universität Karlsruhe (1982)
18. Kuribayashi, E., Tatsuoka, F.: History of earthquake-induced soil liquefaction in Japan. No. 38 in Bulletin of Public Works Research Institute (1977)
19. McKenzie, D.: The generation and compaction of partially molten rock. Journal of Petrology **25**(3), 713–765 (1984)
20. Morsch, O.: Die Physik der Körner. Neue Züricher Zeitung, 13.02.2008 (2008)
21. Muir Wood, D.: The magic of sands — the 20th Bjerrum Lecture presented in Oslo, 25 November 2005. Can. Geotech. J. **44**, 1329–1350 (2007)
22. Nübel, K.: Experimental and Numerical Investigation of Shear Localisation in Granular Material. No. 159 in Veröffentlichungen des Institutes für Bodenmechanik und Felsmechanik der Universität Karlsruhe (2002)
23. Palmer, A., Pearce, J.: Plasticity theory without yield surface. In: A. Palmer (ed.) Symposium on the Role of Plasticity in Soil Mechanics, Cambridge (1973)
24. Pestana, J., Whittle, A.: Compression model for cohesionless soils. Géotechnique **45**(4), 611–631 (1995)

25. Ramberg, H.: Gravity, deformation and the earth's crust. Academic Press, London (1967)
26. Revuzhenko, A.: Mechanics of Granular Media. Springer (2006)
27. Sokoutis, D., Corti, G., Bonini, M., Brun, J., Cloetingh, S., Mauduit, T., Manetti, P.: Modelling the extension of heterogeneous hot lithosphere. Tectonophysics **444**, 63–79 (2007)
28. Topolnicki, M.: Observed stress-strain behaviour of remoulded saturated clay and examination of two constitutive models. No. 107 in Veröffentlichungen des Institutes für Bodenmechanik und Felsmechanik der Universität Karlsruhe (1987)
29. Truesdell, C., Noll, W.: The Non-Linear Field Theories of Mechanics, 2nd edn. Springer (1992)
30. Twiss, R., Moores, E.: Structural Geology. W.H. Freeman and Company, New York (1992)
31. Verdugo, R., Ishihara, K.: The steady state of sandy soils. Soils and Foundations **2**, 81–91 (1996)

The Physics of Granular Mechanics

Yimin Jiang and Mario Liu

Abstract The *hydrodynamic* approach to a continuum mechanical description of granular behavior is reviewed and elucidated. By considering energy and momentum conservation simultaneously, the general formalism of *hydrodynamics* provides a systematic method to derive the structure of constitutive relations, including all gradient terms needed for nonuniform systems. An important input to arrive at different relations (say, for Newtonian fluid, solid and granular medium) is the energy, especially the number and types of its variables.
Starting from a careful examination of the physics underlying granular behavior, we identify the independent variables and suggest a simple and qualitatively appropriate expression for the granular energy. The resultant hydrodynamic theory, especially the constitutive relation, is presented and given preliminary validation.

1 Introduction

When unperturbed, sand piles persist forever, demonstrating in plain sight granular media's ability to sustain shear stresses – an ability that is frequently considered the defining property of solids. On the other hand, when tapped, the same pile quickly degrades, to form a layer (possibly a monolayer) of grains minimizing the gravitational energy. This is typical of liquids. The microscopic reason for this dichotomy is clear: The grains are individually (and ever so slightly) deformed if buried in a pile, which is what sustains the shear stress. When tapped, the grains jiggle and shake, and briefly loose contact with one another. This is why they get rid of some

Yimin Jiang
Central South University, Changsha 410083, China

Mario Liu
Theoretische Physik, Universität Tübingen, Germany, e-mail: mliu@uni-tuebingen.de

of their deformation – which shows up, macroscopically, as a gradual lost of the static shear stress and a continual flattening of the pile. [1]

When sand is being sheared at a constant rate, both solid and fluid behavior are operative. First, the grains are being deformed, increasing the shear stress as any solid would. Second, the same shear rate also provokes some jiggling, just as if the grains were lightly tapped. [2] This leads to a fluid-like relaxation of the shear stress – the larger the shear rate, the stronger the jiggling, and the quicker the relaxation. Note the reason why loading and unloading give different responses (called *incremental nonlinearity* [1, 2]): When being loaded, the solid part of granular behavior increases the stress, while the fluid part decreases it. During unloading, both work in the same direction to reduce the stress.

This entangled behavior, we suspect, lies at the heart of the difficulty modeling sand macroscopically. In addition, there is a "history-dependence" of granular behavior that, being experimentally obvious but conceptually confused and ill-defined, further perplexes the modeler. Obviously, if sand can be characterized, as do other systems, by a *complete set of state variables*, any history-dependence only indicates that the experiments were run at different values of these variables. This is what we believe happens.

The *hydrodynamic theory* is a powerful approach to continuum-mechanical description (or macroscopic field theory), pioneered by Landau [3] and Khalatnikov [4] in the context of superfluid helium. Bei considering energy and momentum conservation simultaneously, and combining both with thermodynamic considerations, this approach is capable of cogently deducing, among others, the proper constitutive relation. Hydrodynamics [5] has since been successfully employed to account for many condensed systems, including liquid crystals [6, 7], superfluid ^{3}He [8, 9, 10], superconductors [11, 12, 13], macroscopic electro-magnetism [14, 15, 16] and ferrofluids [17, 18, 19, 20]. Transiently elastic media such as polymers are under active consideration at present [21, 22, 23].

Two steps are involved in deriving the theory hydrodynamically, the first specifies the theory's *structure*: Being a function of the state variables, the energy itself is not independent. Nevertheless, the form of the energy density $w(s,\rho)$ is left unspecified in this first step, and the differential equations are given in terms of the energy density w, its variables and conjugate variables. [Conjugate variables are the derivatives of the energy with respect to the variables, say temperature $T(s,\rho) \equiv \partial w/\partial s$ and chemical potential $\mu(s,\rho) \equiv \partial w/\partial \rho$ for s, ρ, the entropy and mass density]. In a continuum theory, a number of transport coefficients [such as the viscosity $\eta(s,\rho)$

---

[1] In the usual picture, force chains consisting of infinitely rigid grains is what sustains shear stresses. This does not contradict the above scenario, it is just a different description of the same circumstance: With contacts that are Hertzian (or Hertz-like), grains are infinitely soft at first contact, irrespective of how stiff the bulk material is, since very little material is then being deformed. There is therefore, realistically speaking, always some granular deformation and elastic energy present in any force chain. Now, because grains get rapidly stiffer when being compressed further, the displacement is small, and infinite rigidity is frequently a good approximation. Yet it is the loss of this tiny deformation by tapping that is the cause for the flattening of the pile.

[2] "Jiggling" is used throughout, for random motion of the grains, large or small, that occurs when they are rearranging. Sometimes, words such as wiggle, creep, or crawl may be more appropriate.

or the heat diffusion coefficient $\kappa(s,\rho)$] are needed to parameterize dissipation and entropy production. Neither is their functional dependence specified.

A theory is unique and useful, of course, only when its energy and transport coefficients are made specific, in a second step. This division is sensible, because the first step is systematic, the second is not. The first starts with clearly spelt-out assumptions based on the basic physics of the system at hand, which is followed by a derivation that is algebraic in nature, and hence rather cogent. The second step is a fitting process – one looks for appropriate expressions, by trial and error, for a few scalar functions that, when embedded into the structure of the theory, will yield satisfactory agreement with the many experimental data.

Starting from the physics of granular deformation and its depletion by jiggling, we have identified the variables and derived the structure of the equations governing their temporal evolution [24], calling it GSH, for granular solid hydrodynamics. But our second step is not yet complete, and some proposed functional dependencies are still tentative. The expression for the energy appears quite satisfactory, but our notion of the transport coefficients are still vague. Our final goal is a transparent theory with a healthy mathematical structure that is capable of modeling sand in its full width of behavior, from static stress distribution, via elastoplastic deformation [25, 26], to granular flow property at higher velocities [27, 28, 29, 30].

2 Granular State Variables

In this section, we determine the complete set of granular variables starting from the elementary physics of granular deformation and its depletion by jiggling.

2.1 The Elastic Strain

If a granular medium is sheared, the grains jiggle, roll and slide, in addition to being deformed. Only the latter leads to a reversible energy storage. Therefore, the strain $\varepsilon_{ij} = u_{ij} + p_{ij}$ has two parts, the elastic and plastic one, with the first *defined* as the part that changes the energy. Hence the energy density $w(u_{ij})$ is a function of the elastic strain u_{ij}, which alone we identify as a state variable. For analogy, think of riding a bike on a snowy path, up a steep slope. The rotation of the wheel, containing slip and center-of-mass motion, corresponds to the total displacement d. The gravitational energy $w(d_t)$ of the cyclist and his bike depends only on the center-of-mass movement d_t, the "elastic" or energy-changing portion here. And the gravitational force on the center of mass is $f_g = -\partial w/\partial d_t$. Similarly, the elastic stress is $\pi_{ij} = -\partial w(u_{ij})/\partial u_{ij}$. When grains jiggle, granular deformation relax, hence

$$\partial_t u_{ij} = \mathrm{v}_{ij} - u_{ij}/\tau, \tag{1}$$

with the usual elastic term $v_{ij} \equiv \frac{1}{2}(\nabla_i v_j + \nabla_j v_i)$, and a relaxation term $-u_{ij}/\tau$ that accounts for plasticity. [Note because the total strain obeys $\partial_t \varepsilon_{ij} = v_{ij}$, the evolution of the plastic strain $p_{ij} \equiv \varepsilon_{ij} - u_{ij}$ is also fixed by Eq (1), and given as $\partial_t p_{ij} = u_{ij}/\tau$.] To understand how plasticity comes about, consider first the following scenario with τ = constant. If a granular medium is deformed quickly enough by an external force, leaving little time for relaxation, $\int (u_{ij}/\tau)\,dt \approx 0$, we have $u_{ij} \approx \varepsilon_{ij} = \int v_{ij} dt$ and $p_{ij} = 0$ right after the deformation. The built-up in elastic energy and stress π_{ij} is maximal. If released at this point, the system snaps back toward its initial state, as prescribed by momentum conservation, $\partial_t (\rho v_i) + \nabla_j \pi_{ij} = 0$, displaying an elastic, reversible behavior. But if the system is being held still ($\partial_t \varepsilon_{ij} = v_{ij} = 0$) long enough, the elastic strain u_{ij} will relax, $\partial_t u_{ij} = -u_{ij}/\tau$, while the plastic strain grows accordingly, $\partial_t p_{ij} = u_{ij}/\tau$. When u_{ij} vanishes, elastic energy $w(u_{ij})$ and stress π_{ij} are also gone, implying $\partial_t (\rho v_i) = 0$. The system now stays where it is when released, and no longer returns to its original position. This is what we call plasticity.

However, $1/\tau$ is not a constant in sand: It grows with the jiggling of the grains (as the deformation is lost more quickly) and vanishes if they are at rest. If we quantify the jiggling by the associated kinetic energy, or (via the gas analogy) by a granular temperature T_g, we could account for this by assuming $1/\tau \sim T_g$.

As discussed above, a shear rate would jiggle the grains, giving rise to T_g. For a constant rate, an expression of the form $T_g \sim \sqrt{v_{ij} v_{ij}} \equiv ||v_s||$ is appropriate [see Eq (13) below]. Inserting $1/\tau = \Lambda ||v_s||$ (with Λ the proportionality coefficient) into Eq (1), we obtain the rate-independent expression, $\partial_t u_{ij} = v_{ij} - \Lambda u_{ij} ||v_s||$. Being a function of u_{ij}, the stress $\pi_{ij}(u_{ij})$ therefore obeys the evolution equation,

$$\partial_t \pi_{k\ell} = M_{k\ell ij} \partial_t u_{ij} = M_{k\ell ij}(v_{ij} - \Lambda u_{ij} ||v_s||), \qquad (2)$$
$$M_{k\ell ij} \equiv \partial \pi_{k\ell}/\partial u_{ij} \equiv \partial^2 w/\partial u_{ij} \partial u_{k\ell},$$

which clearly possesses the structure of hypoplasticity [1, 2], a state-of-the-art engineering model originally adopted because sand is incrementally nonlinear, and responds with different stress increases depending on whether the load is being increased ($v_{ij} > 0$, $||v_s|| > 0$) or decreased ($v_{ij} < 0$, $||v_s|| > 0$). It is reassuring to see that the realism of hypoplasticity is based on the elementary physics that granular deformation is depleted if the grains jiggle; and it is satisfying to realize that the complexity of plastic flows derives from the simplicity of stress relaxation.

Under cyclic loading of small amplitudes, because the shear rate is not constant, T_g oscillates and never has time to grow to its stationary value of $T_g \sim ||v_s||$. Therefore, the plastic term $u_{ij}/\tau \sim T_g$ remains small, and the system's behavior is rather more elastic than rendered by Eq (2).

The complete equation for u_{ij} is in fact somewhat more complex, [3]

$$d_t u_{ij} = (1-\alpha) v_{ij} - u^*_{ij}/\tau - u_{\ell\ell}\delta_{ij}/\tau_1, \qquad (3)$$
$$1/\tau = \lambda T_g, \quad 1/\tau_1 = \lambda_1 T_g, \qquad (4)$$

[3] even assuming hard grains, with an elastic strain u_{ij} that is typically tiny, of order 10^{-4}

where u^*_{ij} is the deviatoric (or traceless) part of u_{ij} and $\mathrm{d}_t \equiv \partial_t + \mathrm{v}_k\nabla_k$. The modifications are: (1) The relaxation time for u^*_{ij} and $u_{\ell\ell}$ are different. (2) A shear rate v_{ij} yields an elastic deformation rate $\mathrm{d}_t u_{ij}$ that is smaller by the factor of $(1-\alpha)$.

In contrast to strain relaxation $\sim u_{ij}/\tau$ that is irreversible, α accounts for reversible processes (such as rolling). Without relaxation, elastic and total strain are always proportional, and for say $\alpha = 2/3$, u_{ij} is a third of ε_{ij}. Circumstances are then reversible and quite analogous to a solid – aside from the fact that one needs to move three times as far to achieve the same deformation. So the mechanisms accounted for by α are akin to a lever. [This is also the reason why the stress, or counter-force, is smaller by the same factor, see Eq(19).] Note since any granular plastic motion such as rolling and slipping, be it reversible or irreversible, become successively improbable when the grains are less and less agitated, we expect

$$\alpha(T_g) \to 0, \quad \text{for} \quad T_g \to 0, \tag{5}$$

implying granular media are fully elastic at vanishing granular temperature.

2.2 Mass, Entropy and Granular Entropy

The energy density $w_0(s,\rho)$ of a quiescent Newtonian fluid depend on the entropy density s and mass density ρ, both per unit volume. Defining the temperature and chemical potential as $T \equiv \partial w_0/\partial s|_\rho$ and $\mu \equiv \partial w_0/\partial \rho|_s$, we note that they can be computed only if the functional dependence of $w_0(s,\rho)$ is given. The pressure, a prominent quantity in fluid mechanics, is also a conjugate variable, as it is given by $P \equiv \partial \bar{w}/\partial v$ at constant sv, where $v \equiv 1/\rho$ is the specific volume, $\bar{w} \equiv w_0 v$ the energy per unit mass. Again, P is given once $w_0(s,\rho)$ is. (Note it is not independent from μ and T, since it may be written as $P = -w_0 + Ts + \mu\rho$.)

The conserved energy w depends also on the momentum density $g_i = \rho \mathrm{v}_i$, and is generally given as $w = w_0 + g^2/2\rho$. So the complete set of variables is given as s,ρ and g_i, and the hydrodynamic theory of Newtonian fluids consists of five evolution equations for them. Being a structure of an actual theory, these equations contain w_0, P, also $T, \mu, \mathrm{v}_i \equiv \partial w/\partial g_i$. They are closed only when w_0 is specified. [4]

In continuum-mechanical theories, the entropy s is not always given the attention it deserves. The basic facts underpinning its importance are: The conserved energy w is, in equilibrium, equally distributed among all degrees of freedom, macroscopic ones such as ρ, g_i, and microscopic ones such as electronic excitations or phonons (ie, short wave length sound waves). The entropy s is the macroscopic degree of freedom that subsumes all microscopic ones (typically of order 10^{23}), and accounts for the energy contained in them. Off equilibrium, energy is more concentrated in a few degrees of freedom, typically the macroscopic ones. The one-way, irreversible transfer of energy from the macroscopic to the microscopic ones – in fluid mechan-

[4] Frequently, it is enough to know w_0 in a small environment around given values of s and ρ, or equivalently, of T and P, if these are taken as the independent variables.

ics from ρ, g_i to s – is what we call dissipation, and the basic cause for irreversibility. A proper account of dissipation must consider the variable s, its conjugate variable T, and the entropy production R [with R/T denoting the rate at which entropy is being increased, see Eq (9)]. This remains so for systems (such as granular media) that typically execute isothermal changes.

The energy density of a solid depends on an additional tensor variable, the elastic strain $u_{ij} = \varepsilon_{ij}$, which in crystals is very close to the total strain. The associated conjugate variable $\pi_{ij} \equiv -\partial w_0/\partial u_{ij}$ is the elastic stress – where linear elasticity, or $\pi_{ij} \sim u_{ij}$, represents the simplest case. The hydrodynamic theory of solids consists of eleven evolution equations, for the variables s, ρ, g_i, u_{ij}, which in their structure contain the conjugate variables $T, \mu, \mathrm{v}_i, \pi_{ij}$.

Displaying solid and liquid behavior, granular media have the same variables – in addition to the one that quantifies granular jiggling, for which a scalar should suffice if the motion is sufficiently random. We call it *granular entropy* s_g, and define it to contain all inter-granular degrees of freedom: the stochastic motion of the grains (in deviation from the smooth, macroscopic velocity) and the elastic deformation resulting from collisions. We divide all microscopic degrees of freedom contained in s into the [5] inner- and inter-granular ones, $s - s_g$ and s_g, with the conjugate variables $T \equiv \partial w_0/\partial(s - s_g)$ and $T_g \equiv \partial w_0/\partial s_g$. Equilibrium is established, when both temperatures are equal, and s_g vanishes. (There are overwhelmingly more inner than inter granular degrees of freedom. When all degrees have the same amount of energy, there is practically no energy left in s_g) The equilibrium conditions are:

$$s_g = 0, \quad \bar{T}_g \equiv T_g - T = 0. \tag{6}$$

As zero is the value s_g invariably returns to if unperturbed, it is an energy minimum. Expanding the s_g-dependent part of the energy $w_2 \equiv w - w(s_g = 0)$, we take [6]

$$w_2(s, \rho, s_g) = s_g^2/(2\rho b), \quad \bar{T}_g \equiv \partial w_2/\partial s_g|_s = s_g/\rho b, \tag{7}$$

with $b(s, \rho) > 0$. So the twelve independent variables are: $s, s_g, \rho, g_i, u_{ij}$ and the hydrodynamic theory consists of evolution equations for them all, of which six are given by Eq (3). The rest will be given in section 3. These equations will contain w_0 and the conjugate variables: $T, \bar{T}_g, \mu, \mathrm{v}_i, \pi_{ij}$, also the pressure, given as

$$P_T \equiv -\partial \bar{w}_0/\partial v \equiv -w_0 + \mu\rho + sT + s_g\bar{T}_g, \tag{8}$$

with the derivative taken at constant $sv, s_g v$ and u_{ij}. As we shall see in Eq (24), this is the pressure that accounts for the contribution of agitated grains.

[5] Typical inner granular degrees of freedom are again phonons and electronic excitations.

[6] With $\mathrm{d}w_2 = T\mathrm{d}(s - s_g) + T_g \mathrm{d}s_g = T\mathrm{d}s + \bar{T}_g \mathrm{d}s_g$, we have $T_g \equiv \partial w_2/\partial s_g|_{s-s_g}$ and $\bar{T}_g \equiv \partial w_2/\partial s_g|_s$.

2.3 History Dependence and Fabric Anisotropy

Finally, some remarks about the special role of the density in granular behavior. First, it is quite independent of the compression $u_{\ell\ell}$: Plastic motion rearranges the packaging and change the density by up to 20%, without any elastic compression. Second, the local density only changes if there is some jiggling and agitation of the grains, $\bar{T}_g \neq 0$. Even when non-uniform, a given density remains forever if the grains are at rest. So, if a pouring procedure produces a density inhomogeneity, this will persist as long as the system is left unperturbed, providing an explanation for the history dependence of static stress distribution. Sometimes, these density inhomogeneities have a preferred direction, say, a density gradient along $\hat{x}$. With density-dependent elastic coefficients, the system will then mimic fabric anisotropy, displaying a stress-distribution reminiscent of an anisotropic medium – even when it consists of essentially round grains and the applied stress is isotropic. Our working hypothesis, given a preliminary validation in section 4.1.3, is that both effects are covered by density inhomogeneities. The static stress of a sand pile is calculated there and compared to experiments for two densities, the first uniform and the second with a reduced core density, which we argue is a result of different pouring procedures, being rain-like and funnel-fed, respectively.

3 Granular Solid Hydrodynamics (GSH)

This section presents the remaining six evolution equations. They will be explained but not derived, see [24] for more details and the complete derivation.

3.1 Entropy Production

The evolution equation for the entropy density s is

$$\partial_t s + \nabla_i (s\mathrm{v}_i - \kappa \nabla_i T) = R/T, \tag{9}$$

$$R = \eta \mathrm{v}^*_{ij} \mathrm{v}^*_{ij} + \zeta \mathrm{v}^2_{\ell\ell} + \kappa (\nabla_i T)^2 \tag{10}$$
$$+ \gamma \bar{T}_g^2 + \beta (\pi^*_{ij})^2 + \beta_1 \pi^2_{\ell\ell}.$$

Eq (9) is the balance equation for the entropy s. It is (with R unspecified) quite generally valid, certainly so for Newtonian fluids and solids. The term $s v_i$ is the convective one that accounts for the transport of entropy with the local velocity, and $\kappa \nabla_i T$ is the diffusive term that becomes operative in the presence of a temperature gradient. $R/T > 0$ is the source term. It vanishes in equilibrium, and is positive-definite off it, to account for the fact that the conserved energy w always goes from the macroscopic degrees of freedom to the microscopic ones, $w \longrightarrow s$.

The functional dependence of R changes with the system. In liquids, R is fed by shear and compressional flows, and by temperature gradients [3], as depicted by the first line of Eq (10). In equilibrium, we have $v_{ij}, \nabla_i T = 0$; off it, the quadratic form with positive shear and compressional viscosity, $\eta, \zeta > 0$ and heat diffusion coefficient, $\kappa > 0$, ensures that the entropy s can only increase. In fact, the terms of the first line are, in an expansion of R, the lowest order positive ones that are compatible with isotropy.

The second line of Eq (10), with $\gamma, \beta, \beta_1 > 0$, displays the additional dissipative mechanisms relevant for granular media. As discussed in the introduction, a finite $\bar{T}_g$ or π_{ij}, indicating some jiggling or deformation of the grains, will both relax and give rise to entropy production. Since granular stress π_{ij} will not dissipate for $\bar{T}_g = 0$, we require $\beta, \beta_1 \to 0$ for $\bar{T}_g \to 0$.

Being part of the total entropy, the granular entropy s_g obeys a rather similar equation, though it needs to account for a two-step irreversibility, $w \to s_g \to s$, the fact that the energy goes from the macroscopic degrees of freedom to the mesoscopic, inter granular ones of s_g, and from there to the microscopic, inner granular ones of s, never backwards,

$$\partial_t s_g + \nabla_i (s_g v_i - \kappa_g \nabla_i \bar{T}_g) = R_g / \bar{T}_g, \tag{11}$$

$$R_g = \eta_g v^*_{ij} v^*_{ij} + \zeta_g v^2_{\ell\ell} + \kappa_g (\nabla_i \bar{T}_g)^2 - \gamma \bar{T}^2_g. \tag{12}$$

Eq (11) has the exact same form as Eq (9), so do the first three terms of R_g. But R_g also has a negative contribution. The three positive ones, with $\eta_g, \zeta_g, \kappa_g > 0$, account for $w \to s_g$, how shear and compressional flows, and gradients in the granular temperature produce s_g, the jiggling of the grains. The negative term $-\gamma \bar{T}^2_g$ accounts for $s_g \to s$, how the jiggling turns into heat. There is the same term, though with negative sign, in R, because the same amount of energy arriving at s must have left s_g. As emphasized, all transport coefficients $\eta, \eta_g, \zeta, \zeta_g, \kappa, \kappa_g, \gamma, \beta, \beta_1$ are functions of the state variables (which may alternatively be taken as $T, \bar{T}_g, \rho$, $\pi_{\ell\ell}$ and $\pi^2_s \equiv \pi^*_{ij} \pi^*_{ij}$).

In the stationary and uniform limit, for $R_g = 0$ and $\nabla_i T_g = 0$, macroscopic flows produce the same amount of granular entropy as is leaving, implying

$$\gamma \bar{T}^2_g = \eta_g v^*_{ij} v^*_{ij} + \zeta_g v^2_{\ell\ell}. \tag{13}$$

This is the relation employed to arrive at Eq (2), showing that hypoplasticity holds in the limit of stationary shear rates. Given a shear rate, part of its energy will turn into s_g, which in turn will leak over to s. At the same time, some of the flow's energy will heat up the system directly, with the ratio of the two dissipative channels parameterized by η/η_g and ζ/ζ_g. In dry sand, η, ζ are probably negligible und shall be neglected below – though they should be quite a bit larger in sand saturated with water: A macroscopic shear flow of water implies much stronger microscopic ones in the fluid layers between the grains, and the dissipated energy contributes to R.

Finally, we consider the $\bar{T}_g$-dependence of η_g, ζ_g, γ. Expanding them,

$$\eta = \eta_0 + \eta_1 \bar{T}_g, \quad \zeta_g = \zeta_0 + \zeta_1 \bar{T}_g, \quad \gamma = \gamma_0 + \gamma_1 \bar{T}_g, \tag{14}$$

we shall assume $\eta_0, \zeta_0 = 0$, because

- R_g then stays well defined for $\bar{T}_g \to 0$, see Eq (12);
- Viscosities typically vanish with temperature;
- This fits the Bagnold scaling;
- For $\gamma_0 \gg \gamma_1 \bar{T}_g$ and $\gamma_0 \ll \gamma_1 \bar{T}_g$, respectively, we have from Eq (13), for $v_{\ell\ell} = 0$,

$$\bar{T}_g = (\eta_1/\gamma_0)\,|v^*_{ij}|^2, \quad \bar{T}_g = \sqrt{\eta_1/\gamma_1}\,|v^*_{ij}|. \tag{15}$$

This ensures the existence of an elastic regime at vanishing $\bar{T}_g$, see section 4.2.2.

3.2 Conservation Laws

The three evolution equations left to be specified are conservation laws, for mass, energy and momentum,

$$\partial\rho + \nabla_i(\rho v_i) = 0, \quad \partial w + \nabla_i Q_i = -\rho v_i \nabla_i \phi, \tag{16}$$

$$\partial_t(\rho v_i) + \nabla_i(\sigma_{ij} + \rho v_i v_j) = -\rho \nabla_i \phi, \tag{17}$$

where ϕ is the gravitational potential (on the earth surface, we have $-\nabla_i\phi = G_i$, the gravitational constant pointing downwards). Without specifying the fluxes Q_i, σ_{ij}, these equations are always valid, quite independent of the system, and express the simple fact that being locally conserved quantities (in the absence of gravitation), energy, momentum and mass obey continuity equations. The basic idea of the hydrodynamic theory is to require the structure of the fluxes Q_i, σ_{ij} to be such that, with the temporal derivatives of the variables given by Eqs (3,9,11,16,17), the thermodynamic relation

$$\begin{aligned}\partial_t w(s, s_g, \rho, g_i, u_{ij}) &= (\partial w/\partial s)\partial_t s + (\partial w/\partial s_g)\partial_t s_g + (\partial w/\partial \rho)\partial_t \rho \\ &\quad + (\partial w/\partial g_i)\partial_t g_i + (\partial w/\partial u_{ij})\partial_t u_{ij} \\ &= T\partial_t s + \bar{T}_g \partial_t s_g + \mu \partial_t \rho + v_i \partial_t g_i - \pi_{ij}\partial_t u_{ij}\end{aligned}$$

is identically satisfied, irrespective of w's functional form. This is a rather confining bit of information, enough to uniquely fix the two fluxes as

$$Q_i = (w + P_T)v_i + \sigma_{ij}v_j - \kappa T \nabla_i T - \kappa_g \bar{T}_g \nabla_i \bar{T}_g, \tag{18}$$

$$\sigma_{ij} = (1-\alpha)\pi_{ij} + (P_T - \zeta_g v_{\ell\ell})\delta_{ij} - \eta_g v^*_{ij}, \tag{19}$$

with P_T given by Eq (8), and v^*_{ij} being the deviatory (or traceless) part of v_{ij}. (For details of derivation see [24].) Although now specified to fit granular physics as codified in Eqs (3,9,11), these are still fairly general results, valid irrespective what concrete form w assumes. Moreover, they also nicely demonstrate the dependence on the number and types of variables: Eliminating s_g, or equivalently, taking $\bar{T}_g = 0$, in Eqs (8,18,19), one obtains the solid hydrodynamics. [7] Further eliminating u_{ij} by taking $\pi_{ij} = 0$ leads to the fluid hydrodynamics.

Focusing on the plastic motion, the standard approach (especially the thermodynamic consideration by Houlsby and coworkers, [40]) employs the plastic strain $p_{ij} \equiv \varepsilon_{ij} - u_{ij}$ as the independent variable. Although this starts from the same insight about plastic motion, the connection between elastic strain, stress and energy, so similar in solids and granular media, with formulas that hold for both systems, is lost – or at least too well hidden to be useful, see also the discussion in section 4.3.1.

Enforcing a velocity gradient v_{ij}, the rate of work being received by the system is $-\sigma_{ij}\mathrm{v}_{ij} = -[(1-\alpha)\pi_{ij} + P_T\delta_{ij}]\mathrm{v}_{ij} + [\zeta_g \mathrm{v}_{\ell\ell}\mathrm{v}_{\ell\ell} + \eta_g v^*_{ij} v^*_{ij}]$, see Eq (18). Of these, the terms in the first square brackets, being proportional to the velocity and hence odd under time inversion, are reactive; while those $\sim \mathrm{v}^2$ in the second bracket are even and dissipative. Work received via an odd term will leave if its sign is changed by inverting time's direction; work received via an even term stays, as happens only with dissipative processes. The reappearance of the same factor $(1-\alpha)$ as in Eq (3) is not an accident, but required by energy conservation. If the same velocity leads to an elastic deformation that is smaller by $(1-\alpha)$, then just as with a lever, the force counteracting this deformation $\sigma_{ij} = (1-\alpha)\pi_{ij} + \cdots$ is smaller by the same factor.

This concludes the derivation of the structure of GSH, or granular solid hydrodynamics, given by Eqs (3,8), (9,10,11,12) and (16,17,18,19).

4 Validation of GSH

The advantage of GSH is two-fold, its clear connection to the elementary granular physics as spelt out in the introduction, and more importantly, the stringency of its structure. It cannot be changed at will to fit experiments, without running into difficulties with general principles. The only remaining liberty is the choice of the functional dependence for the energy and some transport coefficients. As this implies much less wiggle room than with typical continuum-mechanical models, any agreement with experimental data is less designed, "hand-crafted," and more convincing, especially with respect to the starting physics.

In what follows, we shall fist examine granular statics, for a medium at rest, $T_g = 0$, then go on to granular dynamics, with enforced flows or stress changes, and some accompanying jiggling, $T_g \neq 0$. An expression for the conserved energy w will be proposed that, in spite of its relative simplicity, reproduces many important granular features when embedded into GSH. As discussed above Eq (6), we divide w into three parts: the micro-, macro- and mesoscopic ones,

$$w = w_0(s,\rho) + [w_1(u_{ij},\rho,g_i) + g^2/2\rho] + w_2(s_g,\rho). \tag{20}$$

[7] In solids, density change and compression are not usually independent. We may account for this by formally setting $P_T = 0$.

The first [8] accounts for the inner-granular degrees of freedom, all subsumed as heat into the true entropy s. We take $w_0 = \langle E(s)/m\rangle\rho$, where $E(s)$ is the energy of a grain, m its mass, and $\langle\rangle$ denotes the average. The second consists of the contributions from the macroscopic variables of momentum density g_i and the elastic strain u_{ij}, where w_1 is given by Eq (21) below. The third, $w_2(s_g,\rho)$ of Eq (7), is further specified in section 4.2.1. It accounts for the inter-granular degrees of freedom, the mesoscaled, strongly fluctuating elastic and kinetic contributions.

4.1 Granular Statics, $\mathbf{T_g = T}$

Given an energy $w_1(u_{ij})$, we can use the stress $\pi_{ij}(u_{ij}) \equiv -\partial w_1/\partial u_{ij}$ and $u_{ij} = \frac{1}{2}(\nabla_i U_j + \nabla_j U_i)$ to close the stress balance $\nabla_j \pi_{ij}(r_i) = \rho G_i$, and determine $\pi_{ij}(r_i)$ with appropriate boundary conditions. As this is done without any knowledge of the plastic strain, we may with some justification call this *granular elasticity* [31].

The relation $u_{ij} = \frac{1}{2}(\nabla_i U_j + \nabla_j U_i)$ remains valid because of the following reasons: In an elastic medium, the stressed state is characterized by a displacement field from a unique reference state, in which the elastic energy vanishes. Because there is no plastic deformation U_i^p, the total displacement is equal to the elastic one. Circumstances appear at first quite different in granular media. Starting from a reference state, a stressed one is produced by the displacement $U_i + U_i^p$, with typically $U_i^p \gg U_i$. Due to sliding and rolling, U_i^p is highly discontinuous, but U_i remains slowly varying, because the cost in elastic energy would otherwise be prohibitive. Fortunately, U_i^p is quite irrelevant: We have innumerable reference states, all with vanishing elastic energy and connected to one another by purely plastic deformations. As a result, we can, for any given displacement $U_i + U_i^p$, switch to the reference state that is separated from the original one by U_i^p, and to the stressed one by U_i. Now, the circumstances are completely analogous to that of an elastic medium.

4.1.1 Yield Surfaces

An important aspect of granular behavior, in the space spanned by the variables, is the existence of yield surfaces. We take them to be the divide between two regions, one in which stable elastic solutions are possible, the other in which they are not – so the system must flow and cannot come to rest. A natural and efficient way to account for yield is to code it into the energy, a scalar. Given the stress balance, the energy is extremal [31] – minimal if convex and maximal if concave. Having the energy being convex within the yield surface, and concave beyond it, any elastic solution that is stable within the surface, will be eager to get rid of the excess energy and become unstable against infinitesimal perturbations beyond it.

[8] Assuming that only w_0 depends on s neglects effects such as thermal expansion which, however, can be easily included if needed.

4.1.2 The Elastic Energy w_1

Our present choice for the elastic energy is [24, 33, 34, 35],

$$w_1(\rho, u_{ij}) = \mathscr{B}\sqrt{\Delta}\left(2\Delta^2/5 + u_s^2/\xi\right), \tag{21}$$

where $\Delta \equiv -u_{\ell\ell}$, $u_s^2 \equiv u_{ij}^* u_{ij}^*$. The energy w_1 is convex only for $u_s/\Delta \le \sqrt{2\xi}$, or equivalently $\pi_s/P_\Delta \le \sqrt{2/\xi}$ (where $P_\Delta \equiv \frac{1}{3}\pi_{\ell\ell}$, $\pi_s^2 \equiv \pi_{ij}^*\pi_{ij}^*$), which coincides with the Drucker-Prager condition. [9] Taking $\xi = 5/3$ gives a friction angle of about 28°. We further take $\mathscr{B} = \mathscr{B}_0\mathscr{B}_1(\rho)\mathscr{C}(\rho, u_{ij})$, where $\mathscr{B}_0$ is a constant, and

$$\mathscr{B}_1 = \left[(\rho - \rho_{\ell p}^*)/(\rho_{cp} - \rho)\right]^{0.15}, \tag{22}$$
$$2\mathscr{C} = 1 + \tanh[(\Delta_0 - \Delta)/\Delta_1]. \tag{23}$$

The coefficient $\mathscr{B}_1$ diverges for the "random closed-pack" density, ρ_{cp}, and is convex only between ρ_{cp} and the "random loose pack" density $\rho_{\ell p}$. [$\rho_{\ell p}^*$ is a constant chosen to yield the right value for $\rho_{\ell p}$ with the relation $\rho_{\ell p} \equiv (11\rho_{cp} + 9\rho_{\ell p}^*)/20$.] It accounts for (1) the lack of elastic solutions for $\rho < \rho_{\ell p}$, when the grains loose contact with one another; (2) the stiffening of granular elasticity with growing density, until it (as an approximation for becoming very large) diverges at ρ_{cp}.

With Δ_0, k_1, k_2, k_3 being constants, and $\Delta_0 = k_1\rho - k_2 u_s^2 - k_3$, we have $\mathscr{C} = 1$ for $\Delta \ll \Delta_0$, and $\mathscr{C} = 0$ for $\Delta \gg \Delta_0$. It changes from 1 to 0 in a neighborhood of Δ_1 around Δ_0, destroying the energy's convexity there. Taking Δ_0 to grow with the density and fall with u_s^2 limits the region of stable elastic solutions to sufficiently small Δ-values, reproducing the virgin consolidation curve and the so-called caps at varying void ratios e, see Fig 1.

4.1.3 Stress Distribution for Silos, Sand Piles and Point Loads

Three classic cases, a silo, a sand pile and a granular sheet under a point load, are solved employing the stress expression derived from the energy of Eq (21), producing rather satisfactory agreement with experiments.

Silos For tall silos, the classic approach is given by Janssen, who starts from the assumption that the ratio between the horizontal and vertical stress is constant, $k_J = \sigma_{rr}/\sigma_{zz}$. Assuming in addition that σ_{zz} only depends on z, not on r, Janssen finds the vertical stress σ_{zz} saturating exponentially with height – a result well verified by observation. (He leaves σ_{rz} and all three radial components: $\sigma_{\theta\theta}, \sigma_{r\theta}$ and $\sigma_{z\theta}$ undetermined.) Having calculated σ_{zz}, one needs the value of k_J to obtain σ_{rr}, usually provided by $k_J \simeq 1 - \sin\varphi$, with φ the friction angle measured in triaxial tests. This makes φ the only bulk material parameter in silo stress distributions. We shall refer to this as the Jaky formula, although it is also attributed to Kézdi.

[9] Only if an energy expression depends on the third strain invariant, could it possibly contain an instability at the true Coulomb condition.

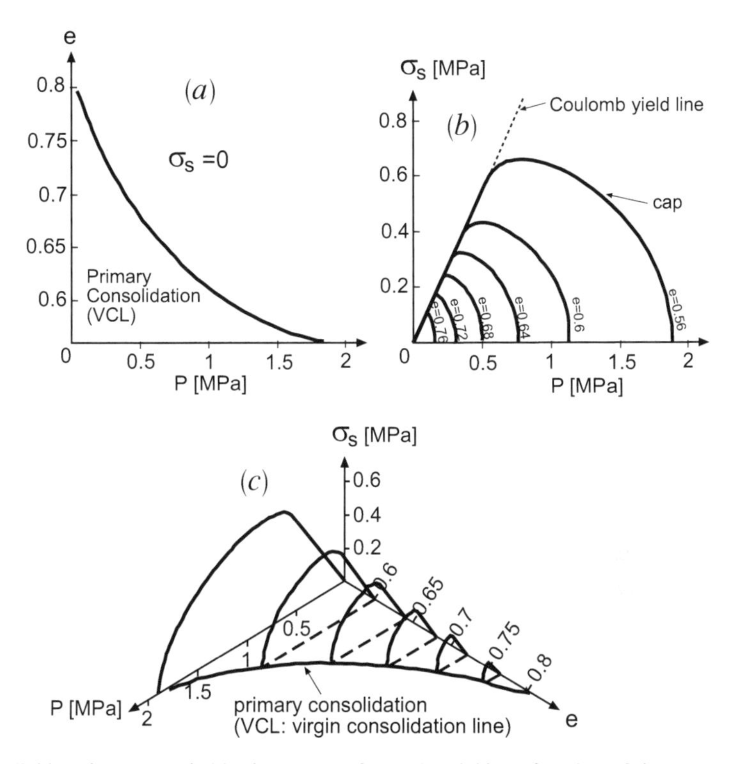

Fig. 1 Yield surfaces as coded in the energy of Eqs (21,22,23), a function of the pressure, shear stress, and void ratio. (a): The *virgin consolidation line*. (b): The bending of the Coulomb yield line, as a function of e. (c): Combination of (a) and (b).

Being important for the structural stability of silos, this formula is (with a safety factor of 1.2) part of the construction industry standard, see eg. DIN 1055-6, 1987. We believe this formula goes well beyond its practical relevance, that it is a key to understanding granular stresses, because it demonstrates the intimate connection between stress distribution and yield, a connection that has not gained the wide attention it deserves. Starting from Eq (21), we calculated [32] all six components of the stress tensor, verifying the Janssen assumptions to within 1%, and found the Janssen constant k_J well rendered by the Jaky formula.

Point Loads The stress distribution at the bottom of a granular layer exposed to a point force at its top is calculated [32] employing Eq (21), without any fit parameter. Both vertical and oblique point forces were considered, and the results agree well with simulations and experiments using rain-like preparation. In addition, the stress distribution of a sheared granular layer exposed to the same point force is calculated and again found in agreement with experimental data, see [32] for more details and references.

Sand Piles The fact that the pressure distribution below sand piles and wedges, instead of always displaying a single central peak, may sometimes show a dip, has intrigued and fascinated many physicists, prodding them to think more carefully and deeply about sand. Recent experimental investigations established the following connection: A single peak results when the pile is formed by rain-like pouring from a fixed height; the dip appears when the pile is formed by funneling the grains onto the peak, from a shifting funnel always hovering slightly above the peak. Employing Eq (21) to consider the stress distribution in sand wedges, we found the pressure at the bottom of the pile to show a single central peak if a uniform density is assumed. The peak turns into a pressure dip, if density inhomogeneity, with the center being less compact, is assumed. The two calculated pressure distributions are remarkably similar to the measured ones, see [31]. The nonuniform density, we believe, is a consequence of pile formation using the hovering funnel: Since the funnel is always just above the peak, the grains are placed there with very little kinetic energy, resulting in a center region below the peak that has a low density. Those grains that do not find a stable position roll down the slope and gather kinetic energy. When they crash to a stop at the flanks, they compact the surrounding, achieving a much higher density.

4.2 Granular Dynamics, $\mathbf{T_g \neq T}$

If a granular medium is exposed either to stress changes, or a moving boundary, the grains will flow, displaying both a smooth, macroscopic velocity, $\mathrm{v}_i \neq 0$, and some stochastic jiggling, $s_g \sim \bar{T}_g \neq 0$. Then the following effects will come into play: First, the energy is extended by a s_g-dependent contribution, $w_2(s_g, \rho)$, see Eq (7). Second, the transport coefficients of Eq (14) become finite. Most importantly, third, the relaxation times τ, τ_1 of Eq (3) are no longer infinite, implying the presence of plastic flows.

4.2.1 The $\mathbf{s_g}$-Dependent Part of the Energy

Specifying the expansion coefficient $b(\rho)$ of Eq (7) as $b = b_0(1-\rho/\rho_{cp})^a$, we find

$$P_T = a\rho\, b_0 \bar{T}_g^2 (1-\rho/\rho_{cp})^{a-1}(\rho/2\rho_{cp}) \tag{24}$$

by employing Eq (8). The density dependence of the expansion coefficient $b(\rho)$ is chosen such that it reproduces the observed volume-dilating pressure contribution $P_T \sim f_2/(\rho_{cp}-\rho)$ from agitated grains [36, 37, 38]. However, we cannot take $a = 0$ as it would imply a diverging granular entropy s_g for $\rho \to \rho_{cp}$. Therefore, we take a to be positiv but small, where $a \approx 0.1$ appears appropriate. (Note that with w_0/ρ independent of ρ and $w_1/\rho \sim \Delta^{2.5}$ – where Δ rarely exceeds 10^{-4} – the respective density derivative and pressure contribution is zero and negligibly small.)

4.2.2 The Hypoplastic Regime

We may choose our parameters such that $\bar{T}_g$ is small at typical velocities of elasto-plastic deformations, though large enough to cover both limits of Eq (15). Then the first term of Eq (19) dominates, because all other terms ($\sim P_T, \eta_g, \zeta_g$) are of order $\bar{T}_g^2$. Then we have $\partial_t \sigma_{ij} = (1-\alpha)\partial_t \pi_{ij} = (1-\alpha)M_{ijk\ell}\partial_t u_{ij}$, with $\partial_t u_{ij}$ given by Eq (3). Stress relaxation, the culprit producing irreversible plasticity, is a term $\sim \bar{T}_g$. For very slow shear flows and $\bar{T}_g \sim ||\mathrm{v}_{ij}||^2$ [first of Eq (15)], it is quadratically small and negligible. This is the elastic regime. At somewhat faster shear flows, the relation $\bar{T}_g \sim ||\mathrm{v}_{ij}||$ [second of Eq (15)] renders $\partial_t \sigma_{ij}$ rate-independent, giving it the basic structure of hypoplasticity, Eq (3). Comparing this results to a state-of-the-art hypoplastic model, we found impressively quantitative agreement, see Fig 2. This is remarkable, because the anisotropy of these figures, determined essentially by $M_{ijk\ell}$, is a calculated quantity: $M_{ijk\ell} \equiv \partial^2 w_1/\partial u_{ij}\partial u_{k\ell}$, with w_1 given by Eq (21).

4.2.3 The Butterfly Cycle

Our last example for validation is not a direct comparison of GHD to some experimental data, but rather an examination of what GHD does, unforced and uncrafted, under typical elasto-plastic deformations. It is solved numerically for stress paths in the triaxial geometry (ie. $\sigma_{xx} = \sigma_{yy}$, $\sigma_{ij} = 0$ for $i \neq j$, similarly for u_{ij}), including all energy terms given above, except $\mathscr{C}$ of Eq (23) that is set to 1 (assuming the yield surface is sufficiently far away). All transport coefficients depend on T_g as specified, but are otherwise constant, independent of stress and density. Also, all variables are taken to be spatially uniform, reducing a set of partial differential equations to or-

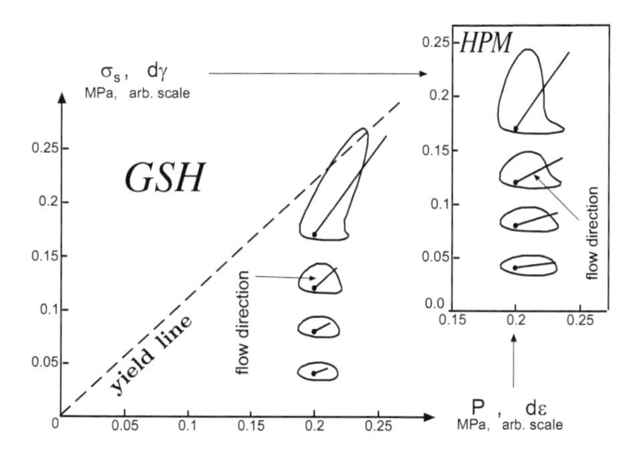

Fig. 2 The change in strain $d\gamma \equiv (\mathrm{v}_{11} - \mathrm{v}_{33})dt, d\varepsilon \equiv -(2\mathrm{v}_{11} + \mathrm{v}_{33})dt$ for given stress rate starting from different points in the stress space, spanned by σ_s, P, as calculated employing (1) GSM, the present theory (taking $1-\alpha = 0.22$, $\tau/\tau_1 = 0.09$, $\zeta_g/\eta_g = 0.33$, $\lambda\sqrt{\eta_g/\gamma} = 114$), and (2) HPM, a typical hypoplastic model, see [39] for more figures and details.

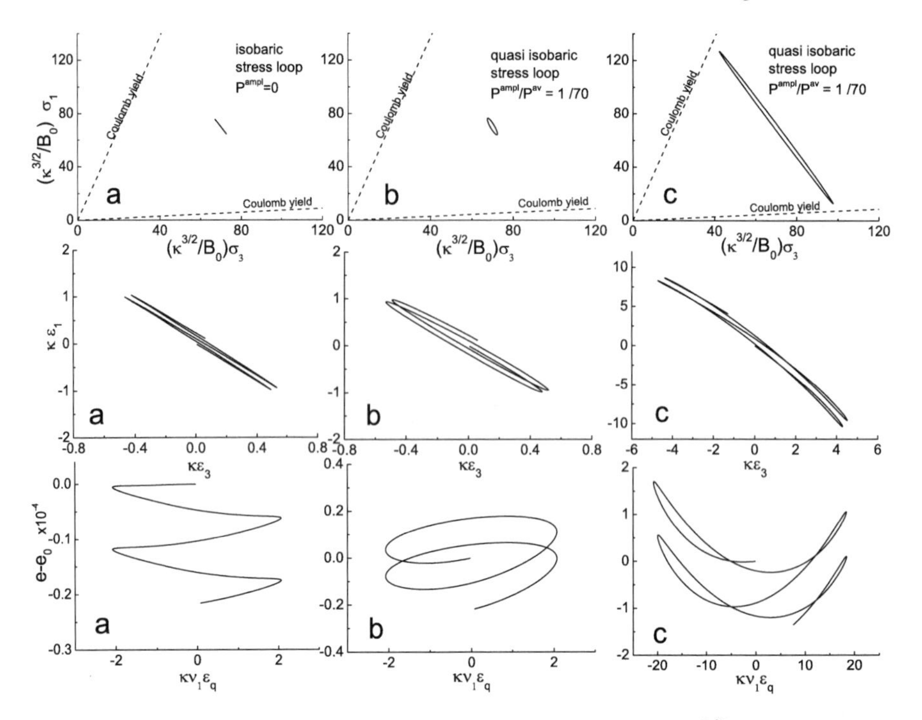

Fig. 3 Upper row: radial stress σ_1 versus axial stress σ_3, rescaled by $B_0\kappa^{-3/2}$. Middle row: radial strain $\varepsilon_1 = \int v_{xx}dt$ versus axial strain $\varepsilon_3 = \int v_{zz}dt$. Lower row: $e - e_0$ (with e_0 the initial void ratio) versus shear strain $\varepsilon_q = \int (v_{zz} - v_{xx})dt$, rescaled by $v_1\kappa$. The stress loads are isobaric for (a) and quasi-isobaric for (b,c); the cyclic amplitude is small for (a,b) and large for (c). The associated strain loci and void ratio are: sawtooth-like for (a), coil-like for (b), butterfly-like (or double-looped) for (c). [The large-amplitude, isobaric plot is quite similar to (c).]

dinary ones in time. In spite of these major simplifications, the results as rendered in Fig 3 display such uncanny realism that it seems obvious GSH has captured some important elements of granular physics. We consider a test with the stress given as

$$P = P^{\text{av}} + P^{\text{ampl}} \cos(2\pi ft), \quad q = q^{\text{ampl}} \cos(2\pi ft + \varphi). \tag{25}$$

Numerical solutions were computed for isobaric test with $P^{\text{ampl}} = 0$ (ie. $P =$ constant) and quasi-isobaric test, with $P^{\text{ampl}} << P^{\text{av}}$ (ie. $P \approx$ constant). The results are shown in Fig 3. they are obtained using the dimensionless parameters: $\kappa \equiv \sqrt{\zeta_1\gamma_1}/\rho b = 18257$, $(\gamma_0/\gamma_1)^2\,(\rho_{cp}b^2\kappa^{3/2}/\mathscr{B}_0 b_0) = 1.07\times 10^{-6}$, $\lambda_1/\lambda = 0.09$, $v_1^2/2 \equiv \eta_1/3\zeta_1 = 1$, $\lambda\sqrt{\eta_1/\gamma_1} = 114$, $\alpha = 0$. The initial conditions are: $e_0 = 0.68085$ (or $\rho_0 = 0.94\rho_{cp}$), v_{ij}, $\bar{T}_g$, $\partial_t\bar{T}_g$, $\partial_t\rho$, $\partial_t u_{ij} = 0$. The averaged pressure $P^{\text{av}} \equiv \sigma_{ii}/3$ is $P^{\text{av}} = 70\mathscr{B}_0\kappa^{-3/2}$, and the amplitude $q^{\text{ampl}} \equiv \sigma_3 - \sigma_1$ is $5v_1\kappa^{3/2}q^{\text{ampl}}/6\mathscr{B}_0 = 10$ for (a,b) and 100 for (c,d). The frequency of $P^{\text{ampl}}, q^{\text{ampl}}$ is $f = 12(\gamma_0/b\rho)$, and the phase lag between them is $\varphi = 58^\circ$.

4.3 Competing Concepts and Misconceptions

Finally, we revisit two previous approaches to come to terms with granular behavior, *granular thermodynamics* by Houlsby et al [40], and *granular statistical mechanics* by Edwards et al [41]. We shall compare GSH to both assuming at most superficial familiarity with them. Also, we refute some misconceptions that have become unfortunately widespread, especially the one about energy not being conserved in sand [*sic*]. These are at best a nuisance in exchanges with referees; and at worst actual obstacles in the progress of our coming to grips with granular modeling.

4.3.1 Granular Thermodynamics

Although considerable work and thoughts have gone into applying thermodynamics to granular media and plastic flow, especially from Houlsby and Collins [40], its basic points are clear and easy to grasp. Taking the entropy production as

$$R = \pi_{ij}\partial_t p_{ij} \tag{26}$$

(where p_{ij} denotes, as before, the plastic strain), it is obvious that the usual linear Onsager force-flux relation, $\partial_t p_{ij} \sim \pi_{ij}$, hence $R \sim \pi_{ij}^2$, does not give a rate-independent R. Therefore, Houlsby, Collins and coworkers consider instead

$$R = \sqrt{\chi_{ijk\ell}\partial_t p_{ij}\partial_t p_{k\ell}} = (\chi_{ijk\ell}\partial_t p_{ij}\partial_t p_{k\ell})/\sqrt{\chi_{ijk\ell}\partial_t p_{ij}\partial_t p_{k\ell}},$$

a rate-independent expression. Equating it to Eq (26), with $\pi_{ij} = -\partial F/\partial p_{ij}$, and F being the free energy density, one then solves for the plastic strain p_{ij} with a given F. One example gives $\partial_t p_{ij} \neq 0$ on a yield surface, characterized by some components of π_{ij} being constant, and $\partial_t p_{ij} = 0$ off it.

GSH starts with the same R, but possesses the additional variable T_g, for which $\bar{T}_g \sim ||\mathrm{v}_s||$ frequently holds, see Eq (13). The linear Onsager force-flux relation

$$\partial_t p_{ij} = \beta\pi_{ij} \quad \text{with} \quad \beta \sim \bar{T}_g, \tag{27}$$

therefore suffices to yield an rate-independent $R \sim \bar{T}_g\pi_{ij}^2$. Note Eq (27) leads directly to the relaxation term: Because $\partial_t u_{ij} + \partial_t p_{ij} = \mathrm{v}_{ij}$, we have $\partial_t u_{ij} - \mathrm{v}_{ij} = -\beta\pi_{ij} = -u_{ij}/\tau$, with $1/\tau \sim \bar{T}_g$. (The last equal sign holds because π_{ij}, β, τ are all functions of u_{ij}, with β, τ as yet unspecified.)

Summarizing, without the variable $\bar{T}_g$, Houlsby and Collins needed to go beyond the well-verified and -substantiated procedure of linear Onsager force-flux relation to maintain rate-independence, obtaining a plastic flow that is confined to the yield surface. In GSH, rate-independence arises naturally within the confines of linear Onsager relation, producing a plastic flow that is as realistic as hypoplasticity, and finite also off the yield surface.

4.3.2 Granular Statistical Mechanics

Generally speaking, it is important to remember that of all microscopic degrees of freedom, the inner-granular ones are many orders of magnitude more numerous than the inter-granular ones. It is the former that dominate the entropy and any entropic considerations. When revisiting granular statistical mechanics, especially the Edwards entropy, it is useful to keep this in mind.

Taking the entropy $S(E,V)$ as a function of the energy E and volume V, or $\mathrm{d}S = (1/T)\mathrm{d}E + (P/T)\mathrm{d}V$, the authors of [41] argue that *a mechanically stable agglomerate of infinitely rigid grains at rest* has, irrespective of its volume, vanishing energy, $E \equiv 0$, $\mathrm{d}E = 0$. The physics is clear: However we arrange these rigid grains that neither attract nor repel each other, the energy remains zero. Therefore, $\mathrm{d}S = (P/T)\mathrm{d}V$, or $\mathrm{d}V = (T/P)\mathrm{d}S \equiv X\mathrm{d}S$. The entropy S is obtained by counting the number of possibilities to package grains for a given volume, and taking it to be e^S. Because a stable agglomerate is stuck in one single configuration, some tapping or similar disturbances are needed to enable the system to explore the phase space.

In GSH, the present theory, grains are neither infinitely rigid, nor always at rest, hence the energy contains both an elastic and a s_g-dependent contribution. [10] And the question is whether granular statistical mechanics is a legitimate limit of GSH. We are not sure, but a yes answer seems unlikely, as both are conceptually at odds in two points, the first more direct, the second quite fundamental: (1) Because of the Hertz-like contact between grains, very little material is being deformed at first, with the compressibility diverging at vanishing compression. This is a geometric fact independent of how rigid the bulk material is. Infinite rigidity is therefore not a realistic limit for sand. (2) As emphasized, the number of possibilities to arrange grains for a given volume is vastly overwhelmed by the much more numerous configurations of the inner granular degrees of freedom, especially phonons. Maximal entropy S for given energy therefore realistically implies minimal macroscopic energy, such that a maximally possible amount of energy is in S (or heat), equally distributed among the inner-granular degrees of freedom. Maximal number of possibilities to package grains for a given volume is a very different criterion.

4.3.3 Energy Conservation

Stemming ultimately from a loose vocabulary, some alleged difficulties to model sand are based on fallacies that need to be refuted here.

The essential difference between granular gas and ideal (atomic or molecular) gas is that the particles of the first undergo non-elastic, dissipative collisions. As a result, their kinetic energy is not conserved, and the velocity distribution typically lacks the time to arrive at the equilibrium Gaussian form. Quantifying the kinetic energy as a granular temperature T_g, it is therefore hardly surprising that the fluctuation-dissipation theorem (FDT), formulated in terms of T_g, is frequently

[10] That grains neither attract nor repel each other is accounted for by the stress vanishing if s_g and u_{ij} do. Then $w_1, w_2 = 0$ and $w_0 \sim \rho$, see Eq (20), implying $\sigma_{ij} = \partial(w_0/\rho)/\partial(1/\rho)\delta_{ij} = 0$.

violated. These are sound results, obtained from a healthy but truncated model that takes the grains as the basic microscopic entity with no heat content. However, some of the further conclusions are deduced forgetting this simplification, rendering them patently absurd. These, and their *[refutation in italic]*, are listed below:

- As the energy is not conserved in sand, neither thermodynamics nor the hydrodynamic method are valid. *[Only the kinetic energy dissipates in granular media, not the total energy. The latter, including kinetic, elastic and heat contributions, remains conserved – as it is in any other system. And only the conservation of total energy is important for thermo- and hydrodynamics.]*
- FDT, along with other general principles either derived from it or in its conceptual vicinity (such as the Onsager reciprocity relation) are all violated. *[There are two versions of* FDT, *only the one given in terms of T_g is violated, not the one in terms of the true temperature T. The latter is a general principle and always valid. For instance, the volume fluctuation is given as $\langle \Delta V^2 \rangle = T(\partial^2 F/\partial V^2)^{-1}$, with F the associated free energy, for a copper block, a single grain, and a collection of grains. If the grains in the collection are jiggling, there is an extra contribution $\sim T_g^2$ in F, see Eq (7), that considerably increases the value of $\langle \Delta V^2 \rangle$. The Onsager relation remains valid because the true* FDT *holds.]*
- The Onsager relation is also violated because the microscopic dynamics, the collision of the grains, is dissipative and hence irreversible. *[The true microscopic dynamics is that in terms of atoms and molecules, the building blocks of the grains. Their dynamics is, as in any other system, reversible.]*

References

1. D. Kolymbas, *Introduction to Hypoplasticity*, (Balkema, Rotterdam, 2000).
2. D. Kolymbas, also W. Wu and D. Kolymbas, in *Constitutive Modelling of Granular Materials* ed D. Kolymbas, (Springer, Berlin, 2000), and references therein.
3. L. D. Landau and E. M. Lifshitz, *Fluid Mechanics* (Butterworth-Heinemann, Oxford, 1987) and *Theory of Elasticity* (Butterworth-Heinemann, Oxford, 1986)
4. I.M. Khalatnikov, *Introduction to the Theory of Superfuidity*, (Benjamin, New York 1965).
5. S. R. de Groot and P. Masur, *Non-Equilibrium Thermodynamics*, (Dover, New York 1984).
6. P.G. de Gennes and J. Prost, *The Physics of Liquid Crystals* (Clarendon Press, Oxford 1993).
7. M. Liu, *Hydrodynamic theory of biaxial nematics*, Phys. Rev. **A 24**, 2720 (1981).
8. D. Vollhardt and P. Wölfle, *The Superfluid Phases of Helium 3*, Taylor and Francis, London (1990).
9. M. Liu, *Hydrodynamics of ^{3}He near the A-Transition,* Phys. Rev. Lett. **35**, 1577 (1975).
10. M. Liu, *Relative Broken Symmetry and the Dynamics of the A_1-Phase,* Phys. Rev. Lett. **43**, 1740 (1979).
11. M. Liu, *Rotating Superconductors and the Frame-independent London Equations,* Phys. Rev. Lett. **81**, 3223, (1998).
12. Jiang Y.M. and M. Liu, *Rotating Superconductors and the London Moment: Thermodynamics versus Microscopics,* Phys. Rev. **B 6**, 184506, (2001).
13. M. Liu, *Superconducting Hydrodynamics and the Higgs Analogy,* J. Low Temp. Phys. 126, 911, (2002)
14. K. Henjes and M. Liu, *Hydrodynamics of Polarizable Liquids,* Ann. Phys. **223**, 243 (1993).

15. M. Liu, *Hydrodynamic Theory of Electromagnetic Fields in Continuous Media,* Phys. Rev. Lett. **70**, 3580 (1993).
16. Y.M. Jiang and M. Liu, *Dynamics of Dispersive and Nonlinear Media,* Phys. Rev. Lett. **77**, 1043, (1996).
17. R.E. Rosensweig, *Ferrohydrodynamics*, (Dover, New York 1997).
18. M. Liu, *Fluiddynamics of Colloidal Magnetic and Electric Liquid,* Phys. Rev. Lett. **74**, 4535 (1995).
19. O. Müller, D. Hahn and M. Liu, *Non-Newtonian behaviour in ferrofluids and magnetization relaxation,* J. Phys.: Condens. Matter 18, 2623, (2006).
20. S. Mahle, P. Ilg and M. Liu, *Hydrodynamic theory of polydisperse chain-forming ferrofluids,* Phys. Rev. **E 77**, 016305 (2008).
21. H. Temmen, H. Pleiner, M. Liu and H.R. Brand, *Convective Nonlinearity in Non-Newtonian Fluids,* Phys. Rev. Lett. **84**, 3228 (2000).
22. H. Pleiner, M. Liu and H.R. Brand, *Nonlinear Fluid Dynamics Description of non-Newtonian Fluids*, Rheologica Acta **43**, 502 (2004).
23. O. Müller, *Die Hydrodynamische Theorie Polymerer Fluide*, PhD Thesis University Tübingen (2006).
24. Y.M. Jiang, M. Liu, *Granular Solid Hydrodynamics*, Grannular Matter,**11-3**, 139 (2009) [DOI 10.1007/s10035-009-0137-3].
25. R.M. Nedderman, *Statics and Kinematics of Granular Materials* (Cambridge University Press, Cambridge, 1992).
26. A. Schofield, P. Wroth, *Critical State Soil Mechanics* (McGraw-Hill, London, 1968).
27. P. K. Haff, *Grain flow as a fluid-mechanical phenomenon*, J.
28. J. T. Jenkins and S. B. Savage, *A theory for the rapid flow of identical, smooth, nearly elastic particles*, J. Fluid Mech. **130**, 187(1983).
29. GDR MiDi, *On dense granular flows*, Eur. Phys. J. **E 14**, 341 (2004).
30. P.Jop, Y. Forterre, O. Pouliquen, *A constitutive law for dense granular flows*, Nature **441**, 727, 2006.
31. D.O. Krimer, M. Pfitzner, K. Bräuer, Y. Jiang, M. Liu, *Granular Elasticity: General Considerations and the Stress Dip in Sand Piles,* Phys. Rev. **E74**, 061310 (2006).
32. K. Bräuer, M. Pfitzner, D.O. Krimer, M. Mayer, Y. Jiang, M. Liu, *Granular Elasticity: Stress Distributions in Silos and under Point Loads,* Phys. Rev. **E74**, 061311 (2006);
33. Y.M. Jiang, M. Liu, *Granular Elasticity without the Coulomb Condition,* Phys. Rev. Lett. **91**, 144301 (2003).
34. Y.M. Jiang, M. Liu, *Energy Instability Unjams Sand and Suspension,* Phys. Rev. Lett. **93**, 148001(2004).
35. Y.M. Jiang, M. Liu, *A Brief Review of "Granular Elasticity",* Eur. Phys. J. **E 22,** 255 (2007).
36. L. Bocquet, J. Errami, and T. C. Lubensky, *Hydrodynamic Model for a Dynamical Jammed-to-Flowing Transition in Gravity Driven Granular Media*, Phys. Rev. Lett., **89**, 184301 (2002).
37. W. Losert, L. Bocquet, T. C. Lubensky, and J. P. Gollub, *Particle Dynamics in Sheared Granular Matter*, Phys. Rev. Lett., **85**, 1428 (2000);
38. L. Bocquet, W. Losert, D. Schalk, T. C. Lubensky, and J. P. Gollub, *Granular shear flow dynamics and forces: Experiment and continuum theory*, Phys. Rev., E **65**, 011307 (2002);
39. Y.M. Jiang, M. Liu, *From Elasticity to Hypoplasticity: Dynamics of Granular Solids,* Phys. Rev. Lett. **99**, 105501 (2007).
40. I. F. Collins and G. T. Houlsby, *Application of thermomechanical principles to the modelling of geotechnical materials*, Proc. R. Soc. Lond. A **453**, 1975, (1997).
41. S.F. Edwards, R.B.S. Oakeshott, *Theory of powders*, Physica A **157**, 1080 (1989); S.F. Edwards, D.V. Grinev, *Statistical Mechanics of Granular Materials: Stress Propagation and Distribution of Contact Forces*, Granular Matter, **4**, 147 (2003).

Are we there yet? Following the energy trail in cohesionless granular solids

Antoinette Tordesillas and Robert P. Behringer

Abstract We review recent developments in experiment, simulation and theory with an eye towards highlighting salient aspects of shearing and failure of granular materials under quasi-static loading. The exchanges and flow of energy within the deforming medium manifest themselves in the evolving kinematics and contact force distribution. As such, we focus on particle motion and the evolution of force chain networks, particularly within shear bands and under conditions that generate stick-slip behavior. A common goal is to reconcile theoretical descriptions of these dissipative processes at different length scales and to relate these to simulation and experimental observations. Implications and challenges brought to the fore for micromechanical constitutive theory are centered around the dominance of nonaffine deformation and force chain evolution across multiple length scales, from the particle to the bulk.

1 Introduction

Tracking the energy trail through space and time can be a useful way to unravel crucial aspects of the rheology of dense, cohesionless granular materials under quasistatic loading. In this context, any attempt to trace the exchanges and flow of energy within the deforming medium must necessarily begin at the source – the force chains. Specifically, force chains carry the largest forces in a granular system, and,

Antoinette Tordesillas
Department of Mathematics and Statistics,
University of Melbourne, Victoria 3010, Australia
e-mail: atordesi@ms.unimelb.edu.au

Robert P. Behringer
Department of Physics and Center for Nonlinear and Complex Systems,
Duke University, Durham, North Carolina 27708, USA
e-mail: bob@phy.duke.edu

hence, they store the largest amount of potential energy. Failure of a force chain results in a collective release of energy accumulated at the chain contacts in the period prior to its collapse and thus dictates the mechanism for energy flow. In the absence of plastic deformation at the particle level (e.g. wear of surface asperities and particle crushing), the frictional rearrangements that result from force chain failure form the predominant mechanism for plastic deformation. In this paper, we review recent developments from our combined studies on force chain evolution, covering the three fronts of experiment, modeling and simulation. Indeed, there is perhaps no other topic in the broad science of granular media that draws together the two disciplines of physics and mechanics more than that of force transmission. It is from these two perspectives that the following review has been cast.

The experiments we describe here exploit the special properties of photoelasticity to obtain forces and other relevant properties in two-dimensional granular systems. We specifically focus on two experiments that highlight the important aspects of shearing and failure of granular materials. The first involves studies of particle motion within the shear band of a 2D Couette flow. The second focuses on the force network, and the way in which the force chains of this network fail when sheared under conditions that generate stick-slip behavior or fluctuations in the macroscopic stress.

The theoretical developments we present here focus on the characterization and modeling of force chain structures and their evolution, and the incorporation of new insights from these studies into constitutive theory. Past literature on force transmission in granular systems have highlighted two crucial aspects of force chain evolution: (a) the connection between the force chains of the discrete medium and the trajectory of the major compressive principal stress in the corresponding continuum idealisation; and (b) the role of the surrounding weak network contacts (e.g. Oda and co-workers [1, 2, 3, 4], Radjai et al. [5], Antony [6], Majmudar and Behringer [7], Lee et al. [8]). Force chains have been observed to preferentially align in the direction of the most compressive principal stress, with the surrounding weak network particles providing lateral stability to the chains under axial compression. While these properties seem obvious from a descriptive analysis of data from physical and numerical experiments, the quantitative characterization and modeling of force chain evolution requires development from three fronts: (a) an objective description of what constitutes a force chain, (b) an analysis of the kinematics and statics within a force chain column and its laterally supporting neighbors, and (c) methods which can unravel the essential aspects of the interplay between the micro or contact mechanisms of failure, i.e. frictional sliding and rolling, and the mesoscopic structural mechanism of force chain failure via confined buckling. We discuss recent attempts at addressing these issues.

The remainder of the paper is arranged in four parts. The first part is devoted to experimental studies and is presented in Section 2. The second part focuses on the characterization and theoretical modeling of force chains and is presented in Section 3. The third part discusses implications of these studies for constitutive theory and is given in Section 4. Concluding remarks are given in Section 5. The projects discussed here, as well as other closely related work, involve a number of collaborators, including: Dapeng Bi, Bulbul Chakraborty, Eric Clement, Karen Daniels, Somayeh Farhadi, Junfei Geng, Isaac Goldhirsch, Robert Hartley, Silke Henkes, Dan Howell, Giles Hunt, Paul Johnson, Lou Kondic, Qun Lin, Gregg Lois, Stefan Luding, Trush Majmudar, Chris Marone, Hans Muhlhaus, Maya Muthuswamy, Corey OHern, John Peters, Jie Ren, Guillaume Reydellet, Trevor Shannon, Jingyu Shi, Matthias Sperl, Tim Tshaikivsky, Brian Utter, David Walker, and Stuart Walsh, Peidong Yu, and Jie Zhang.

2 Experiments

The plastic deformation of disordered solids is a topic of considerable recent interest [10, 11, 12, 13, 14, 9, 15, 16]. In particular, the phenomena appear across a large collection of materials, including glasses, foams, colloids and granular materials, which we consider here. For instance, at low temperatures, i.e. in the absence of thermal noise, plastic behavior in molecular materials [11, 12, 13, 14] may be similar to what occurs in a granular materials. Both molecular and granular materials can exist in disordered jammed states. When sheared, they deform irreversibly at the microscopic level. In the granular case, force chains, long filamentary structures, form in resistance to the shear, as shown in Figure 1.

Here we discuss two types of experiments that highlight the important aspects of shearing and failure of granular materials. The first involves studies of particle motion within the shear band of a 2D Couette flow. The second focuses on the force network and the way in which the force chains of this network fail when sheared under conditions that generate stick-slip behavior.

2.1 Nonaffine deformation in granular shear

A key issue is the relation between microscopic phenomena and larger scale plasticity, that is, how can one relate the smaller scale irreversible motion to the large-scale deformation, described in terms of homogenized strain fields. The former is referred to as affine motion, and the latter, nonaffine motion. Falk and Langer (FL)[9] have developed models of elasto-plastic deformation of amorphous solids and these have been applied recently to granular materials [15, 16, 17]. FL [9] consider the deformation of small clusters of N particles. They fit local deformations over a time Δt to an affine deformation. The residual variance between the fit and the detailed particle motion, D^2_{min}, characterizes the local affine deformation. Also part of this approach is the idea of shear transformation zones, local structural rearrangements that characterize the shear. Here, however, we focus just on the nonaffine motion.

Specifically, we characterize particle motion by determining affine and nonaffine components in a 2D granular Couette (shear) flow. The experimental apparatus [18], is a 2D Couette shearing experiment where $\sim 45,000$ bidisperse photoelastic disks

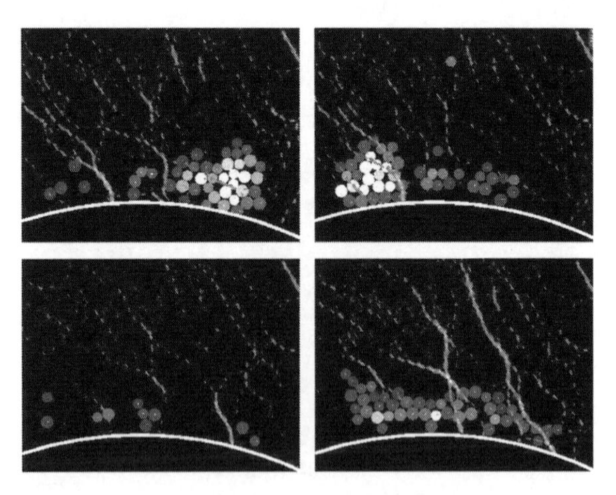

Fig. 1 Time sequence of images emphasizing the spatially inhomogeneous and temporally intermittent nature of force chains (bright lines) and active areas of nonaffine deformation (shaded disks). Each disk shows a local instantaneous value for D^2_{min} (see text), where brighter corresponds to larger D^2_{min}. Images are separated in time by $5\Delta t$ (see text), and run left-to-right, top-to-bottom. The shearing wheel, located near the bottom of each image, rotates counterclockwise. Normally black, the edge of the wheel is highlighted for visability.

lie on a flat horizontal surface and are bounded by a slowly rotating inner shearing wheel (radius $20.5cm$) and an outer concentric ring (ID= 51cm). The wheel rotates at a rate f = inverse rotation period = $1mHz$, leading to quasistatic shear. As described in [18, 20], we track particle rotation, displacement, and force.

Motion occurs primarily in the shear band of width less than 10 grain diameters, near the shearing wheel. The mean tangential velocity $v_\theta(r)$ and local shear rate $\partial v_\theta/\partial r$ decrease roughly exponentially with radial distance from the wheel, where we choose $r = 0$ to be the edge of the shearing wheel.

We characterize localized, mesoscopic regions within $2.2d$ of a reference particle in the shear band. Here d is the mean particle diameter, and there are $N \simeq 15$ particles in such a neighborhood. Over a time, $\Delta t = 1.85s$, which is short relative to $1/f$, we follow the motion of the particles in these clusters. We determine the displacement of each particle relative to the cluster center of mass (CoM), which removes the larger scale flow. We then least-squares fit the cluster deformation, thus determining the affine deformation. Specifically, we least squares fit data for the the initial locations relative to the CoM, r_i, and final locations relative to the CoM, r'_i, for the particles in a cluster to the form $r'_i = \mathbf{E}r_i$. $\mathbf{E}$ is a 2×2 matrix. The variance from this fit, $D^2_{min} = \Sigma(r'_i - \mathbf{E}r_i)^2$ is essentially the parameter of FL. Since $\mathbf{E}$ is in general not symmetric, its eigenvalues may be (and often are) complex. We then characterize $\mathbf{E}$ by both a symmetric tensor, $\mathbf{F}$, and a rotation, given by a rotation matrix, $\mathbf{R}_\theta$: $\mathbf{E} = R_\theta \mathbf{F}$. Inverting the order of $\mathbf{F}$ and $\mathbf{R}_\theta$ does not change θ or the eigenvalues of $\mathbf{F}$, and θ is uniquely determined if we restrict $-\pi/2 < \theta \leq \pi/2$. We write the deformation in terms of the eigenvalues, ε_i, of $\varepsilon = \mathbf{F} - \mathbf{I}$, where $\mathbf{I}$ is the unit tensor. Thus, θ, and the ε_i characterize the affine part of the local evolution, whereas D^2_{min} characterizes the extent of nonaffine motion. In general, all measures of the deformation are spatially and temporally intermittent, as seen for instance in Figure 1. Hence, we present data for probability distribution functions (PDF's) of various measures as functions of r.

Figure 2 gives PDF's for the eigenvalues of ε, in the form $\delta\varepsilon = \varepsilon_2 - \varepsilon_1$ and $2\bar{\varepsilon} = \varepsilon_1 + \varepsilon_2$, where by definition $\varepsilon_2 \geq \varepsilon_1$. Different data sets, i, are binned over radial widths of size d, corresponding to a series of radial distances d apart, starting from the shearing wheel, where the first series is centered at $r = d/2$. Note that as one might expect, $\bar{\varepsilon}$ has zero mean, since it corresponds to the dilation of the local patches, and $\delta\varepsilon$ which measures shear deformation, set by the velocity profile, has non-zero mean and a typical shape that also applies to the data for D^2_{min}. PDF's for θ are similar to those for $\delta\varepsilon$, with mean values of typically a few degrees or less.

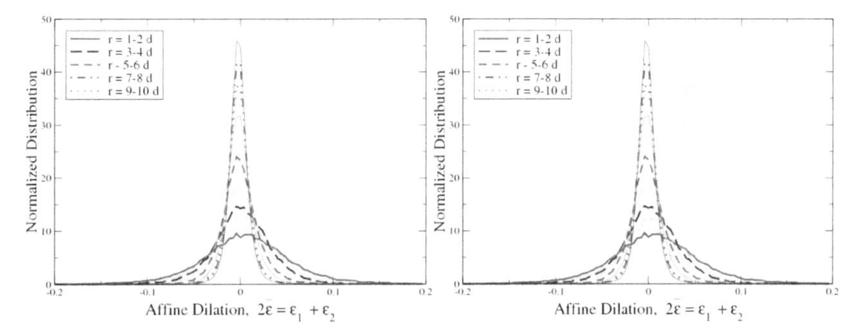

Fig. 2 PDF's of the affine dilation, $\varepsilon_1 + \varepsilon_2$ (top), and affine shear deformation, $\varepsilon_2 - \varepsilon_1$ (bottom). Each series corresponds to a radial bin of width d. The notation $r = n - (n+1)d$, etc. indicates a bin centered at $(2n+1)d/2$

Figure 3(top) gives PDF's of D^2_{min}. Roughly, the peak in these PDF's moves to larger values of D^2_{min} for r closer to the shearing wheel, although the position of the peak does not vary strictly monotonically with r, due to boundary effects. The various PDF's are well fitted by the form $f(x) = C_1 x^{C_2} e^{-x/C_3}$, although the exponent C_2 and the scaling factor C_3 vary from fit to fit.

If, as in Figure 3(bottom), we rescale each of the data sets in the vertical direction by the peak fit value and in the horizontal direction by the mean $\langle D^2_{min}\rangle$, we find that the data for different r collapse fairly well onto a common curve. Averages $\langle D^2_{min}\rangle$ (inset), are set by the local shear rate $\dot{\gamma}$ within the shear band (inset, bottom), which resembles what we have previously observed for diffusivities, i.e. $D \propto \dot{\gamma}$ [18].

D^2_{min} is mesosopic in the sense that it is a locally averaged measure of a grain-scale process. We suggest that the scaling for the PDF's of D^2_{min} is the result of an underlying microscopic phenomena. Interestingly, the PDF's $P_1(\delta\mathbf{r_i})$ of the individual nonaffine particle displacements, $\delta r_i = r'_i - \mathbf{E}r_i$ in Figure 4 (left), are approximately gaussian, $P_1(\delta r_i) \simeq A\exp[-(\delta r_i/\Delta r)^{\alpha}]$, where $\alpha \leq 2$. (See e.g. Figure 4 (right), $\ln(-\ln(P_1(\delta r_i)))$ as a function of $\ln(\delta r_i)$, with P_1 normalized so that $P_1(0) = 1$.)

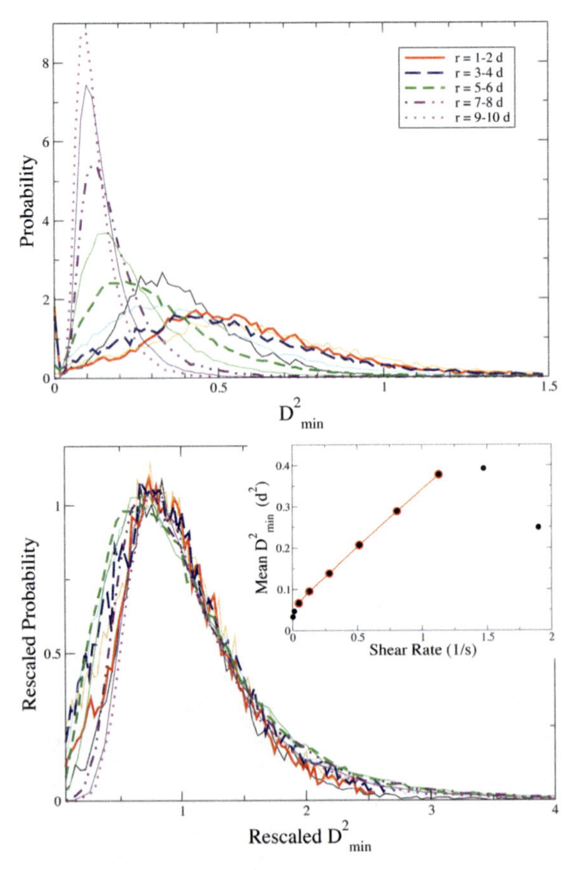

Fig. 3 Top: PDF's of D^2_{min} (units of d^2) at various distances r from the shearing surface (in units of d) for 1080 images. Bottom: Rescaled data from top. Each data set is fitted and rescaled by the peak magnitude and mean value of D^2_{min}. Inset: $\langle D^2_{min}\rangle$ vs. local shear rate. Note the drop in $\langle D^2_{min}\rangle$ (departure from red line) at high $\dot{\gamma}$, i.e. near wheel. The notation $r = n-(n+1)d$, etc. indicates a bin centered at $(2n+1)d/2$.

This form for $P_1(\delta r_i)$ suggests an explantion for the form of the PDF's for D^2_{min}. We can write the PDF for D^2_{min} as

$$P(D^2_{min}) = \int P_N(\delta r_1,..\delta r_N)\delta(D^2_{min} - \Sigma(\delta r_i)^2)d\delta r_1...d\delta r_N, \qquad (1)$$

where P_N gives the PDF for the N nonaffine particle displacements in a cluster, and where with $D^2_{min} = (\Sigma_N \delta r_i^2)$. If we assume that the individual particle displacements are uncorrelated and gaussian-distributed: $P_1(\delta r_i) = B\exp[-(\delta r_i/\Delta r)^2]$, then

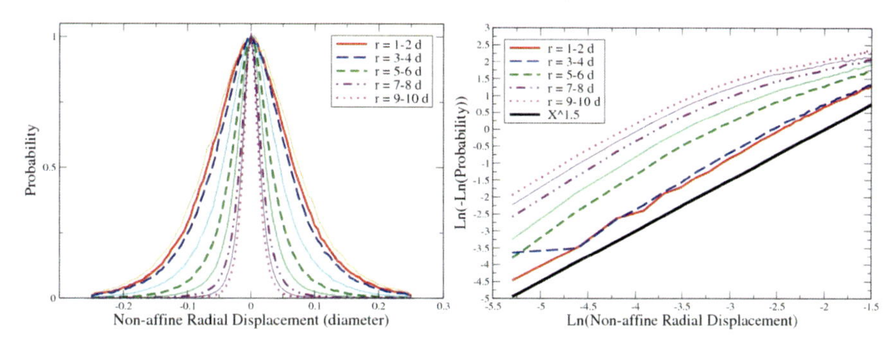

Fig. 4 Data for individual particle nonaffine radial displacements, $P_1(\delta r_i)$. Top: on linear scales, Bottom: on ln | ln | vs. ln scales (solid black line shows slope 3/2). PDF's for the nonaffine azimuthal displacements are similar. The notation $r = n-(n+1)d$, etc. indicates a bin centered at $(2n+1)d/2$.

the computation of the PDF for D^2_{min} follows. The assumptions of independent displacements with guassian distributions are roughly correct, although not rigorous. The integral yields $P(D^2_{min}) = C_1(D^2_{min})^{N-1}\exp(-D^2_{min}/C_2)$, where the C_i are constants. This expression is roughly consistent with the experimentally obtained results (Figure 3).

When we compare various mesoscopic and microscopic measures, such as $\langle D^2_{min}\rangle$ and the radial and tangential diffusivities, D_{rr} and $D_{\theta\theta}$, we find that all can be rescaled to a common curve, within experimental error. In Figure 5, we have renormalized each quantity so that it is 1 at $r = 2d$. As one might expect, diffusion coefficients follow the mean measure of local plastic rearrangements $\langle D^2_{min}\rangle$. Note that on dimensional grounds, D^2_{min} has units of x^2 and is measured over a time interval Δt. Diffusivity D is given by $D \sim \langle(\Delta x)^2\rangle/\Delta t$. In these experiments, displacements scale as d, times scales as the local inverse shear rate $\dot{\gamma}^{-1}$, so $D \sim \dot{\gamma}d^2$ and $D^2_{min} \sim \dot{\gamma}\Delta t d^2$.

It is interesting that the scales for the nominally macroscopic quantities, the mean azimuthal flow, the more 'mesoscopic' affine displacements, and the roughly microscopic nonaffine displacements are all of comparable size. For example, the lengths used to scale the quantities in Figure 5 to unity at $r = 2d$ are all roughly similar. These are: (i) for the displacement due to mean azimuthal flow during Δt scale factor is $0.21d$. (ii) For the affine strains, the relevant length is the size of the region under consideration. Multiplying this length, $2.2d$, times, respectively, the value of $\varepsilon_2 - \varepsilon_1$ at the peak of the distribution, and the width of this distribution, we obtain $0.103d$ and $0.088d$. Scaling the width of the distribution for $\varepsilon_1 + \varepsilon_2$, we obtain $0.11d$. (iii) The 'microscopic' length scales are $[D_{rr}/\dot{\gamma}]^{1/2} = 0.095d$ and $[D_{\theta\theta}/\dot{\gamma}]^{1/2} = 0.135d$ for diffusion. (iv) The 'mesoscopic' length scale for D^2_{min} is $[D^2_{min}/N]^{1/2}$ (N = typical number of particles in a cluster), which at $r = 2d$ is $0.159d$. (v) Finally, the rms widths of the azimuthal and radial components of the distributions of the δr_i are $0.072d$ and $0.071d$, respectively.

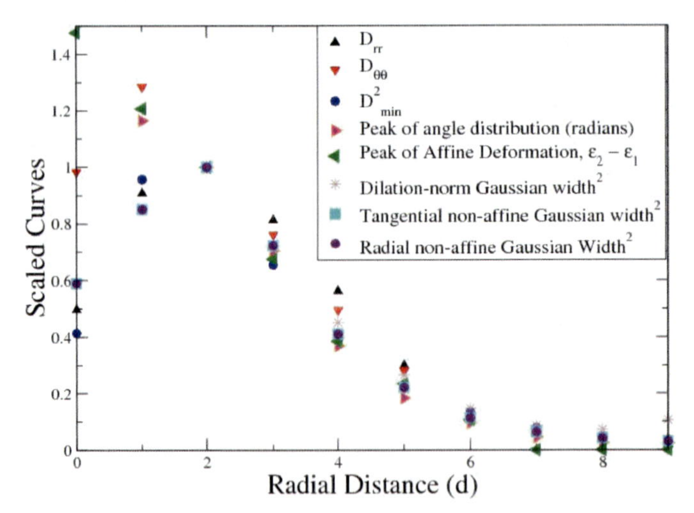

Fig. 5 Rescaled $\langle D^2_{min} \rangle$, widths of nonaffine PDF's, $P_1(\delta \mathbf{r_i})$, diffusivities, and peak values of $P(\theta)$ and $P(\varepsilon_2 - \varepsilon_1)$ versus radial distance from the shearing wheel. All quantities are scaled to 1 at $r = 2d$. For reference, at $r = 2d$, $\dot{\gamma} = 0.60 s^{-1}$, $D_{rr} = 0.0054 d^2/s$, $D_{\theta\theta} = 0.011 d^2/s$, $< D^2_{min} >= 0.38 d^2$, $\theta_{peak} = 0.045 rad$, $(\varepsilon_2 - \varepsilon_1)_{peak} = 0.047 d$. Square width for $P(\varepsilon_2 - \varepsilon_1)$–$0.0016 d^2$; Square widths for distributions of δr_i in respectively radial and azimuthal directions: $0.0050 d^2$ and $0.0053 d^2$.

In summary, our analysis which separates affine, nonaffine and mean motions within a granular shear band, indicates that all three types of motion contribute in nearly equal amounts to the complex motion within a shear band. All quantities from the present analysis are highly fluctuating, necessitating a statistical distribution. PDF's for D^2_{min}, a measure of the nonaffine and/or diffusive properties associated with the flow, have a characteristic shape which we can understand in terms of N weakly correlated nonaffine particle displacements, δr_i, for particles within a cluster. The distributions for the δ_i have a quasi-gaussian form, but with an exponent of roughly 1.5 rather than 2, within the exponential.

2.2 *Granular friction and stick-slip*

Here we seek to understand the grain-scale physical mechanisms for granular friction and in particular, friction between a solid surface and a granular material. This is of interest both in regard to stick-slip behavior in fault zones [24], and more generally in regard to the frictional properties of solids [25]. These experiments consist of a solid object, the 'slider' being pulled across a granular surface at constant speed through a spring. We focus on the low-speed limit where stick-slip dynamics apply.

Recently, various studies have probed granular friction [26, 27, 28, 29, 30, 31]. A classic experiment, as in Nasuno et al., is to consider the motion of a slider driven over a granular surface. But, in the Nasuno et al. experiments, the connection be-

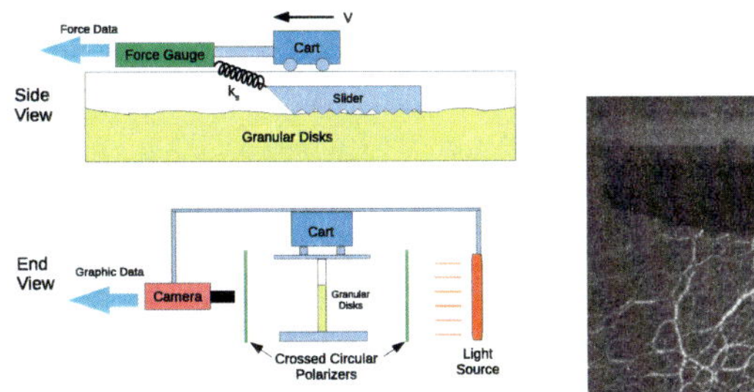

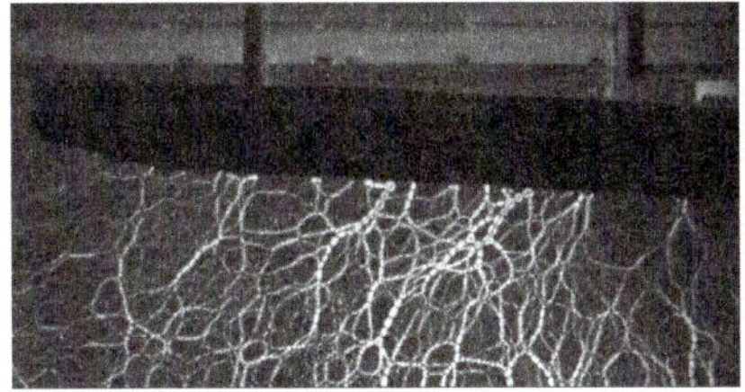

Fig. 6 Left: sketch of apparatus. Right: Typical force structure image. The dark structure is the metal slider, moving to the left. Bright regions below the slider are particles experiencing a large force. These tend to be organized along force chains oriented diagonally down and left, along the compressive direction for the shear.

tween the slider and the internal granular dynamics was not accessible. Other recent work involves the use of photoelastic particles [30, 31]. A key feature of this work is that by using photoelastic techniques, it yields simultaneous information on the slider dynamics and the granular behavior. In general, friction in granular and atomic systems show a range of complex and interesting behavior [32, 33, 34, 35, 36, 25].

We use as a backdrop to the present experiments a very simple model of a solid block of mass M sliding on a horizontal plane. The block is pulled through a horizontal spring of force constant k_s, and the end of the spring opposite the block is pulled at a constant speed, V. The two interact by ordinary solid-on-solid friction. The normal and tangential forces, F_n and F_t are due to gravity (Mg), and friction. Static friction supports tangential forces when $|F_t| < \mu_s|F_n|$. When $|F_t|$ exceeds $\mu_s|F_n|$, the frictional force switches instantaneously to a retarding kinetic frictional force of $\mu_k F_n$. The frictional force reverts instantaneously to static friction when the relative velocity between the objects returns to zero, provided that $|F_t| < \mu_s|F_n|$.

The behavior of the block is simple. We imagine that at $t = 0$, the block is a rest, the spring is neither stretched nor compressed, and that the puller starts to move at speed V. Initially, the block remains at rest (stick), until F_p exceeds $\mu_s Mg$. For $F_p \geq \mu_s|F_n|$ the block slides, and kinetic friction provides a drag force $\mu_k Mg < \mu_s Mg$. Sliding occurs for a time Δt, and during that time, the slider moves a distance Δx, after which the block returns to rest, and a state of static friction. Newton's equations of motion for the slip phase yield the displacement, $x(t)$ of the slider relative to an appropriate reference position x_o: $x(t) = A\cos(\omega t + \phi) + Vt + x_o - \mu_k Mg/k_s$. Various quantities in this expression include: $A = -(V_o^2 + V^2)^{1/2}/\omega$, $\tan(\phi) = -V/V_o$. $V_o = (\mu_s - \mu_k)g/\omega$ is a natural velocity scale, $\omega^2 = k_s/M$, $\Delta t = (\pi - 2\phi)\omega$, and $\Delta x = (2V_o + \omega \Delta t V)/\omega$. As noted above, in the present experiments, we operate in the stick-slip regime, and $V_o >> V$. Note that $\omega^{-1} = (M/K_s)^{1/2}$ is a natural time scale for the motion, and when V is small, slip occurs for roughly half a period, $2\pi/\omega$. It is also possible to show that $\mu_s - \mu_k = (\omega^2/2g)(\Delta x - \Delta t V)$. Note that because F_p/Mg at failure determines μ_s, a determination of Δx and Δt provides μ_k.

We show a sketch of the apparatus, which is similar in function to the block-plane-spring model, and a typical photoelastic image in Figure 6 (left/right). The basic mechanism involves a slider of mass M which is pulled via a spring over a quasi-2D granular surface. The pulling force is F_p. The pulling is carried out via a cart that is advanced at a constant driving/pulling speed, V by lead screw driven by a stepper motor screw. Attached to the cart is a spring of force constant k_s, whose other end is attached to the slider. The cart also carries a force gauge which measures the pulling force, F_p. The analogue output of the gauge is sampled by an A/D converter. The cart also carries a high-speed video camera whose function we describe below.

The granular system in these experiments is 2D. It consists of photoelastic disks (PSM, Vishay Measurements) which reside in a channel formed by two vertically oriented Plexiglas sheets. We obtain video images of the photoelastic response of the granular layer, measurements of pulling forces, and the slider motion. The channel is 1.5m long, by 15.5 cm tall, and has a gap slightly greater than the thickness of the particles. It is filled to a depth of about 20 mean grain diameters with about about 10^4 bidisperse disks. These have diameters of 4mm and 5mm, and thickness 3.2mm, and an average particle size $d = 4.37mm$. The sliders are made of brass or aluminum, and have a series of half-round cut-outs, of the mean particle diameter across the bottom. The sliders vary in length from 30 d to 60 d, and they have masses between 50 to 250 g.

As noted above, the cart carries both the force gauge and the photoelastic imaging system. The latter consists of a circularly polarized light source, a video camera operating up to frame rates up to 500 Hz, and a second crossed circular polarizer. The video and force measurements of F_p are synchronized. We note that the photoelastic technique, which produces images such as that in Figure 6 (Right), allows the determination of quantitative particle-scale force information [37, 38, 39].

The stick-slip process adheres to the following scenario. In a stick state, the slider remains essentially at rest as the cart advances. During this phase, the pulling force, F_p, builds up at a uniform rate set by V, as in Figure 7. When the slip threshold is reached, F_p decays quickly, and the slider advances rapidly. These events are non-periodic. The values of F_p at the beginning and ending of slip are broadly distributed. In order to analyze our results, we define a stick (slip) event, Figure 7, as any increase (decrease), ΔF_p, of F_p with time that is larger than a threshold slightly larger than the force gauge resolution. The resulting properties that we obtain from our data, such as the probability distribution functions (PDF) of relevant quantities are insensitive to modest increases of this threshold for slip or slip events. Below, we consider PDF's for F_p at the beginning (end) of slip, F_{start} (F_{stop}); the duration over which stick (slip) occurs, Δt; and the energy changes of the pulling spring, ΔE_p, and the difference between static and kinetic friction coefficients.

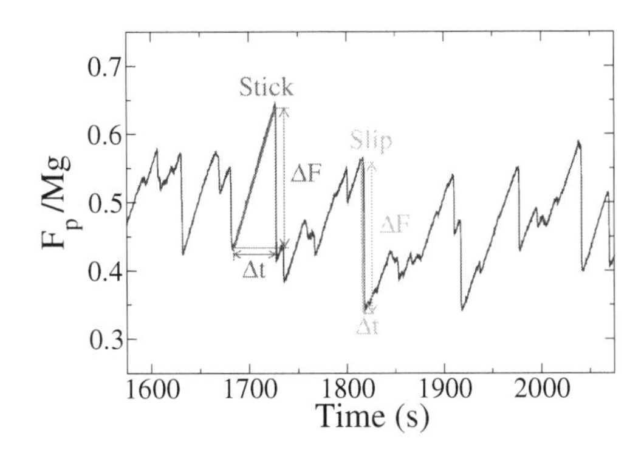

Fig. 7 Typical time series for the pulling force, F_p, with $M = 219.0g, V = 0.033cm \cdot s^{-1}, k_s = 34.3N \cdot m^{-1}$.

PDF's of F_{start}/Mg, which effectively defines $\mu_s = F_{start}/Mg$, are broad, nearly overlapping, and roughly gaussian (see also Albert et al. [40]) with means close to 0.5 and variances of ~ 0.07. Interestingly, the PDF's for F_{stop}/Mg are very similar, with the only substantive difference being a slight leftward shift relative to the PDF's for F_{start}/Mg. These PDF's do not change significantly for different V and k_s in the stick-slip regime.

We next consider the energy decrease following a slip event, which is $\Delta E = \Delta E_p + \Delta E_g$. Here, ΔE_g is the potential energy stored inside the granular disks, and $E_p = F_p^2/(2(k_s)$. Because the grains are stiff relative to the pulling spring, E_p dominates E_g by over an order of magnitude, and $\Delta E = (F_{start}^2 - F_{stop}^2)/(2(k_s)$ provides a good measure of energy lost in a slip event.

These data are particularly interesting [31], because they bear some resemblance to data for earthquake magnitudes, m, the well known Gutenberg-Richter (GR) power-law distribution, $N(m)$, and models for self-organized criticality [41]. The GR relation deserves some comment. It is a cumulative probability distribution function (CDF) as compared to the data that we present here. In fact, the PDF for m, $P(m)dm$, is related to the CDF by $N(m) = \int_m^\infty P(m')dm'$. As such, a PDF is a more sensitive test for power-law behavior, than a CDF. The GR CDF is $\log_{10} N(m) = a - bm$, where $b \simeq 1$, from the global range of earthquake data. The earthquake magnitude, m, can be related to the seismic energy E_s by $\log_{10}(\Delta E_s) = 1.5m + 4.8$. The PDF of ΔE_s is related to the PDF of m by $P(\Delta E_s) = P(m)dm/d(\Delta E_s)$. Hence, the PDF for ΔE_s should be $P(E_s) \propto \Delta E_s^{-\varepsilon}$ where $\varepsilon = 1 + 2b/3$. A value of $b \simeq 1$ implies that $\varepsilon \simeq 5/3$.

PDF's of ΔE_p of slip events from this exeriment for various parameters can be described by a common power law with exponent -1.2 ± 0.1. Deviations from a power law at the lower and higher ends of the data are caused by measurement resolution and finite slider mass respectively.

We next consider the PDF's for slip time, Δt. In the ideal block model above, we would expect a delta function distribution peaked at $\Delta t \simeq \tau = \pi/\omega = \pi(M/k_s)^{1/2}$. The experimentally determined PDF's for $\Delta t/\tau$ are nearly exponentially distributed, $P = \exp(-C\Delta t/\tau)$, where $C = 0.2$ is roughly constant over all parameters. We do not obtain a sharp distribution, but τ is the right time scale for the slip events. We note too, that for large enough τ or possibly V, the distribution falls below an exponential.

The last distribution that we consider is for $\mu_k - \mu_s = \Delta\mu = \mu_s - (\omega^2/2g)(\Delta x - \Delta t V)$. Here, we use the experimentally determined data for Δx and Δt, and put these into the above expression, as an effective definition of $\Delta\mu$ and hence μ_k. The PDF's for $\Delta\mu$ are roughly power laws We understand this by noting that for slow V, $\Delta\mu$ is dominated by Δx, which in turn is tied to the behavior of ΔE. The distributions for $\mu_k = \mu_s + \Delta\mu$ exhibit Gaussian-like PDF's similar to those of μ_s because PDF's for $\Delta\mu$ are concentrated near $\Delta\mu \simeq 0$.

In order to probe the obvious complexity of this system, we track the slider motion at frame rates of 500Hz, fast enough to capture many details of the dynamics. Specifically, over the time leading up to and during a slip event, we track the slider position, $x(t)$. We then numerically compute $v = \dot{x}$ and $a = \dot{v}$, and the frictional force $f_g = F_p - Ma$. We have carried out such studies for a considerable number of cases. Several features are of interest. For instanced, slip is preceded by a slow creep-like displacement. In addition, we find that the slider motion during slip has detailed structure. In this, as in other large events, there are multiple roughly harmonic cycles, where additional cycles begin before the previous one is complete. In addition, f_g fluctuates about and closely tracks F_p.

In order to further unfold the complexity of slip events, we use high speed photoelastic imaging (frame rate $= 500Hz$). We focus on the region below the slider, since this is where there is significant evolution of the force network. In order to pull out thc changcs in the force network, we digitally subtract sequential images, and then enhance the contrast. We begin imaging very close to the maximum of F_p, and the images span most of the event. From this analysis, we identify several important processes. As F_p builds up, the grains move very little, but the force carried by the network strengthens. The first significant change in the network occurs via small adjustments in the network, where a force chain has failed. In some cases, small failures of the force network, such as this one, may lead to a very small event, or may also be part of a larger event. Thus, they may end without further slip/failure. However, such a failure may be a precursor to a larger event. In this case, the initial network failure leads to forward motion of the slider and additional load on the remaining network. The impact on the the remaining force chains can have several different forms. Some chains may be further compressed, increasing their load, while others may be restructured, for instance, as one of its particles shifts forward to form part of a new force chain. Other force chains may fail via several identifiable mechanisms. Particles in a chain may slip, and this seems to occur most often near the boundary with the slider. Alternatively, a chain may buckle, or a chain may deform so that its upper part rotates forward. Both processes tend to weaken a chain's mechanical strength and its ability to resist the horizontal force applied through the spring. All of these processes lead to forward motion of the slider. This slip may be arrested after a few chains fail. However, the failure of one or more chains may cause a cascade of multiple chain failures. That is, force chains appear to have a threshold of force beyond which they fail. When some chains fail, additional load can be placed on the remaining chains. Note that forward motion of the slider has two effects. First, it reduces the pulling force, but it may also weaken the remaining force chains. Thus, the reaction to the failure of one force chain depends on the balance between any reduction in the pulling force, changes in the strength of the force network, and the possibility that some force chains will be pushed past their thresholds of failure. The typical scenario for a large event is that one chain fails, and that, through the processes above, other chains fail consequently. Eventually, a new force-balanced state forms. Some or all of the original force chains may have disappeared, to be replace by a new chains which are able to support the load from the pulling spring. Since the slider moves forward much faster than the pulling speed, V, the pulling force is reduced at the end of an event. For a large event, the pulling force is substantially smaller that its value just before slip. Cascades occur as a result of the fact that breaking or buckling of one chain applies more load to the remaining chains, which may buckle or rotate towards the vertical, weakening their resistance to horizontal load. The data for slider velocity and acceleration suggest that the slider begins to return to rest following an initial slip, but then undergoes additional slip associated with further failure of the network. This multi-chain failure process underlies the complexity of the dynamics, and also explains why a simple block model can provide only a heuristic description of events.

3 Theory

The experiments discussed above, along with other earlier and more recent investigations (e.g. [18, 20, 30, 44, 39, 45, 46]) have provided observations and data that have guided and stimulated the theoretical advances described here. Specifically, analytical methods developed to quantitatively characterize and model force chain evolution and associated nonaffine deformation form the foci of this section.

3.1 Characterization of force chains and their evolution

Here we present techniques developed to quantitatively characterize force chains and their evolution. All of these techniques rely on data at the level of particles, e.g. contact forces, contact locations, and the displacements and velocities of particle centers. While in the past, this level of information has been confined only to particle based simulations, the breakthrough by Majmudar and Behringer [7], have made possible for this data to be generated from photoelastic disk experiments, at least for two-dimensional assemblies. Indeed, some of the results described in what follows have arisen from analysis of data from these experiments (e.g. [44]).

3.1.1 Identifying force chains

The first step in extending the practical applicability of existing modeling tools, from the perspectives of the discrete element method and continuum theory, is the development of an algorithm for identifying force chains. This algorithm, developed originally in Peters et al. [47] and subsequently improved in Muthuswamy and Tordesillas [48], is based on the concept of a force chain being that of a quasilinear arrangement of three or more particles through which above average loads are transmitted. A force moment tensor s_{ij} is used to calculate an estimate of the load carried by a particle:

$$s_{ij} = \sum_{c=1}^{N} f_i^c r_j^c \tag{2}$$

where N is the number of contacts of the particle and f_i^c and r_j^c represent, respectively, the components of the contact force and the unit normal vector from the particle's center to the point of contact. The most compressive principal force moment, i.e. the largest eigenvalue of s_{ij}, represents the *load* carried by the particle, while its corresponding eigenvector represents the direction of force transmission. Groups of at least three particles whose eigenvectors are aligned within a predefined tolerance angle, and whose loads are each above the global average constitute a force chain. The procedure was incorporated into an algorithm which takes contact force data as the known input, and provides the force chains and thus the force chain network as the output. This procedure has been recently adapted to assemblies consisting of irrregularly shaped particles by Pena et al. [49].

3.1.2 Properties of force chains

Using the method in Section 3.1.1, we have characterized force chains in various granular systems under biaxial compression and punch indentation tests. Analysis is undertaken from both a geometrical and structural mechanics perspective.

Studies in [47, 48] suggest that force chain length, expressed in terms of the number of particles in the chain, exhibits an exponential frequency distribution, consistent with the experiments in Howell et al. [46] and, more recently, in Sanfratello and Fukushima [50]. In comparing dense versus loose assemblies, results suggest that, on average, dense systems favor shorter chains. In this context, materials with shorter force chains may be macroscopically stronger than those with longer chains, in agreement with [46] and [51]. Interestingly, in order to explain observed macroscopic differences in sheared fault gouges, Anthony and Marone [52] conjectured that shorter chains can support greater stress since there are fewer potential failure points. Indeed, intuitively, one would expect longer chains to be less stable.

In addition to length, a material's ability to form straighter force chains also influences its macroscopic strength or load-bearing capacity. Chains that are able to sustain higher loads are straighter [46]. In [48], a quantity called *curvature*, c, is used to measure the departure from a perfectly linear or straight chain configuration. This can be expressed as the ratio of the sum of the magnitudes of branch vectors linking each particle to the next in the chain to the magnitude of the vector linking the first to last particle, i.e.

$$c = \frac{\sum_{p=1}^{n-1} |\mathbf{x}^{p+1} - \mathbf{x}^{p}|}{|\mathbf{x}^{n} - \mathbf{x}^{1}|}; \tag{3}$$

here n is the number of particles in the chain, and $\mathbf{x}^p$ is the position vector of particle p. Thus $c \geq 1$, where the lower bound $c = 1$ corresponds to a perfectly straight force chain. This measure is similar to the 'end-to-end' vector in polymer rheology [53]. Results suggest that increasing interparticle friction, packing density and degree of polydispersity – properties that are known to enhance material stability and strength – promote the formation of straighter force chains.

3.1.3 Properties of the force chain network

The geometry of individual force chains alone is not the only factor at play. The topological properties of the whole force chain network are also important. For example, in [48], it was shown that a force chain network with more branch-points generally means a greater capacity for the material to support more load, since there are more pathways available for stress transmission. Once an existing pathway for stress transmission is no longer viable, force may be simply transferred to a different branch of the force chain network, without the need for particles to rearrange. This finding is consistent with observations from experiments that examined the effect of packing density: the force chain network was observed as changing from "intermittent, long, radial chains" for less densely packed, to being "more tangled, with many shorter intersecting chains" for more densely packed (Howell et al. [46]), where "tangled" is quantified here by the degree of branching.

3.1.4 Confined force chain buckling

The most obvious starting point for the characterization of force chain evolution is to simply examine the evolution with strain of the geometrical properties described above in Sections 3.1.2-3.1.3. However these properties alone do not reveal the whole story on why and how force chains evolve to accommodate the changes in the applied loads to the boundary of the material. Accordingly, more recent studies take the next steps for development by focussing on the kinematics, stability and failure of force chains as a structure [44, 54, 55, 56, 57, 58, 59, 60, 61, 62, 63, 64]. In these studies, the kinematics and statics of these structures are simultaneously accounted for and *with respect to their surrounding neighbors*. Methods of this kind are necessary if we are to gain insights into the concurrent and interrelated mechanisms (e.g. dilatation) that are known to significantly affect the stability and lifetime of force chains.

To facilitate these studies, an algorithm was first developed for the purposes of identifying force chains or parts thereof that have undergone buckling, i.e. buckled force chain segments (Tordesillas [54]). A strain interval of interest, $[\varepsilon_A, \varepsilon_B]$, is chosen: for example, this may span a drop in stress ratio (an 'unjamming event') or a single time step in the DEM simulation. A set of three filters is then sequentially applied: (a) eliminate all particles not in force chains at ε_A; (b) out of those remaining, eliminate those which have not decreased in potential energy during $[\varepsilon_A, \varepsilon_B]$; (c) out of those remaining, identify and isolate all 3-particle force chain segments which have buckled during $[\varepsilon_A, \varepsilon_B]$. To determine if a segment has buckled during $[\varepsilon_A, \varepsilon_B]$, we consider the angle between the branch vectors from the central particle to the two outer particles. The decrease in this angle over the interval in question is defined as being twice the buckling angle, θ_b. A buckled force chain segment, i.e. a BFC, is simply one where $\theta_b > 0$. The set of particles remaining after all three filters have been applied is the set consisting of all BFC segments and is referred to hereafter as BFCs. A CBFC corresponds to a particle cluster involving a BFC and its confining first ring neighbors, as determined from a Delaunay triangulation.

To study the evolution of force chain buckling, we used data from a series of biaxial compression tests: under constant confining pressure in [54, 55, 56, 60], and under constant volume in [44]. The latter, [44], involved data from both DEM simulation and experiments on photoelastic disk assemblies. In all cases, buckling events were found to initiate just prior to peak shear stress. Moreover, these particle clusters (i.e. CBFCs) were confined to the shear bands, consistent with earlier findings: see, for example, those by Oda and co-workers [1, 2, 3, 4] and, more recently, by Rechenmacher [65]. Reminiscent of the "slip-stick" phenomenon described in Section 2.2, distinct fluctuations in the shear stress can be observed. Successive cycles of a decrease (unjamming or slip event) followed by an increase (jamming or stick event) in the shear stress (or stress ratio) were observed in the post peak, strain-softening and critical state regimes. These fluctuations have been observed in numerous experimental and numerical studies, including those which focus on geological fault gouge processes (e.g. [66, 67, 68, 69] and references cited therein). Note that the use of the terms 'unjamming or slip events' and 'jamming or stick events' in what follows is strictly in reference to these strain intervals in the post peak regimes. Sudden bursts or peaks in the population of buckling events or BFCs coincide with the drops in the stress ratio or unjamming periods in the post-peak regimes [44]. In agreement with the observations in [30], and as discussed earlier in Section 2.2, the local failure of the force chain network, here via buckling of a few force chains, triggers the onset of macroscopic slip or unjamming. While initially local, force chain buckling quickly spreads along the shear band, before inducing elastic unloading of the regions exterior to the band, and ultimately precipitating the sudden drop in the macroscopic shear stress.

Next the kinematics of confined force chain buckling was quantified in [44, 54, 55, 56, 60], using new measures of local micropolar deformation developed in [55]. These measures were borne out of a thermomechanical analysis, the same framework used in the derivation of the constitutive law to be discussed in upcoming sections. This consistency in the definition of deformation is crucial if we are to robustly integrate information derived from analysis of particle data into micromechanical continuum theory. The new deformation measures account for particle rotations, and are based on a local measure of micropolar strain whose volumetric average matches the macroscopic strain experienced by the whole assembly. The local micropolar strain, experienced by a single particle relative to its first ring neighbors, is

$$\varepsilon_{ij} = \frac{1}{2A} \sum_{c \in B} (p_i^c + p_i^{c+1}) e_{jk3} (l_k^{c+1} - l_k^c), \tag{4}$$

where the sum is taken anticlockwise over the set of branch vectors B associated with the particle, l_i^c is a branch vector between the reference particle and its neighbor 'c', e_{jk3} is the permutation symbol, p_i^c is a vector describing the relative motion of the particle c from the frame of the reference particle, and A is the area of the Delaunay polygon for the particle [55]. This polygon is made up of the triangles derived from a Delaunay triangulation of the particle and its first ring of neighbors. The vector p_i^c is defined as

$$p_i^c = u_i^c - u_i + e_{ij3} l_j^c \omega, \tag{5}$$

where u_i and u_i^c denote respectively the displacement of the reference particle and its neighbor since the start of loading, and ω is the rotation of the reference particle. A more detailed account of this strain definition is presented in [55]. The measure of the local particle strain rate is thus given by

$$\dot{\varepsilon}_{ij} = \frac{1}{2A} \sum_{c \in B} (\dot{p}_i^c + \dot{p}_i^{c+1}) e_{jk3} (l_k^{c+1} - l_k^c), \tag{6}$$

from which the volumetric strain rate $\dot{\varepsilon}_{vol} = \frac{1}{2}(\dot{\varepsilon}_{11} + \dot{\varepsilon}_{22})$ can be computed.

Similarly, the particle curvature rate may be expressed in terms of the rate of relative rotation $\dot{\Phi}^c = \dot{\omega}^c - \dot{\omega}$ at a branch vector as

$$\dot{\kappa}_i = \frac{1}{2A} \sum_{c \in B} (\dot{\Phi}^c + \dot{\Phi}^{c+1}) e_{ij3} (l_j^{c+1} - l_j^c). \tag{7}$$

Local measures of the nonaffine deformation can be computed by comparing the relative motion predicted by the local measures of strain and curvature to the actual relative motion between particle contacts [56]. The nonaffine strain rate is:

$$\dot{\Delta}^{\varepsilon} = \frac{1}{2A} \sum_{c \in B} |\Delta \dot{\mathbf{p}}^c| (|\mathbf{l}^{c+1} - \mathbf{l}^c| + |\mathbf{l}^c - \mathbf{l}^{c-1}|), \tag{8}$$

where $\Delta \dot{p}_i^c$ represents the difference between the relative motion of the particle c from the frame of the reference particle and the motion implied by the particle strain, namely

$$\Delta \dot{p}_i^c = \dot{p}_i^c - \dot{\varepsilon}_{ij} l_j^c . \tag{9}$$

Similarly, if $\Delta \dot{\Phi}^c = \dot{\Phi}^c - \dot{\kappa}_i l_i^c$, then the nonaffine curvature rate is given by

$$\dot{\Delta}^{\kappa} = \frac{1}{2A} \sum_{c \in B} |\Delta \dot{\Phi}^c| (|\mathbf{l}^{c+1} - \mathbf{l}^c| + |\mathbf{l}^c - \mathbf{l}^{c-1}|) . \tag{10}$$

Following [18, 20], we performed a detailed examination of the development of nonaffine deformation and kinematic diffusion within the shear band. The results of this study are reported in [56] and indeed many aspects of the study emulate those presented in Section 2.1. However, space limitations confine the discussion here to only some specific aspects of the kinematics of confined force chain buckling.

Certain trends prevail in all of the tests in [44, 54, 55, 56, 60]. We show in Figure 8 the strain evolution of the macroscopic stress ratio, $\sin\phi = (\sigma_{22} - \sigma_{11})/(\sigma_{22} + \sigma_{11})$, and the rate of energy dissipation for the system examined in [60]. A strong temporal correlation is evident between the peaks in the dissipation rate and the peaks in the global averages of deformation of the following types: (a) local nonaffine strain $<\Delta^{\varepsilon}>$, (b) local nonaffine curvature $\bar{R}<\Delta^{\kappa}>$, (c) rate of dilatation $\dot{\varepsilon}_v$, and (d) rate of curvature $|\dot{\kappa}|$. We quantified the correlation between the total amount of energy dissipated during unjamming periods and the corresponding total amount of deformation for the various types (a)-(d) and found that the squares of the Pearson's product moment correlation coefficient (r^2) are, respectively: (a) 0.566, (b) 0.675, (c) 0.937, and (d) 0.850. Similarly, the correlation between the amount of energy dissipated and the corresponding decrease in the potential energy in the unjamming periods is r^2=0.969. Spatially, these quantities are all strongly correlated as was shown in [60]. In particular, dissipation is mainly confined to the shear band, as the regions outside sustain relatively little deformation. Thus, altogether, these findings suggest that the mechanism that is chiefly responsible for dissipation is confined to the band, is associated with high levels of nonaffine deformation, dilatation and relative rotations – and occurs during periods of unjamming. To determine whether this mechanism is confined force chain buckling, we first measured the correlation between the amount of energy dissipated and the corresponding population of BFCs during unjamming. A strong correlation was found: r^2=0.859. Next we performed a detailed examination of force chain failure in [55, 56], with a view toward quantifying the associated nonaffine deformation, dilatation and relative rotations developed during buckling. Here we found that, temporally and spatially, the dissipative mechanism of force chain buckling is indeed a major contributor to deformation on the mesoscopic and macroscopic scales. Peaks in the volumetric strain, micropolar curvature, nonaffine strain and nonaffine curvature from CBFCs all occur during unjamming or drops in stress ratio. Moreover, despite CBFC branch vectors (i.e. branch vectors between particles belonging to the same CBFC) constituting a relatively small fraction of the total number of branch vectors in the system, CBFCs contribute a disproportionately large amount to nonaffine deformation. Finally, the highest correlation levels between the two measures of local nonaffine deformation, i.e. nonaffine strain and nonaffine curvature, can be observed inside the persistent shear band, where the buckling of force chains is the dominant mechanism [56]. This confirms the key role of particle rotations in force chain buckling; recall that particle rotation is the common term in these nonaffine measures (equations (10) and (8)).

We have also examined in [64] the evolution of the stability of force chains, both prior to and after buckling, in the aforementioned DEM simulations. This employs a procedure for computing the limiting upper and lower bounds for the stability of a particle cluster based on a stiffness matrix, a method that we have recently developed by extending the structural mechanics approach of Bagi [70]. Once these limiting values are established, the stability of the cluster can be estimated by combining these values with the relative populations of plastic and elastic contacts in the cluster under consideration. Stabilities of other emergent mesoscopic structures are also reported in Tordesillas and Lin [64].

Techniques from graph theory and complex networks are also being explored in [61, 62, 63]. In the main, a unique aspect of these studies is that they are designed to characterize the *evolution* of contact and contact force networks – specifically with respect to the stability and failure by buckling of laterally confined force chains. To illustrate, we highlight two example networks from these studies: L-expansion network and K-core network. For the biaxial test in [54], the majority of particles in low order cycles, as determined from the conditional L-expansion network where $L = 2,3$, are also members of the force chain network: see Figure 9(a). L-expansion networks act as a network filter: by design, the resulting networks only contain $L+1$-cycles, i.e. closed self-avoiding path of length $L+1$ in a graph [71]. We investigated the *coevolution* of these low order cycles and force chains. The degradation of 3-cycles, and not of 4-cycles, among force chains was found to be that which results in the loss of stability of force chains: force chains rich in 3-cycles are relatively stable and more resistant to buckling than those deplete of 3-cycles [63]. K-core networks are also revealing of the loss of internal connectivity and associated spread of instability in the deforming material. This network filter is obtained by pruning all the vertices (particles) and their respective edges (contacts) with degree less than K until all that is left is the core network of particles with contacts greater than or equal to K [72]. The 3-core network for the same system examined in [54] at the strain state just after the peak shear stress (i.e. $|\varepsilon_{yy}| = 0.0361$) is shown in Figure 9(b), together with all the CBCFs that have emerged in the period leading up to and including this strain state. Interestingly, this strain state is the very first point in the loading history of the material when the 3-core network completely separates into two distinct networks. Based on other indicators of shear band evolution, as discussed in [54, 56], the 3-core network captures remarkably well the full development of the shear band and the onset of the critical state regime.

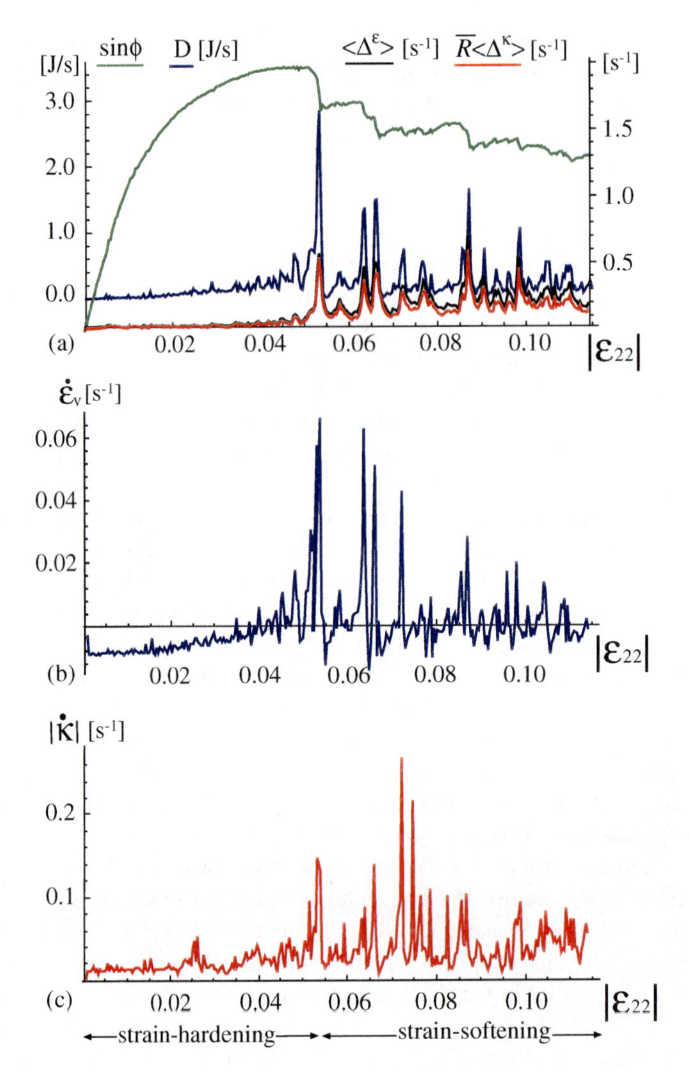

Fig. 8 (color online) (a) Strain evolution of macroscopic stress ratio sinϕ and global energy dissipation rate D together with the global averages of local nonaffine motion $<\Delta^{\varepsilon}>$ and $\bar{R}<\Delta^{\kappa}>$, where $\bar{R}$ and $<.>$ denote average particle radius and the ensemble average operator. Peak sinϕ is 0.66 and occurs at $|\varepsilon_{22}| = 0.05$. Strain evolution of the (b) global volumetric strain rate $\dot{\varepsilon}_v$, and (c) magnitude of the global curvature rate $|\dot{\kappa}|$. All rates are measured over strain intervals $|\Delta\varepsilon_{22}| = 3.85 \times 10^{-4}$.

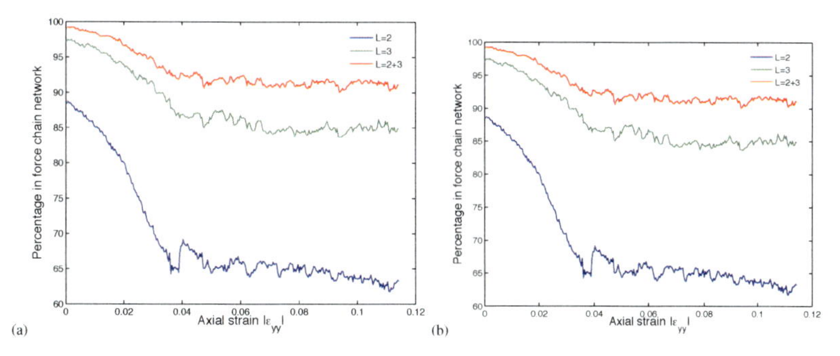

Fig. 9 (color online) (a) The evolution with axial strain $|\varepsilon_{yy}|$ of the percentage of particles in 3,4-cycles that also belong to the force chain network. The peak shear stress is at around $|\varepsilon_{yy}| = 0.034$. (b) The 3-core network (green) at the onset of critical state when the shear band has just become fully developed. The spatial distribution of all the CBFCs that have emerged prior to this point is also shown: red particles are the buckled force chains (buckling threshold angle used is 1°) and blue particles are their confining neighbors.

3.2 Modeling of force chain evolution

3.2.1 Localized buckling of a force chain

When analyzed within the framework of structural mechanics, the response of a constrained granular medium under load bears a number of familiar characteristics to those of typical engineering structures. In particular, loss of initial stability can often be captured at a point of bifurcation. This renders such systems open to study from the point of view of nonlinear bifurcation theory, for example, as applied to highly-nonlinear plate and shell problems and stemming notably from the work of Koiter [73]. In Hunt et al. [58], we employ these techniques to model the elastic buckling of a force chain consisting of uniformly sized circular particles. Resistance to relative rotation and sliding between force chain particles operates at the contacts. Lateral support is also provided to each particle. Examination of the critical buckling load shows that force chains buckle initially into periodic shapes, the wavelengths of which depend on the lateral support. A parallel analysis on rigid struts highlights similarities that suggest the possibility for this buckling mode to evolve into a localized form, extending over a finite number of particles, under increasing load. It is conjectured that this localization represents an evolutionary route to the shear band in granular materials.

3.2.2 Initial force chain buckling and onset of shear banding

The connection between force chain buckling and shear banding is examined in more detail in Tordesillas et al. [59], with particular attention paid to the hypothesis made by Thornton & Zhang [74] that the *"the initiation of shear bands is an elastic buckling problem*". Accordingly, a linearised elastic buckling analysis is employed to determine the minimum critical buckling load and the corresponding buckling mode m_{min} for a force chain of $N+2$ particles. We quantify the dependence of the half wavelength of this buckling mode, $(N+1)/m_{min}$ particles, and the minimum critical load on material properties, boundary conditions and the number of particles in the chain. In particular, we explore a possible link between the half wavelength of this buckling mode and the thickness of the shear band by imposing kinematical conditions known to apply at the boundaries of and inside the shear band. Two types of boundary conditions for the force chain are investigated, as illustrated in Figure 10: pin-ended condition in which the center of the particle at one end is fixed while that of the particle at the other end is confined to vertical translation only (C1 condition); particles at both ends are prevented from rotating (C2 condition). Figure 11 illustrates a possible connection between the buckled configuration of the force chain for C2 and the shear band thickness. Note that we observe force chain configurations of the forms depicted in Figure 10(b) and (c) in DEM simulations (e.g. Tordesillas [54]).

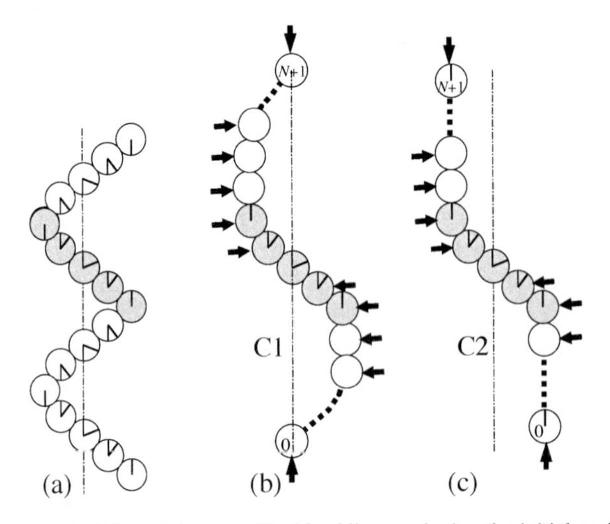

Fig. 10 (a) A segment of the minimum critical buckling mode, i.e. the initial periodic buckling profile of the force chain. Possible corresponding localized elastic buckling configurations for two force chain boundary conditions (b) C1 and (c) C2. End particles 0 and $N+1$ do not move laterally in C1, and do not rotate in C2.

The minimum critical buckling load provides a measure of the load-carrying capacity of the force chain. In general, as can be seen in Figure 12, this load decreases as the contact stiffness parameters decrease. These results also suggest that the degree of influence of the stiffness parameters on the critical buckling depend on the wave number. For those buckling modes with small wave numbers, the critical buckling load is mainly influenced by the stiffness of the lateral support k^s; the lower the stiffness of this support, the wider the shear band (Figure12(a)). For those buckling

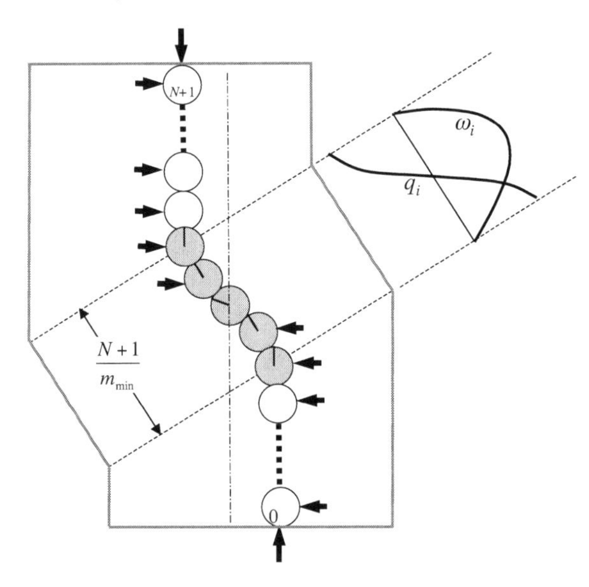

Fig. 11 Depiction of elastic localized buckling of a force chain of particles $i = 0, \ldots\ldots, N+1$ with localized buckled segment spanning shear band width. Boundaries of the shear band lie along particle centres. Included is a schematic of the rotation ω_i and displacement q_i distributions across the band.

modes with large wave numbers, the critical buckling load is greatly influenced by the stiffness of the tangential contact force k^t (Figure 12(b)); the lower the stiffness of the tangential force between particles, the thinner the shear band. The effect of rolling stiffness k^r on the critical buckling load resides mainly on those modes with intermediate wave numbers (Figure 12(c)); however, its influence on band thickness is more complex and is not monotonic. Predicted band thicknesses, according to the half wavelength of the minimum critical buckling mode, i.e. $(N+1)/m_{min}$ particles, were found to be consistent with those observed in DEM simulations for the same parameter values. Figure 13 shows the generally weak effect of the total number of force chain particles on the minimum critical load. For C1, the minimum critical buckling load remains essentially invariant with respect to the total number of particles in the force chain; a slightly greater influence by the total number of particles on this critical load is evident for C2 conditions. We emphasize here that this work is in its nascent stages and the trends reported here are valid only for the specific range of parameters examined thus far. Work is ongoing to determine how robust these trends are, and with an eye to broadening the scope of application of these models.

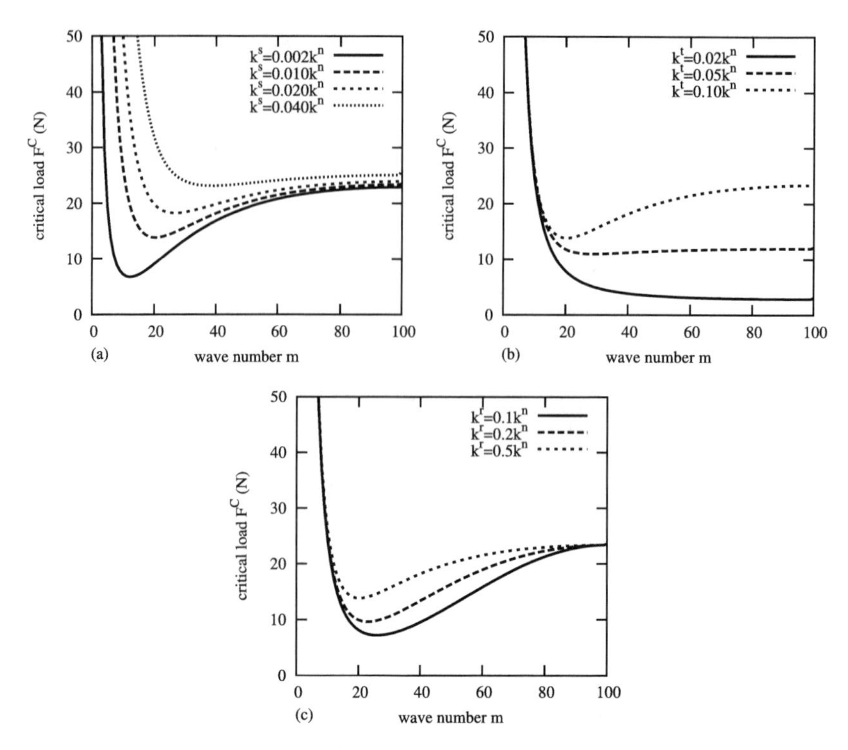

Fig. 12 Variation of critical loads with wave number of buckling modes for different parameters (a) k^s, (b) k^t, (c) k^r, for C1 boundary conditions and 101 particles. Reference values of parameters: $R = 0.00114m$, $k^s = 0.01k^n$, $k^t = 0.1k^n$, $k^r = 0.5k^n$ with $k^n = 1.0 \times 10^5 (N/m)$.

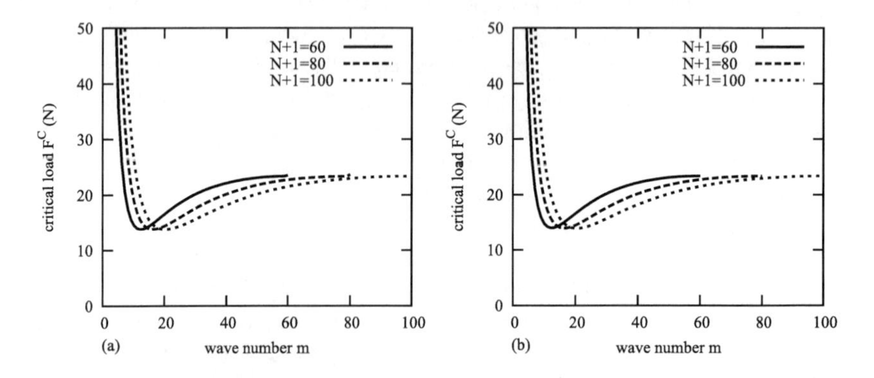

Fig. 13 Variation of critical loads with wave number of buckling modes for two different types of boundary conditions: (a) boundary conditions C1, (b) boundary conditions C2, for force chains with different numbers $(N+1)$ of particles. Values of parameters: $R = 0.00114m$, $k^s = 0.01k^n$, $k^t = 0.1k^n$, $k^r = 0.5k^n$ with $k^n = 1.0 \times 10^5 (N/m)$.

3.2.3 Elastic-plastic buckling of a force chain

The analysis discussed in Sections 3.2.1 and 3.2.2 is confined to elastic buckling and give valuable new insights into the possible mechanics of the initiation of shear bands. We now turn our attention to the elastic-plastic buckling of force chains which dominates rheological behavior in the ensuing regime – from the initiation of the shear band through to the critical state (Tordesillas and Muthuswamy [57]).

In [57], the force chain is considered to be loaded under a controlled vertical displacement (compression). The force chain is represented by a columnar structure consisting of three circular, deformable particles of radius R. The generalized case of a force chain of arbitrary length is the subject of an upcoming paper and will not be reported on here due to space limitations. There are several reasons for considering the 3-particle force chain structure and its confining neighbors: we discuss these in detail in [57]. Among these is its consistency with the size of the representative volume element used in constitutive formulations to be discussed in Section 4. In accordance with the experiments and DEM simulations in [44, 54], sliding at contacts between force chain particles during buckling is assumed to be insignificant. Thus, there are two sources of energy dissipation: frictional rolling at the contacts along the force chain, and all possible modes of frictional dissipation in the laterally supporting weak network. To account for the former, we assume that the three force chain particles interact via contact laws consisting of: linear springs, for the normal and tangential forces, and a spring-slider for the contact moment, where the slider is used to represent the plastic response of the contact moment. To account for the latter, we model all possible sources of confining support from the weak network to the force chain as a single effective lateral supporting force, again modeled as a spring-slider to account for energy dissipation. The sources of confining support encompass the applied confining pressure, normal and tangential forces, and moments operating at contacts between weak network and between weak network and force chain particles.

By varying particle-scale properties, we shed new light on the interconnections between material behavior at three length scales: the microscopic or particle scale, the mesoscopic event of confined buckling of force chains, and the observed behavior on the macroscale. Using this elastic-plastic buckling model, we have shown explicitly and quantitatively the effect of resistance to relative particle rotations on macroscopic strength. The introduction of rolling resistance serves to stabilize force chains, both directly and indirectly. It increases the stability of the force chain structure directly, by providing greater resistance to relative rotation which is an underpinning mechanism of buckling. Thus, it enables force chains to withstand more load and store more energy, before the onset of instability or critical buckling load: see Figure 14. Rolling resistance also enhances force chain stability via the lateral supporting force. For very small rolling resistance, the plastic buckling regime is reached rapidly, resulting in a force-displacement response exhibiting a snapdown path. However, the addition of an initial imperfection smooths this path, and results in the peak stress increasing with increasing rolling friction. These findings elucidate why, as observed in DEM simulations of biaxial tests where force chains are rarely if ever perfectly straight, both the macroscopic peak shear stress and the steady critical-state shear stress increase with increasing rolling friction [54]. Moreover, the model can reliably reproduce the trends observed in DEM on the progression of plastic buckling. As rolling friction is increased, the origin of plastic buckling switches from the force chain column to the laterally supporting weak network contacts. This is consistent with DEM data which show that, as rolling friction is increased, so does the population of buckling segments where the first contact to reach plastic threshold belongs to the laterally supporting weak network [60].

The model in [57] has also shed light on the controversial issue of the use of rolling resistance or contact moment in DEM simulations involving circular or spherical particles [2, 3]. In the absence of rolling resistance in DEM simulations, the tangential forces within the force chain and surrounding weak network particles, along with the confining pressure, are the elements that provide the resistance to force chain buckling. This explains why shear bands can still form in these systems. This study has quantified the tremendous influence that the lateral support, provided by the weak network particles, has on the stability of force chains. The degree of lateral support determines the extent to which force chains can persist before buckling: the greater this support, the higher are the critical buckling load and the displacement at the critical buckling load. Moreover, a higher plastic threshold for the lateral force means less plastic displacement and dissipation, and hence a relatively higher post-critical load-carrying capacity. These results thus yield new insights into the underlying causes of several well-known trends observed under increasing confining pressure: peak shear stress, strain at peak shear stress, steady critical-state shear stress all increase while the rate of softening decreases (e.g. [75, 76]).

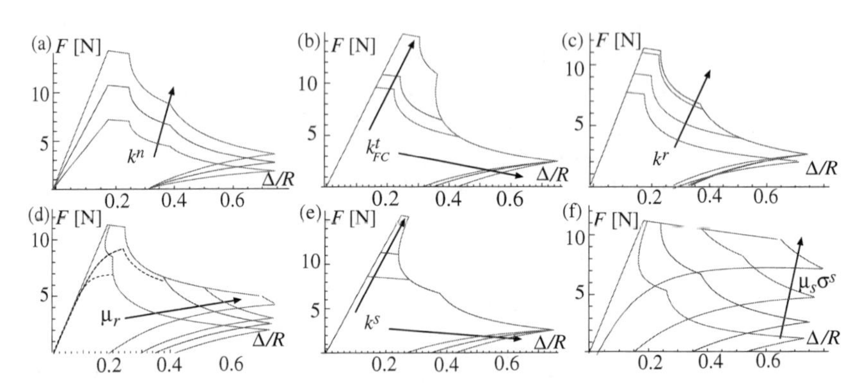

Fig. 14 Effect on the force-displacement curve, F versus Δ/R, of varying each parameter. Arrows point in direction of increasing parameter value. Dotted lines in (d) show model with initial imperfection angle of 2°. See [57] for a complete list of parameter values.

Finally, above and beyond the establishment of fundamental interconnections in material behavior across multiple length scales, the analysis presented in Tordesillas and Muthuswamy [57] has yielded a missing key element in micromechanical constitutive theory: a completely micromechanical formalism for the so-called internal variables and their evolution laws. In what follows, the significance of the studies presented here and in the previous section is discussed from the vantage point of constitutive theory.

4 Implications for constitutive theory

The gauntlet thrown down to practitioners of the continuum micromechanics of granular materials has been the development of constitutive laws, without need for any phenomenological parameters above the particle length scale. From the standpoint of mathematical modeling alone, this challenge is fraught with many difficulties, not least of which is that mechanisms governing bulk behavior are inherently multiscale. Thus, if and when such mechanisms are identified and quantitatively characterized, there arises an even greater problem of establishing a purely micromechanical formulation that engenders 'physical transparency' in the transition from the particle to the bulk. Key to this problem is the formulation for energy dissipation or plastic response of the material [54]. Unlike classical plasticity theory in which dissipative mechanisms are all lumped into one variable, i.e. the plastic strain, a purely micromechanical formalism demands that the fundamental origins of dissipation be explicitly identified in order for internal variables to be given clear physical meaning. In Tordesillas and Muthuswamy [57, 60], the kinematics of key internal events that involve loss of connectivity in the local material domain, i.e. rearrangement events, is explored as a means to establish a micromechanical formalism for internal variables.

In deforming granular materials, the rotational degrees of freedom of particles lead to a richness in kinematics that significantly complicates rheological behavior (e.g. [2, 77, 78, 79]). For these systems, no clear rearrangement event prevails throughout deformation. However, as demonstrated previously in Sections 2 and 3, experiments and discrete element (DEM) studies lend considerable credence to Oda's hypothesis that confined buckling of force chains is a governing event for dissipation and failure of dense, cohesionless granular systems. Accordingly, the new approach for constitutive modeling expounded below embodies *internal variables that represent the nonaffine deformation of the dissipative mechanism of confined force chain buckling*.

4.1 Thermomicromechanics of a Cosserat Continuum

The thermomicromechanical formulation in [81] was developed for a Cosserat or Micropolar continuum representation of a granular material. Its underpinning homogenization scheme is on the scale of a particle and its first ring of neighbors. The formulation proceeds in three steps, as illustrated in Figure 15. Step 1 re-

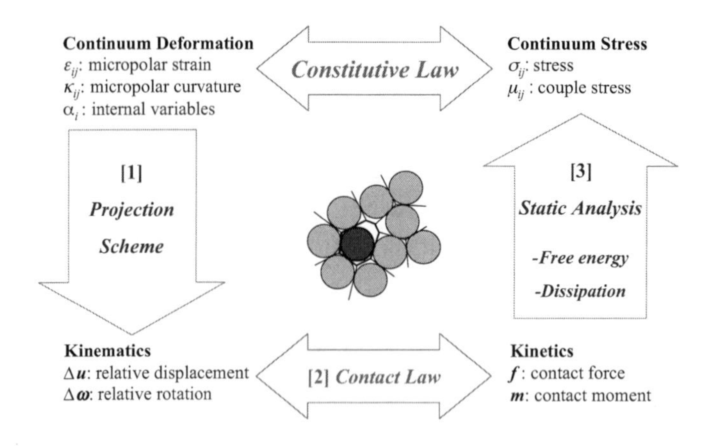

Fig. 15 (color online) The thermomicromechanical approach to constitutive modeling. Inset figure shows the scale of the underpinning homogenization scheme, and associated Voronoi tessellation, i.e. a particle and its first ring of neighbors. Resulting constitutive law is of a micropolar type.

lates the continuum micropolar deformation quantities of strain, ε_{ij}, and curvature, κ_{ij}, to the motions of the particle centers. Step 2 links the particle motions to the forces and moments at their contacts via a contact law. Step 3 connects these forces and moments to the stress, σ_{ij}, and couple stress, μ_{ij}, via the stored free energy $\psi = \psi(\varepsilon_{ij}, \kappa_{ij}, \alpha^1, ..., \alpha^n)$: $\sigma_{ij} = \partial\psi/\partial\varepsilon_{ij}; \mu_{ij} = \partial\psi/\partial\kappa_{ij}$ where $\alpha^1, ..., \alpha^n$ are a set of internal variables (tensors, vectors, or scalars). The α^i represent dissipative mechanisms responsible for the loss of stored free energy. To close the governing system of equations, each requires an evolution law that satisfies $\partial\psi/\partial\alpha^i \otimes \delta\alpha^i \leq 0$, in accordance with the 2nd Law of Thermodynamics. Note here that this inequality applies to individual internal variables, hence repeated indices do not indicate a summation. Complete details of this approach can be found in [81, 82] and the references cited therein.

The central challenge lies in Step 1. The majority of continuum models employ the assumption of an affine deformation field: the relative contact displacement Δu_i is expressed in terms of a single strain and curvature tensor. In two dimensions, we have

$$\Delta u_i = 2R\varepsilon_{ij}n_j + 2R^2 e_{ij3}\kappa_k n_j n_k, \tag{11}$$

where R is the particle radius, n_i the contact normal vector and e_{ij3} the permutation symbol. However, past studies have shown that this type of deformation is rarely observed in dense granular materials (e.g. [77, 16, 83, 18, 20] and references cited therein). In particular, such a constraint on the kinematics precludes various instabilities that emerge on the mesoscopic domain and, as such, may lead to models that: (a) overestimate the strength and stability of the material, and (b) fail to capture strain-softening under dilatation. One way of capturing nonaffine deformation is to decouple the relative normal and tangential displacements [82]. When combined with rotations, these yield

$$\Delta u^{en} = 2R(\varepsilon_{ij} - \varepsilon_{ij}^{pn})n_i n_j; \tag{12}$$

$$\Delta u^{et} = 2R(\varepsilon_{ij} - \varepsilon_{ij}^{pt})t_i n_j + 2R^2(\kappa_i - \kappa_i^p)n_i; \tag{13}$$

$$\Delta \omega^e = 2R(\kappa_i - \kappa_i^p)n_i \tag{14}$$

where t_i is the contact tangential vector, and the three internal variables introduced are: the normal plastic strain ε_{ij}^{pn}, the tangential plastic strain ε_{ij}^{pt}, and the plastic curvature κ_i^p. These variables represent nonaffine deformation associated with the normal, tangential, and rotational motion, respectively. The nonaffine deformation provides a measure of the fluctuating field, specifically, the deviation of actual particle motions from those dictated by the local measures of strain and curvature. The resulting constitutive laws are:

$$\sigma_{ij} = \frac{2R^2}{V}\{k^n(\varepsilon_{kl} - \varepsilon_{kl}^{pn})\int_\Omega n_i n_j n_k n_l \Phi(\mathbf{n})d\mathbf{n} + k^t(\varepsilon_{kl} - \varepsilon_{kl}^{pt})\int_\Omega t_i n_j t_k n_l \Phi(\mathbf{n})d\mathbf{n}\}, \tag{15}$$

$$\mu_i = \frac{2R^4}{V}\{(k^t + k^r)(\kappa_j - \kappa_j^p)\int_\Omega n_i n_j \Phi(\mathbf{n})d\mathbf{n}\}, \tag{16}$$

where V is the area of the Voronoi polygon for the particle, $\Phi(\mathbf{n})$ is the contact density distribution function with $\mathbf{n}$ being the contact normal vector, k^n, k^t, k^r are spring stiffness constants associated with normal and tangential forces and contact moment (rolling resistance) respectively, and δ_{ij} is the Kronecker delta [82]. The quantity $\Phi(\mathbf{n})d\mathbf{n}$ gives the probability of finding a contacting neighbor within $d\mathbf{n}$ of $\mathbf{n}$, and the range of integration Ω in equations (15) and (16) is taken over all possible orientations of the contact normal $\mathbf{n}$ around a particle. Evolution of contacts is accounted for based solely on the contact laws, in accordance with [84]. Thus, with the exception of the internal variables and their corresponding evolution laws, which are yet to be defined, the material parameters in the constitutive relations in equations (15) and (16) are entirely expressed in terms of particle scale properties. In [82], the recourse to a curve-fitting analysis of DEM data on bulk properties to obtain relations for the internal variables and their evolution laws leads to a constitutive law that is, strictly speaking, phenomenological. However, the important point is that the deviation from the micromechanical formulation is now isolated and confined to the internal variables and their evolution laws: see Figure 16. We now discuss how this missing piece of the puzzle can be established directly from a structural mechanics analysis of confined force chain buckling.

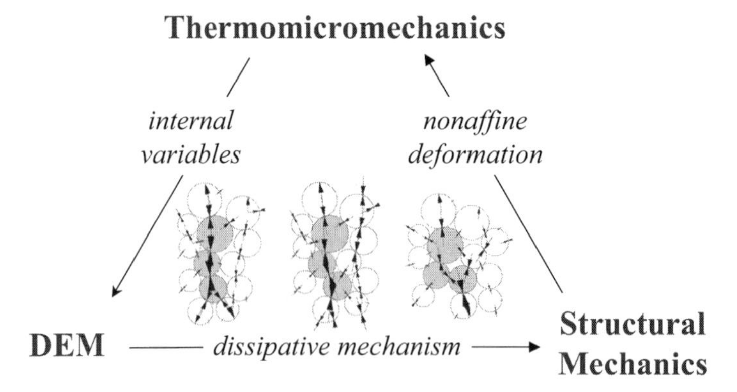

Fig. 16 Theoretical framework for modeling materials in which energy dissipation is primarily governed by structural rearrangements on the mesoscopic scale. Inset shows the buckling of a force chain (grey particles) under lateral confinement from surrounding weak network neighbors, taken from a DEM simulation.

4.2 Constitutive model

Having established a strong correlation between dissipation, nonaffine deformation and buckling events, the kinematics of buckling of force chains under lateral confinement from their supporting neighbors can be now derived using structural mechanics: recall Section 3.2. Internal variables are henceforth given clear physical meaning: the nonaffine deformation of confined buckling of a force chain. In [57], we use a thermodynamics-based formulation to examine the plastic buckling regimes, thereby ensuring a set of evolution laws for the internal variables, that is in compliance with the Second Law of Thermodynamics. The analysis in [57] yielded explicit expressions for the positions and rotations of, and hence contact forces and moments between, the particles, in terms of the degree of buckling, or 'buckling angle' θ that was discussed in Section 3.1.4. As the force chain is axially compressed, four different stages of the elastic-plastic buckling process are encountered: pre-buckling stage prior to critical buckling load, elastic buckling, initial plastic buckling stage during which the contacts in either the force chain column or those in its lateral support remain elastic and, finally, the fully plastic stage when contacts in the lateral support and the force chain column have reached their plastic threshold. The evolution laws for the nonaffine deformation during these stages of buckling embody the coupling between the deviatoric and volumetric strains required to capture dilatancy – the missing micromechanical component of the thermomechanical formulation proposed in [82]. Due to space limitations, the reader is referred to [57] for details of the governing equations for the various stages of buckling, and to [60] for the complete derivation of the resulting thermomicromechanical constitutive law.

4.3 Predictive Capabilities of Model Based on Force Chain Failure

Some preliminary results that demonstrate the potential of the constitutive formulation discussed in the preceding section are now summarised. Particular attention is paid to the following interrelated processes: strain-softening under dilatation, the development of noncoaxiality, and shear band formation and evolution.

The constitutive law successfully captures the defining process of strain-softening under dilatation at both the mesoscopic and macroscopic level. The evolution of this process at the mesoscopic scale is presented in Figure 17(a). The volumetric component of nonaffine deformation (ε_v^{pn}) is shown, demonstrating the model's behavior at all five stages of deformation which includes unloading. The model captures an elastic regime, during the pre-buckling stage in which nonaffine deformation is negligible. Nonaffine deformation essentially takes effect at the onset of buckling.

The model has been implemented to model macroscopic response using bifurcation theory [85] and finite volume method in [87]. In particular, we investigated the development of noncoaxiality between the principal stress and strain rate directions in [87]. At the mesoscopic level of a confined buckling of a force chain, the principal strain rate direction remains relatively unchanged, while the principal stress direction rotates away from this direction as buckling proceeds. Thus the noncoaxility at the mesoscopic scale is due to the rotation of the principal stress direction. At the macroscopic level, the development of noncoaxiality has been examined for: (i) the shear band region in the biaxial compression test, and (ii) the global specimen undergoing simple shear deformation. For case (i), both the directions of principal stress and principal strain rate rotate towards the direction orthogonal to the band. Finite noncoaxiality occurs at the onset of shear banding and the angle of noncoaxiality between the principal stress and strain rate directions during shear band development remains relatively constant. The development of noncoaxiality in a simple shear test, case (ii), is shown in Figure 17(b). The principal stress direction initially rotates rapidly away from the direction of applied shear, and then settles to a near constant value. Similar trends were observed in [74]. The evolution of noncoaxiality on the macroscale reflects the collective response of a complex network of force chains in the system, as demonstrated in [30] and discussed earlier in Section 2.2. This highly ramified network continually evolves with strain, and exhibits strongly intermittent behavior due to the continual collapse of old and generation of new force chains. At any given strain state, the force chains in the system are oriented in different directions, are subject to different loading conditions (some force chains are under axial compression, while those in the direction of shear are under extension), and those under axial compression in the strain-softening regime are at different stages of buckling. The strong primary force chains, which are those that resist and are directed in the opposite direction to the shear, undergo confined buckling during shearing. At least in the systems examined in [87], the rotation of the principal stress direction on the macroscale reflects the collective rotation of the principal stress directions of these primary force chains during buckling.

In [85], the onset and evolution of shear banding has been explored in the context of one-dimensional bifurcation theory in accordance with [86]. As shown in Figure

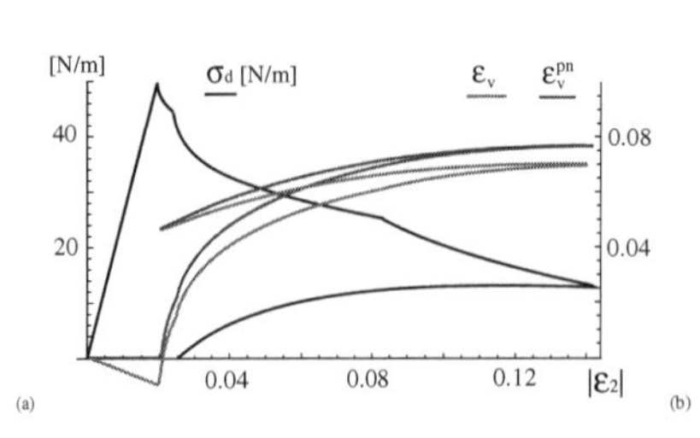

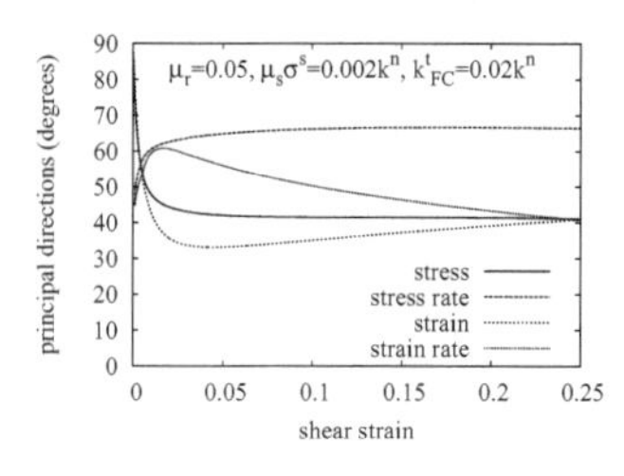

Fig. 17 (color online) (a) Deviatoric stress σ_d and volumetric components of strain ε_v and non-affine strain ε_v^{pn} against $|\varepsilon_2|$. See [60] for parameter values. (b) Variation with shear strain of the principal directions at the centerline for material under simple shear. Applied pressure is 50,000 Pa; inset shows some parameter values and the other parameter values are $R = 0.00114m$, $k^n = 1.05 \times 10^7 N/m$, $k^t = 0.5k^n$, $k^r = 0.5k^n$, $k^s = 0.05k^n$.

18, the band thickness, distribution of rotation and the emergent evolution of normal contact force anisotropy are in good agreement with those from DEM biaxial compression tests displaying a single shear band [54, 2]. Strain softening is achieved on the macroscopic scale, as can be seen from the decrease in maximum normal contact force (Figure 18(c)). Although not shown here, this strain-softening is accompanied by dilatation inside the band, similar to the behavior displayed in Figure 17(a). The direction of rotation of the highly anisotropic normal contact force distribution is consistent with the angle of inclination of the shear band. As seen from Figures 18(b) and (c), buckling results in the rotation of force chains inside the band region, and is thus consistent with the direction of rotation of the normal contact force distribution in a single band, here inclined at 46° to the minor principal stress axis. Compared to the DEM simulation, the bifurcation model reasonably predicts the degree of rotations across the band, except for the centerline. Particle rotations in the DEM are calculated and averaged over all particles at a fixed distance along the axis perpendicular to the band. Here, we observed significant levels of particle rotation along the centerline of the band where the local void ratio is relatively high, especially amongst rattlers (i.e. particles which have at most one contact).

The above exposition describes the role of models of mesoscopic force chain buckling in constitutive theory. However, the merits of introducing these to other approaches for modeling of dense granular materials, e.g. cellular automata [88, 89, 89] have also been probed. This is especially relevant in the context of granular materials under quasi-static shear, such as that discussed in Section 2.2, where the continual failure of old and formation of new force chains clearly govern rheology: see, for example, [66, 68, 74, 77, 78, 80]. Preliminary results, from a lattice model in [92] and a cellular automaton which captures the collective failure of old and formation of new force chains in [91], show promise. The core component of each of these models is the structural mechanics model of force chain buckling described in Section 3.2. While refinements are still ongoing, an important and unique feature of these new breed of discrete models, like that of the thermomicromechanical continuum model in [60], is physical transparency. By design, their formulation enables the explicit interconnections of emergent processes at multiple length scales to be unravelled and understood. Specifically, the cellular automaton in [91] not only reproduces patterns of stick-slip consistent with those observed in recent experiments [66], but also offers the advantage that macroscopic trends can be directly tied to underlying mechanisms on the microscopic and mesoscopic levels.

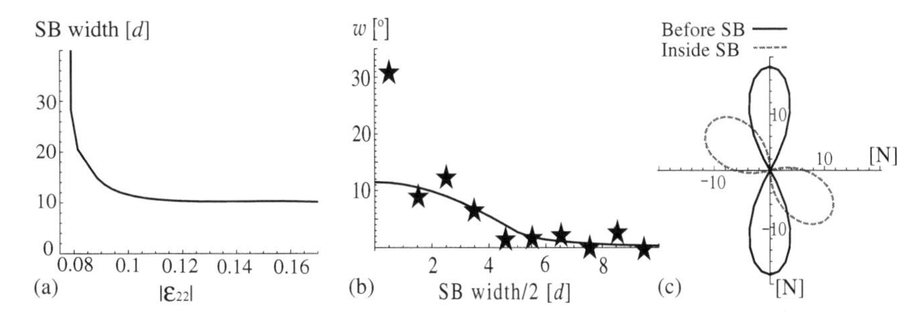

Fig. 18 (a) Shear band (SB) width, measured in particle diameters (d) against axial strain magnitude $|\varepsilon_{22}|$. (b) Rotation, symmetric about the central axis of the SB, with lines and stars respectively showing analytical results and data from DEM simulation with identical parameters [54]. (c) Angular distribution of normal contact force for system in (a), (b). See [60] for a complete list of parameter values.

5 Conclusion

We have reviewed some recent advances in experiment, simulation and theory with an eye towards several key aspects of microscopic and mesoscopic processes emerging in granular shear and failure. We have demonstrated the significance and role of nonaffine deformation and force chain evolution, particularly within shear bands and under conditions that generate stick-slip behavior. Attempts to reconcile theoretical descriptions of these processes at different length scales have been described. While many of these studies are still ongoing, important headways have been achieved in relating the theoretical advances to experimental observations. One approach for micromechanical constitutive development has been discussed, demonstrating how results from experiments and simulation of microscopic and mesoscopic processes can be directly introduced into a thermomechanical framework. By design, resulting models bear physical transparency, allowing explicit interconnections of emergent processes at multiple length scales to be established. Some glimpses of the promise of this approach have been presented in the context of defining aspects of granular rheology: shear bands, strain-softening under dilatation and noncoaxiality. *We are not there yet* – but some pathways forward have been framed.

Acknowledgements AT acknowledges the support of the Australian Research Council (DP0772409) and the US Army Research Office (W911NF-07-1-0370). RPB acknowledges the support of the US Army Research Office (W911NF-07-1-0131), NSF grants DMR-0555431 and DMR-0906908, and LANL Subcontract Number: 64898-001 -08.

References

1. M. Oda and H. Kazama. Microstructure of shear bands and its relation to the mechanisms of dilatancy and failure of dense granular soils. *Géotechnique*, 48(4):465-481, 1998.
2. K. Iwashita and M. Oda. Micro-deformation mechanism of shear banding process based on modified distinct element method. *Powder Technology*, 109:192-205, 2000.
3. M. Oda and K. Iwashita. Study on couple stress and shear band development in granular media based on numerical simulation analyses. *International Journal of Engineering Science*, 38:1713-1740, 2000.
4. M. Oda, T. Takemura and M. Takahashi. Microstructure in shear band observed by micro-focus x-ray computed tomography. *Géotechnique*, 54(8):539-542, 2004.
5. F. Radjai, D.E. Wolf, M. Jean and J.J. Moreau. Bimodal character of stress transmission in granular packings. *Physical Review Letters*, 80:61-64, 1998.
6. S.J. Antony. Evolution of force distribution in three-dimensional granular media. *Physical Review E*, 63(1):1-13 2000.
7. T.S. Majmudar and R.P. Behringer. Contact force measurements and stress-induced anisotropy in granular materials. *Nature*, 435:1079-1082, 2005.
8. J.S. Lee, J. Dodds and J.C. Santamarina. Behavior of rigid-soft particle mixtures. *Journal of Materials in Civil Engineering, ASCE*, 19(2):179-184, 2007.
9. M. Falk and J. Langer, Dynamics of viscoplastic deformation in amorphous solids *Physical Review E* 57:7192–7205 (1998).
10. S. Luding, J. Phys. Anisotropy in cohesive, frictional granular media *Condensed Matter* 17-S2623–S2640 (2005).
11. C. Maloney and A. Lemaître, Subextensive Scaling in the Athermal, Quaistatic Limit of Amorphous Matter in Plastic Shear Flow *Physical Review Letters* 93:016001 (2004).
12. C. Maloney and A. Lemaître, Universal Breakdown of Elasticity at the Onset of Material Failure *Physical Review Letters* 93:195501 (2004).
13. M. J. Demkowicz and A. S. Argon, Autocatalytic avalanches of unit inelastic shearing events are the mechanism of plastic deformation in amorphous silicon *Physical Review B* 72:245206 (2005).
14. F. Leonforte, R. Boissière, A. Tanguy, J. P. Witmer, and J.-L. Barrat, Continuum limit of amorphous elastic bodies. III Three-dimensional systems *Physical Review B* 72:224206 (2005).
15. A. Lemaitre, Origin of a Repose Angle: Kinetics of Rearrangements for Granular Materials *Physical Review Letters* 89:064303 (2002).
16. A. Lemaitre, Rearrangements and Dilatancy for Sheared Dense Materials *Physical Review Letters* 89:195503 (2002).
17. M. L. Falk, M. Toiya, and W. Losert, Shear transformation zone analysis of shear reversal during granular flow *arXiv:0802.0485* (2008).
18. B. Utter and R. P. Behringer, Self-diffusion in dense granular shear flow *Physical Review E* 69:031308 (2004).
19. B. Utter and R. P. Behringer, Transients in sheared granular matter *European Physical Journal E* 14:373–380 (2004).
20. B. Utter and R. P. Behringer, Experimental Measures of Affine and Nonaffine Deformation in Granular Shear *Phys. Rev. Lett.* 100:208302 (2008)
21. F. Rouyer and N. Menon, Velocity Fluctuations in a Homogeneous 2D Granular Gas in Steady State *Physical Review Letters* 85:3676–3679 (2000).

22. W. Losert, D. G. W. Cooper, J. Delour, A. Kudrolli, and J. P. Gollub, Velocity statistics in excited granular media *Chaos* 9:682 (1999).
23. D. Fenistein, J. W. van de Meent, and M. van Hecke, Universal and Wide Shear Zones in Granular Bulk Flow *Phys. Rev. Lett.* 92:094301 (2004).
24. C. Marone, C. B. Raleigh and C. H. Scholz, Frictional Behavior and Constitutive Modeling of Simulated Fault Gouge *J. Geophysical Reserch* 95:7007–7025 (1990).
25. M. H. Müser, M. Urbakh and M. O. Robbins, Statistical mechanics of static and low-velocity kinetic friction *Advances in Chemical Physics* 126:187–272 (2003).
26. S. Nasuno, A. Kudrolli and J. P. Gollub, Friction in granular layers: Hysteresis and Precursors *Phys. Rev. Lett.* 79:949 (1997).
27. S. Nasuno, A. Kudrolli and J. P. Gollub, Sensitive force measurements in a sheared granular flow with simultaneous imaging *Powders & Grains 97* p.329., R. P. Behringer and J. T. Jenkins, eds., Balkema, Rotterdam (1997).
28. W. Losert, J.-C. Geminard, S. Nasuno and J. P. Gollub, Mechanisms for slow strengthening in granular materials *Physical Review E* 61:4060 (2000).
29. S. Siavoshi, A. V. Orpe and A. Kudrolli Friction of a slider on a granular layer: Nonmonotonic thickness dependence and effect of boundary conditions *Physical Review E* 73:010301(R) (2006).
30. Peidong Yu and R. P. Behringer, Granular Friction: A Slider Experiment *Chaos* 15:041102 (2005).
31. K. E. Daniels and N. W. Hayman, Force chains in seismogenic faults visualized with photoelastic granular shear experiments *Journal of Geophysical Research* 113:B11411 (2008).
32. J. H. Dieterich, Time-dependent friction in rocks *J. Geophysical Research* 77:3690–3697 (1972).
33. A. Ruina, J. Slip instatility and slip variable friction laws *Geophysical Research* 88:10359–10370 (1983).
34. T. Baumberger, P. Berthoud and C. Caroli, Physical analysis of the state- and rate-dependent friction law. II. Dynamic friction *Physical Review B* 60:3928-3939 (1999).
35. F. P. Bowden and D. Tabor, *The Friction and Lubrication of Solids*, Oxford University Press (1954).
36. F. Heslot, T. Baumberger, B. Perrin, B. Caroli and C. Caroli, Creep, stick-slip, and dry-friction dynamics: Experiments and a heuristic model *Physical Review E* 49:4973–4988 (1994).
37. T. S. Majmudar and R. P. Behringer, Contact force measurements and stress-induced anisotropy in granular materials *Nature* 435:1079 (2005).
38. T. S. Majmudar, M. Sperl, S. Luding and R. P. Behringer, The jamming transition in granular systems. *Physical Review Letters* 98:058001 (2007).
39. J. Zhang, T. S. Majmudar, A. Tordesillas and R. P. Behringer, *Granular Matter* submitted (2009).
40. I. Albert et al. Stick-slip fluctuations in granular drag. *Physical Review E* 64:031307 (2001).
41. P. Bak, C. Tang and K. Wiesenfeld, Self-Organized Criticality: An Explanation of $1/f$ Noise. *Physical Review Letters* 59:381–384 (1987).
42. R. Burridge and L. Knopoff, *Bulletin Seismological Society of America* 57:3411 (1967).
43. J. M. Carlson, J. S. Langer and B. E. Shaw, Dynamics of earthquake faults. *Reviews Modern Physics* 66:657–670 (1994).
44. A. Tordesillas, J. Zhang and R. Behringer. Buckling force chains in dense granular assemblies: physical and numerical experiments. *Geomechanics and Geoengineering*, 4:3-16, 2009.
45. J. Zhang, J.R. Farhadi, R.P. Behringer, T.S. Majmudar and A. Tordesillas. A Dense 2D Granular Material Subject to Cyclic Pure Shear. *Powders and Grains*, 2009.
46. D.W. Howell, R.P. Behringer and C.T. Veje. Stress Fluctuations in a 2D Granular Couette Experiment: A Continuous Transition. *Physical Review Letters*, 82(26):5241-5244, 1999.
47. J.F. Peters, M. Muthuswamy, J. Wibowo and A. Tordesillas. Characterization of force chains in granular material. *Physical Review E*, 72:041307, 2005.

48. M. Muthuswamy and A. Tordesillas. How do interparticle contact friction, packing density and degree of polydispersity affect force propagation in particulate assemblies? *Journal of Statistical Mechanics: Theory and Experiment*, P09003, 2006.
49. A.A. Pena, H.J. Herrmann, P.G. Lind. Force chains in sheared granular media of irregular particles. *Powders and Grains*, 2009.
50. L. Sanfratello and E. Fukushima. Distribution of Force Chain Lengths: A Useful Statistic that Characterizes Granular Assemblies. *Physical Review E*, submitted, 2008.
51. M. Muthuswamy and A. Tordesillas. Multiscale analysis of the effects of changing gravity on stress propagation in a material subject to an indenting rigid flat punch. *Proc. of 10th ASCE Aerospace Division International Conference on Engineering, Construction and Operations in Challenging Environments (Earth & Space 2006)*, Aerospace Division of the American Society of Civil Engineers, 2006.
52. J.L. Anthony and C. Marone. Influence of particle characteristics on granular friction. *Journal of Geophysical Research*, 110:B08409, 2005.
53. A.D. McNaught and A. Wilkinson. IUPAC Compendium of Chemical Terminology. Royal Society of Chemistry, http://www.iupac.org/publications/compendium/ 1997
54. A. Tordesillas. Force chain buckling, unjamming transitions and shear banding in dense granular assemblies. *Philosophical Magazine*, 87(32):4987-5016, 2007.
55. A. Tordesillas, S.D.C. Walsh and M. Muthuswamy. The effect of local kinematics on the local and global deformations of granular systems. *Mathematics and Mechanics of Solids*, DOI:10.1177/1081286507089844, 2008
56. A. Tordesillas, M. Muthuswamy, and S.D.C. Walsh. Mesoscale measures of nonaffine deformation in dense granular assemblies. *Journal of Engineering Mechanics -ASCE*, 134(12):1095-1113, 2008.
57. A. Tordesillas and M. Muthuswamy. On the modeling of the confined buckling of force chains. *Journal of Mechanics and Physics of Solids*, 57:706-727, 2009.
58. G.W. Hunt, A. Tordesillas, S.C. Green and J. Shi. Force-chain buckling in granular media: a structural mechanics perspective. *Philosophical Transactions*, 2009; in press.
59. A. Tordesillas, G.W. Hunt, J. Shi and J. Hammond. Structural evolution of force chains. (submitted)
60. A. Tordesillas and M. Muthuswamy. A thermomicromechanical approach to multiscale continuum modelling of dense granular systems. *Acta Geotechnica*, 3(3):225-240, 2008.
61. A. Tordesillas, E. Lam and P.T. Metzger. Hyperstaticity and loops in frictional granular packings. arXiv:0906.1703v1 [cond-mat.stat-mech]
62. D.M. Walker and A. Tordesillas. Topological evolution of granular materials: connection to dissipation and stability. (submitted)
63. A. Tordesillas, Q. Lin and D.M. Walker Force cycles and force chains. (submitted)
64. A. Tordesillas and Q. Lin. Conformational structures and cluster rearrangements in dense granular media viewed from the standpoint of structural stability. (in prep)
65. A. L. Rechenmacher. Grain-scale processes governing shear band initiation and evolution in sands. *Journal of the Mechanics and Physics of Solids*, 54(1):22–45, 2006.
66. M. Knuth and C. Marone. Friction of sheared granular layers: Role of particle dimensionality, surface roughness, and material properties. *Geochemistry, Geophysics, Geosystems*, 8(3):Q03012, doi:10.1029/2006GC001327 2007.
67. R.G. Cain, N.W. Page and S. Biggs. Microscopic and macroscopic aspects of stick-slip motion in granular shear. *Physical Review E*, 64:016413, 2001.
68. I. Albert, P. Tegzes, R. Albert, A.L. Barabasi, T. Vicsek, B. Kahng and P. Schiffer. Stick-slip fluctuations in granular drag. *Physical Review E*, 64(3):031307, 2001.
69. D. Volfson, L.S. Tsimring and I.S. Aranson. Stick-slip dynamics of a granular layer under shear. *Physical Review E*, 69:031302, 2004.
70. K. Bagi. On the concept of jammed configurations from a structural mechanics perspective. *Granular Matter*, 9:109-134, 2007.
71. L. da Fontoura Costa. L-percolations of complex networks. *Physical Review E*, 70:056106, 2004.

72. S. N. Dorogovtsev, A. V. Goltsev and J.F.F. Mendes. k-core architecture and k-core percolation on complex networks. *Physica D*, 224:7–19, 2006,
73. W.T. Koiter. On the stability of elastic equilibrium. Ph.D. thesis, Tech- nische Hogeschool, Delft (Technological University of Delft), 1945 Holland. English translation issued as NASA, Tech. Trans., F 10, 833, 1967.
74. C. Thornton and L. Zhang. A numerical examination of shear banding and simple shear non-coaxial flow rules. *Philosophical Magazine*, 86(21-22):3425-3452, 2006.
75. D. Kolymbas and W. Wu. Recent results of triaxial tests with granular-materials. *Powder Technology*, 60(2): 99-119, 1990.
76. J. Desrues and G. Viggiani Strain localization in sand: an overview of the experimental results obtained in Grenoble using stereophotogrammetry. *International Journal for Numerical and Analytical Methods in Geomechanics*, 28: 279-321, 2004.
77. F. da Cruz, S. Emam, M. Prochnow, J.N. Roux and F. Chevoir. Rheophysics of dense granular materials: Discrete simulation of plane shear flows. *Physical Review E*, 72(2):021309, 2005
78. F. Dalton and D. Corcoran. Self-organized criticality in a sheared granular stick-slip system. *Physical Review E*, 63(6):061313, 2001.
79. B.S. Gardiner and A. Tordesillas. Micromechanical constitutive modelling of granular media: evolution and loss of contact in particle clusters. Journal of Engineering Mathematics, 52:93-106, 2005.
80. E. Aharonov and D. Sparks. Stick-slip motion in simulated granular layers. *Journal of Geophysical Research*, 109:B09306, 2004.
81. S.D.C. Walsh and A. Tordesillas. A thermomechanical approach to the development of micropolar constitutive models of granular media. *Acta Mechanica*, 167(3-4):145-169, 2004.
82. S.D.C. Walsh, A. Tordesillas, and J.F. Peters. Development of micromechanical models for granular media: The projection problem. *Granular Matter*, 9(5):337-352, 2007.
83. M.R. Kuhn. Heterogeneity and patterning in the quasi-static behaviour of granular materials. *Granular Matter*, 4(4):155-166, 2003.
84. B.S. Gardiner and A. Tordesillas. Micromechanical constitutive modelling of granular media: evolution and loss of contacts in particle clusters. *Journal of Engineering Mathematics*, 52(1):93-106, 2003.
85. A. Tordesillas and J. Shi. Micromechanical analysis of failure propagation in frictional granular materials. *International Journal for Numerical and Analytical Methods in Geomechanics*, doi: 10.1002/nag.792. 2009.
86. A. Tordesillas, J.F. Peters, and B. Gardiner. Shear band evolution and accumulated microstructural development in Cosserat media. *International Journal for Numerical and Analytical Methods in Geomechanics*, 28:981-1010, 2004.
87. A. Tordesillas, J. Shi, and H. Muhlhaus. Noncoaxiality and force chain evolution. *International Journal of Engineering Science*, doi: 10.1016/j.ijengsci.2008.12.011, 2009.
88. G.W. Baxter and R.P. Behringer. Cellular automata models of granular flow. *Physical Review A*, 42(2):1017-1020 (1990).
89. G.W. Baxter and R.P. Behringer. Cellular automata models for the flow of granular materials. *Physica D*, 51: 465-471 (1991).
90. J. Geng and R.P. Behringer. Slow Drag in 2D Granular Media. *Physical Review E*, 71:011302 (2005).
91. A. Tordesillas, and M. Muthuswamy. Stick-slip and force chain buckling. *Powders and Grains* 2009.
92. A. Tordesillas, J. Shi, and T. Tshaikiwsky. Stress-dilatancy and force chain evolution. (submitted)

Micromechanical alternatives to phenomenological hardening plasticity

Itai Einav and Giang D. Nguyen

Abstract Phenomenology is often applied in science through curve-fitting experimental results. This can be quite a useful exercise as long as it is not possible to derive a theory that can describe the observed results using first principles. Some of the most classical models of soil mechanics are in fact heavily based on curve-fitting procedures. Such models — e.g., critical state soil mechanics and kinematic hardening plasticity — are kept being used albeit the possibility to derive laws that are based on first principles. This may be fine if the mathematical structure of the alternative models is more difficult to apply. Here, however, we demonstrate that this is not always the case. It is possible to derive model alternatives to critical state soil mechanics and kinematic hardening plasticity that are based on first principles and that are equivalently mathematically simple.

1 Introduction

Phenomenological curve-fitting can be very useful when it is not possible to derive a theory that can explain the observed results using first principles. However, curve-fitting is strictly limited to describing only the set of observable experiments. It is then possible to combine the curve-fitting equations with a mathematical framework, in order to predict conditions beyond the experiments. The mathematical framework of plasticity is a convenient way for such an extension for the modelling of materials, which is a popular procedure in the soil mechanics community (both in academia and practice).

While the use of curve-fitting models is still the predominant tool of characterisation in practice, soil mechanics research is gradually learning to accept the role of first principles; this is evident through the increasing popularity of the discrete

Einav, Itai and Nguyen, Giang D.
School of Civil Engineering, The University of Sydney, Sydney, NSW 2006, Australia, e-mail: i.einav@civil.usyd.edu.au

element method (DEM; Cundall and Strack, 1979), i.e., using Newton's laws of motion. Yet it is hard to predict *when* and *if* the use of the discrete simulations could ever be adequate for engineering practice. '*When*' is the question of computer power, which is a particularly relevant when large-scale engineering problems are at stake. The question of '*if*' arises because the discrete element method does not fully resolve the need for constitutive assumptions: instead of using models of representative volume element, here the problem converts into defining particle contact models. Phenomenology then quickly finds its way back, although with less support from experiments that at the micro scale turn to be more complex.

The derivation of effective and simple constitutive models, based on first principles, provides a good reconciliation between the practical requirement for convenient solutions to large-scale problems and the academic pursuit of fundamental understanding. However, deriving effective constitutive models that are both simple and that are based on first principles is easier said than done. As a result, some of the most classical models of soil mechanics are in fact rooted on curve-fitting procedures; often those are structured mathematically using the theory of plasticity. Such models — e.g., critical state soil mechanics and kinematic hardening plasticity — are kept being used for their simplicity and ability to explain some of the observed behaviour, but not necessarily for adequate use of first-principles. Here, we present alternative theories for the kinematic hardening plasticity and critical state models, which are equally simple and that are based on clear first-principles of statistical homogenisation and energy balance.

Mathematical models of non-hardening plasticity involve the definition of a single stationary yield surface in stress-space. Within the yield surface the constitutive behaviour is kept elastic, while the onset of yielding manifests inelastic deformations, accompanied by energy dissipation. These models are mathematically convenient, but geomaterials present a much more complicated behaviour. Stress-strain investigations using elementary material tests reveal the following trends: (1) in cyclic shear tests hysteresis behaviour is observed, and (2) the yield threshold is ever increasing under continuously increasing isotropic pressure. To *implicitly* recover these aspects, traditional constitutive modelling of geomaterials involves, respectively: (1) multi-surface kinematic hardening, and (2) isotropic hardening (as in critical state soil mechanics).

The purpose of this paper is to motivate simple micromechanics explanations to the roots of these phenomena. We present alternative modelling concepts that *explicitly* adopt the micromechanical roots of both the isotropic and kinematic hardening phenomena. The two alternative formulations are equally simple, can describe the same phenomenological aspects *explicitly* rather than *implicitly*, and provide additional information that relates to meaningful microscopically-based variables and physical parameters.

2 Kinematic hardening plasticity and micromechanical roots

2.1 Phenomenological analog model

Consider the observable curve shown in Fig. 1 representing the stress-strain response of an elementary material to a given conceptual mechanical experiment. This example is certainly not new, but it serves an easy entrance into our discussion. In the spirit of phenomenological plasticity let us look at the material as a black box, without questioning what it is actually made of. As we load the material we see a linear stress-strain response, characterised by a slope E, which we call the elastic Young's modulus. However, as we keep loading the material beyond a certain stress threshold, σ_u, we observe a change to the slope from E to K. Upon unloading and reloading, the stress threshold seems to 'kinematically' change its value. In order for the abrupt change in slope, 'yield', to occur in the unloading direction, an opposite increment of stress is needed that equals $2\sigma_u$. We call this effect 'kinematic hardening', and 'feel ready' to propose a phenomenological constitutive model.

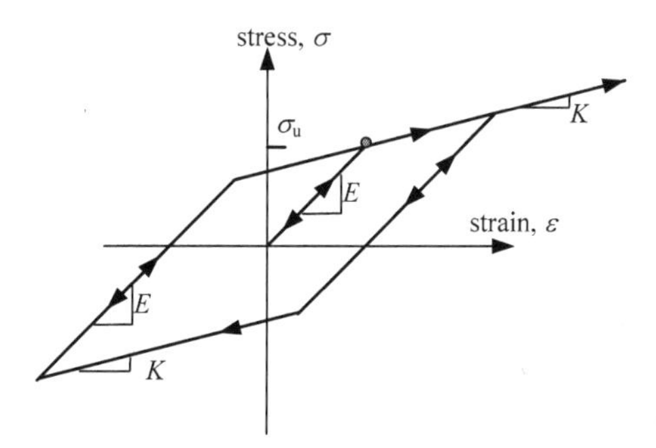

Fig. 1 Stress-strain curve observed as an outcome of a conceptual mechanical experiment.

It is well-known that data of Fig. 1 could be replicated in exact manner by a 1D mechanical analog of a spring with a stiffness modulus H placed in parallel with a slider with stress threshold

$$k = \sigma_u \quad , \tag{1}$$

both of which placed in a series with another spring of modulus E (see Fig. 2).

From geometrical compatibility and force/stress equilibrium it follows that:

$$\varepsilon = \varepsilon_e + \varepsilon_p; \quad \sigma = \sigma_H + \sigma_k \tag{2}$$

where $\varepsilon_e = \sigma/E$ and $\varepsilon_p = \sigma_H/H$ are the elastic deflection/strain within the springs E and H; ε_p is also the macroscopic plastic strain, signifying the overall permanent deformation; σ and ε are the force/stress and the deflection/strain acting externally to the entire mechanical analog; σ_k, the force/stress within the slider, must always satisfy the 'yield criterion':

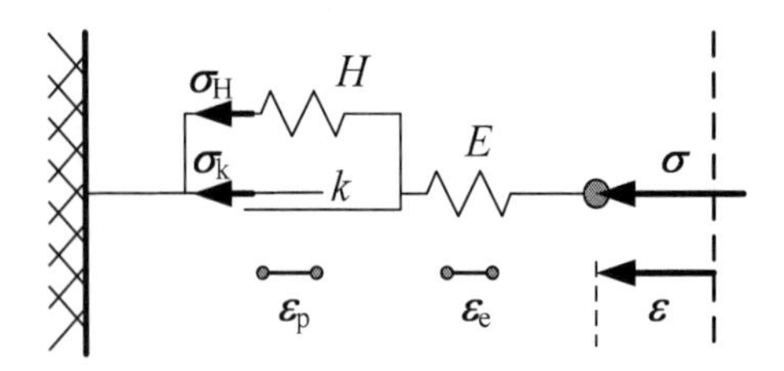

Fig. 2 Mechanical analog that is capable of curve fitting the result shown in Fig. 1.

$$y = \sigma_k^2 - k^2 = (\sigma - H\varepsilon_p)^2 - k^2 \leq 0 \tag{3}$$

Considering eqs. (2) and (3) for the yield situation ($y = 0$), we write $\varepsilon = \sigma/E + (\sigma - k)/H$ (assuming that the element is loaded in the positive direction). Therefore during yield the tangential slope K in Fig. 1 could be captured by setting H to satisfy:

$$K = \frac{EH}{E+H} \tag{4}$$

One can safely conclude that the model above does not violate the fundamental laws of mechanics since it is reproduced directly from a physically-based mechanical analog. It can even be shown that it satisfies the first and second laws of thermodynamics (Maugin, 1992). However, note that the 'real' material is not made out of two springs and a single slider. The analog (or use of thermodynamics) is merely giving us the convenient framework for a consistent mathematical derivation. It tells us nothing about the micromechanics and what the actual material is made of. If microscopic information is not available, then the model is as good as any other model. However, if microscopic information becomes available, we should question whether it is possible to assign physical meaning to the analog's structure and properties.

2.2 Alternative analog model

An alternative 1D analog model defines the mechanical system of two parallel springs, E_i ($i = a,b$; see Fig. 3a), each of which is placed in a series with a corresponding sliding element of threshold k_i ($i = a,b$). Following the logic in the previous section it can be shown that by assigning particular values to the parameters it is possible to deduce exactly the stress-strain curve shown in Fig. 1. For example, we start by requiring that

$$E = E_a + E_b \quad . \tag{5}$$

This condition enables to predict the initial elastic slope in Fig. 1. Equilibrium further suggests that

$$\sigma = \sigma_a + \sigma_b \quad , \tag{6}$$

where σ_i ($i = a, b$) is the local stress in either one of the two analog's legs. Again, the slider must satisfy the 'yielding criterion':

$$y_i = \sigma_i^2 - k_i^2 \leq 0 \quad (i = a, b) \quad . \tag{7}$$

Compatibility requires that $\varepsilon_a = \varepsilon_b = \varepsilon$, with $\varepsilon_i = \varepsilon_{e,i} + \varepsilon_{p,i}$. The local stresses are given by the elastic law of the individual springs, $\sigma_i = E_i \varepsilon_{e,i}$, therefore the yield surfaces can be represented in total strain space using $y_i = (E_i(\varepsilon - \varepsilon_{p,i}))^2 - k_i^2 \leq 0$ $(i = a, b)$. If we further require that one of the slider threshold would never be reached, e.g., say $k_b \rightarrow \infty$, then because $\sigma_b = E_b \varepsilon(\varepsilon_{p,b} = 0)$ and using the equilibrium eq. (6) we confirm that upon yielding ($\sigma_a = k_b$) the tangential stress-strain slope is $\partial\sigma/\partial\varepsilon = E_b$. Therefore, using eq. (5) we can set the stiffness parameters to fit the observed slopes:

$$E_b = H; \quad E_a = E - H \quad . \tag{8}$$

For the case of $k_b \rightarrow \infty$ first yielding (in the other slider) occurs when $\varepsilon_{p,a} = 0$, in which case $(E - H)\varepsilon = k_a(\varepsilon_{e,b} = \varepsilon)$. Therefore, by demanding that the analog provides first yielding according to the observed experiment, $\sigma = \sigma_u$, and noting the equilibrium eq. (6), we use

$$k_a = \frac{E - H}{E}\sigma_u \quad . \tag{9}$$

This shows that the stress-strain curve in Fig. 1 can be replicated by the current analog, exactly as the previous analog managed to do.

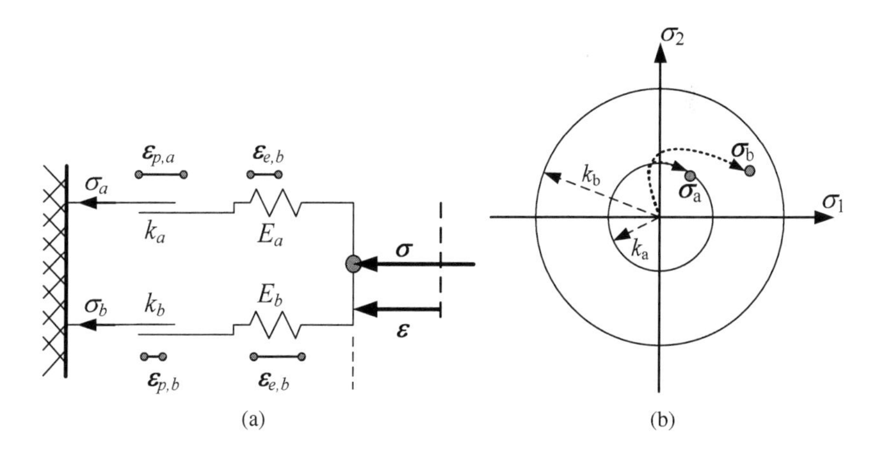

Fig. 3 (a) A second mechanical analog. This analog is also capable of replicating the results in Fig. 1. However, this time it is possible to assign physical meaning to the analog's parameters, as we demonstrate in section 2.3. (b) The extension of the model to account for the more general stress model is possible by replacing the 1D sliders using yield surface (here via a 2D representation).

In the next section we will show that this time (unlike the previous analog) it is possible to assign physical meaning to the analog's parameters. Before we continue to do that, it is useful to note the generalisation of this analog to the more general situation of stress tensors. This is made possible simply by replacing the local stress scalars in the yield equations (7), by local stress tensors:

$$y_i = \frac{1}{2}\boldsymbol{\sigma}_i' : \boldsymbol{\sigma}_i' - k_i^2 \leq 0 (i = a, b) \quad , \tag{10}$$

where the bold notation denotes a tensor (e.g., $\boldsymbol{\sigma}_i$ would be the stress tensor in element i); $\boldsymbol{\sigma}_i'$ is the deviatoric part of $\boldsymbol{\sigma}_i$; the symbol ':' designates dot product between the tensors. The generalisation requires a constitutive evolution equation for the local plastic strain tensors. Thermodynamics is then becoming handy, showing (Einav and Collins, 2008) that the local plastic strain incremental change is then given by the local flow rule:

$$\delta \boldsymbol{\varepsilon}_{p,i}' = \lambda_i \partial y_i / \partial \boldsymbol{\sigma}_i' = \lambda_i \boldsymbol{\sigma}_i' \quad (i = a, b) \quad . \tag{11}$$

For more complete detail on the thermomechanical aspect of this model we recommended the paper by Einav and Collins (2008). At this point, however, what we have shown is sufficient to demonstrate that the extension of the second analog to general stress condition means that the local 1D sliders could now be described by yield surfaces (see Fig 3b for a 2D representation), whose flow rule are associated according to eq. (11). Unlike conventional models of kinematic hardening plasticity, here the yield surfaces are inert in space of unit element forces (here, the forces along the analog's legs) and the hysteresis effects are captured by gradual yielding thanks to the variable nature of the local stresses.

2.3 The micromechanical interpretation of the second analog

We have just managed to demonstrate that from a phenomenological point of view there is no difference between the analogs shown in Figs. 2 and 3a. Both can replicate the stress-strain data, if the model parameters are set as described above. Both satisfy the laws of thermodynamics (see Einav, 2004). The question is then: are they really equally relevant?

To question this, let us now consider a separate scenario: this time we get a 'real' material that is randomly made from two distributed elasto-plastic phases (as shown in Fig. 4a). Assume that in this material:

1. the solid fractions of the distinct phases are ϕ_1 and ϕ_2,
2. the corresponding stiffness of each of the two phases is h_1 and h_2; we set $E_i = h_i \phi_i$ for i = 1 or 2,
3. the corresponding strength of each of the two phases is k_1 and k_2.

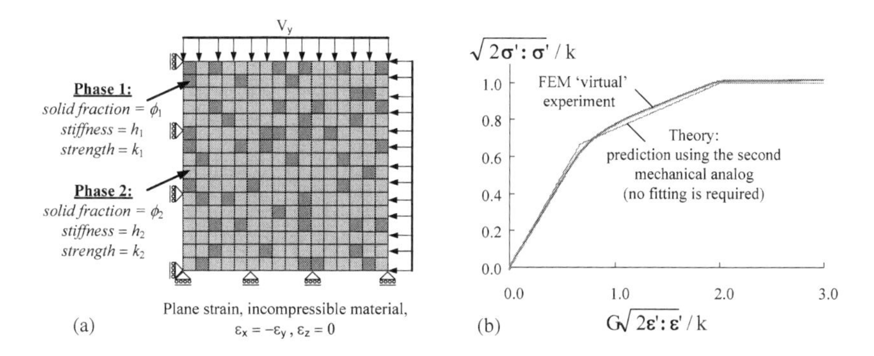

Fig. 4 (a) A random two-phase elasto-plastic material, subjected to an isochoric shear test (as an infinitely large chessboard problem). (b) The shear stress-strain curve of the random material, as extracted from a virtual FEM experiment and predicted (remarkably well) by the simple analog model of section 2.2. Note, the theory predicts the simulation without any form of curve-fitting, only by using the same physical parameters.

Now consider the particular case of a random two-phase elasto-plastic material whose second phase is infinitely strong ($k_2 \to \infty$). Obviously, if the local strains were found to be distributed uniformly across the real material sample, then its effective constitutive response would be that of the second analog, as shown in Fig. 1, with effective macroscopic properties that are given by eqs. (5), (8), and (9). The success of the second analog depends on the crudeness of the strain equivalence assumption. In random two-phase *elastic* material, the strain equivalence is known to provide an upper bound for the true overall stiffness (e.g., Doghri, 2000); in random two-phase *rigid perfectly plastic* material, this assumption provides an upper bound to the true overall collapse stress (e.g., Ostoja-Starzewski and Ilies, 1996).

Let us investigate the practicality of the strain-equivalence assumption in the more general case of a random elasto-plastic material (with all of the coefficients of stiffness and strength being finite), and evaluate the success of the simple analog of section 2.2 to predict experiments without curve-fitting. For that purpose consider the following virtual experiment. Take a two-phase random von-Mises elasto-plastic material, as shown in Fig. 4a. For a given solid fraction of the phases whose stiffness and strength are known, it is possible to compare the prediction of the analog (with its extension to 3D, viz. eq. (10)) against results from a virtual isochoric shear experiment using the finite element method (FEM). The simple analog predicts quite remarkably the experiment, as shown in Fig. 4b. In our view, this demonstration should motivate to seriously consider aborting models that are based on the first analog of section 2.1, and start using the second alternative of section 2.2.

3 Generalisation of the micromechanical analog

3.1 A one-dimensional model for random elasto-plastic material

We extend the logic of the second analog (section 2.2) to that shown in Fig. 5a: the so-called Masing-Iwan model (Masing, 1926; Iwan, 1966). It is well established that this analog results in the typical stress-strain curve presented in Fig. 5b. Hysteresis response is still observable but the stress-strain curve is characterised by a gradually changing slope. This time we do not have a single yield mechanism, but rather have a continuous yielding. Again, by adopting the strain equivalence assumption, this analog can be interpreted for a general random elasto-plastic material, in terms of an infinite number of elasto-plastic phases, i.e., where the stiffness and strength coefficients follow a certain probability density distribution function. For simplicity, in the following we describe only two ideal cases: (1) a constant strength but variable stiffness, and (2) a constant stiffness by variable strength. In order to be consistent with our previous section we adopt a slightly different analysis from that of Einav and Collins (2008). While the concept is similar, it seems that the current analysis is more convenient.

Assume that the strength is constant across the material and equal k, and denote $\phi(\eta)$ as the solid fraction of all of the elements η with stiffness $h(\eta - d\eta/2) < h(\eta) < h(\eta + d\eta/2)$. We can write the following, simply by extending the corresponding equations in section 2.2:

$$E = \int \phi(\eta)h(\eta)d\eta; \quad k(\eta) = k \tag{12}$$

$$\int \phi(\eta)d\eta = 1 \tag{13}$$

$$y(\sigma_\eta, \eta) = \sigma_\eta(\eta)^2 - k^2 \leq 0 \tag{14}$$

$$\sigma_\eta(\eta) = h(\eta)\varepsilon_e(\eta) \tag{15}$$

$$\varepsilon_e(\eta) = \varepsilon - \varepsilon_p(\eta) \tag{16}$$

$$\begin{aligned} \sigma &= \int \phi(\eta)\sigma_\eta(\eta)d\eta = \int \phi(\eta)h(\eta)\varepsilon_e(\eta)d\eta \\ &= E\varepsilon - \int \phi(\eta)h(\eta)\varepsilon_p(\eta)d\eta \quad . \end{aligned} \tag{17}$$

Eq. (12) shows that the average stiffness E is the statistical mean of the distributed phases' stiffness $h(\eta)$, where the solid fraction $\phi(\eta)$ plays the role of a probability density function, according to Eq. (13), in the spirit of homogenisation. Eq. (14) represents the yield criterion of element η, whose local stress is $\sigma_\eta(\eta)$. Eq. (15) denotes the local elastic law of that stress. Eq. (16) arises from compatibility, the decomposition of the total strains into elastic and plastic local strains, $\varepsilon_e(\eta)$ and $\varepsilon_p(\eta)$, and the assumption of strain equivalence $\varepsilon(\eta)$=ε for any η. The last equation, eq. (17), represents equilibrium.

We can conceive that overall failure would occur when all of the slider elements yield, i.e., when the equality in (14) is met for any η. In that case the ultimate stress is given by $\sigma_u = \pm k$.

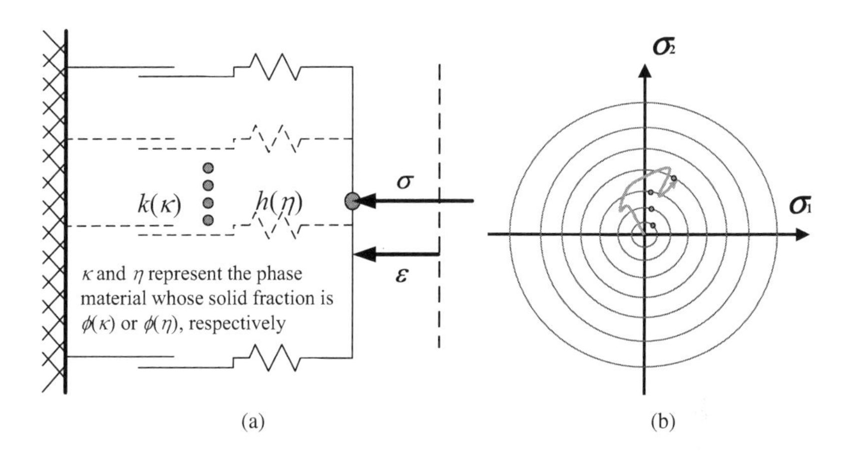

Fig. 5 (a) A distributed parallel spring-slide analog. (b) An extension of the scalar model to account for the more general tensorial case is possible by replacing the 1D sliders using infinite number of yield surfaces (here via a 2D representation.

Now consider the case that the stiffness of all the elements is constant and equal E, and denote $\phi(\kappa)$ as the solid fraction of all of the elements κ with strength in the range $k(\kappa - d\kappa/2) < k(\kappa) < k(\kappa + d\kappa/2)$. We can write the following equations, replacing those in (12-17):

$$\sigma_u = \int \phi(\kappa)k(\kappa)d\kappa; \quad h(\kappa) = E \tag{18}$$

$$\int \phi(\kappa)d\kappa = 1 \tag{19}$$

$$y(\sigma_\kappa, \kappa) = \sigma_\kappa(\kappa)^2 - k(\kappa)^2 \leq 0 \tag{20}$$

$$\sigma_\kappa(\kappa) = E\varepsilon_e(\kappa) \tag{21}$$

$$\varepsilon_e(\kappa) = \varepsilon - \varepsilon_p(\kappa) \tag{22}$$

$$\sigma = \int \phi(\kappa)\sigma_\kappa(\kappa)d\kappa = E\int \phi(\kappa)\varepsilon_e(\kappa)d\kappa$$
$$= E\left(\varepsilon - \int \phi(\kappa)\varepsilon_p(\kappa)d\kappa\right) \quad . \tag{23}$$

The conceptual result is similar to that of the variable stiffness model case in eqs. (12-17).

The tensorial form of the above model was proposed recently by Einav and Collins (2008), proving consistency with the first and second laws of thermodynamics. In the spirit of eqs. (10) and (11), it is sufficient for now to note that the major difference is attributed to the associated flow rules to the local yield surfaces; these two properties (yield criterion and flow rule) are written as follows (e.g., for the constant strength case):

$$y_\kappa(\boldsymbol{\sigma}_\kappa, \kappa) = \frac{1}{2}\boldsymbol{\sigma}'_\kappa(\kappa) : \boldsymbol{\sigma}'_\kappa(\kappa) - k(\kappa)^2 \leq 0 \tag{24}$$

$$\delta\boldsymbol{\varepsilon}'_p(\kappa) = \lambda(\kappa)\partial y_\kappa(\boldsymbol{\sigma}_\kappa, \kappa)/\partial\boldsymbol{\sigma}'_\kappa(\kappa) = \lambda(\kappa)\boldsymbol{\sigma}'_\kappa(\kappa) \quad . \tag{25}$$

Therefore, the extension into the tensorial form entails that the sliders could be replaced by a series of nested and stationary yield surfaces (see Fig. 6a).

3.2 The micromechanical interpretation

In both studied cases (constant stiffness or strength), it can be confirmed that the two most important observable parameters, the overall stiffness E (here constant) and strength σ_u are practically predictable directly from micromechanical information of local parameters, again under the assumption that the strain is uniform across the sample. Whether we should make use of the constant spring model, the constant strength model, or their combination (a possible extension) is not a question of phenomenology but a question of the material at hand. In many materials, the stiffness is fairly constant across the sample, while the strength tends to vary quite significantly (for example, Weibullian strength of brittle particles (Weibull, 1951)). In that respect it is important to mention the contribution of Chaboche (2003) who presented a theory with some similarities to the above, although neglecting the random aspects of the property distributions, connection to phenomenological plasticity, and extensibility to tensorial models.

In accord with the tensorial model, we find that each of the yield surfaces bound a single 'micro' stress $\boldsymbol{\sigma}_\kappa(\kappa)$. Since the macro-stress is the average of the micro-stresses, the overall 'macro' behaviour of the macroscopic stress-strain curve should follow a gradually flattening curve during proportional loading thanks to the gradual yielding of the local stresses. For a Weibull distribution of strengths, the closed form solution of such a stress-strain curve was given by Einav and Collins (2008), based on the spring-slider analog. To demonstrate that this model provides an adequate representation of a general random multi-phase elasto-plastic material, albeit the strain equivalence assumption, a comparison of the theory against virtual FEM simulations (with Weibull statistic of strength) is presented in terms of the stress-strain curve during hysteresis loading. The corresponding strength contours are shown along with the boundary condition of isochoric shear in Fig. 6a, while the prediction of the analog model to the hysteresis cycle is demonstrated in Fig. 6b.

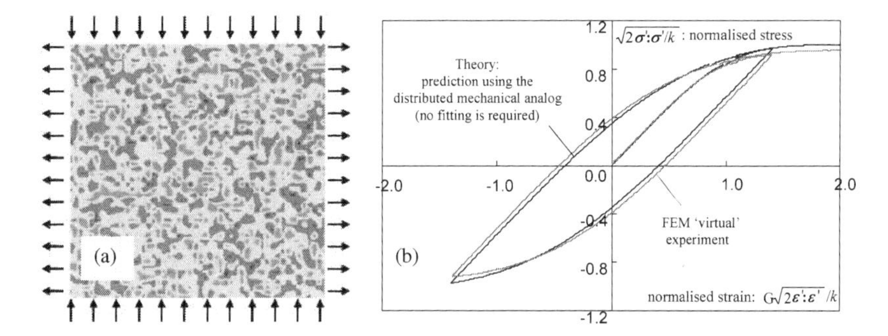

Fig. 6 (a) A random multi-phase elasto-plastic material, subjected to an isochoric shear test (contours represent the strength coefficient). (b) The analog prediction to a virtual experiment of multiphase random elasto-plastic material with variable strength using the FEM (as shown by the inset: taken from Einav & Collins, 2008). No fitting is required since the parameters of the FEM and analog are exactly the same.

3.3 Discussion about the physics of kinematic hardening

Recall the Iwan-Mroz analog model (Iwan, 1967; Mroz, 1967). In this model a single free spring is connected in a series to many individual 'parallel spring-slider elements', each of which is in a series with the next parallel spring-slider element in the line. While the stress is uniform along the various spring-slider elements (and the free spring), it is hard to see how this model could be interpreted for the currently studied random elasto-plastic material, even under an assumption of uniform stress. For example, it is hard to establish to which physical parameter the free spring of the Iwan-Mroz model could be connected. Nevertheless, this analog is commonly playing the starting point of kinematic hardening plasticity models. In the general tensorial case, each of the sliders are being replaced by a yield surface. All of these yield surfaces can translate in the space of sub-element stresses (here, the stress acting on a unit of parallel spring and slider) via the 'translation rule', representing the difference between the stresses within the springs and sliders of those sub-elements.

It seems that the popularity of this 'series' Iwan-Mroz model brought many researchers in the soil mechanics community to adopt the misconception that 'kinematic hardening' is the outcome of translating yield surfaces in stress-space. But this is a rather phenomenological description, and not a micromechanical interpretation; in the previous subsection, however, we did find a way to connect an analog to the physical properties, i.e., via the extension of the 'parallel' Masing—Iwan model. Einav (2004) discusses more about the differences between the thermodynamics potentials of the 'series' Iwan-Mroz and the 'parallel' Masing-Iwan model; it can be shown that the Iwan-Mroz model introduces an undesirable term to the Helmholtz free energy, that introduces the translation rules for the yield surfaces. Following Collins and Einav (2005) it can be supported that this assumption is only phenomenological, and arise directly from heterogeneities of local stresses across the sample. This is, however, being directly captured by the 'parallel' Masing-Iwan model. As motivated before and elaborated by Einav and Collins (2008), the internal variables of the 'parallel' Masing-Iwan model posses a clearer physical meaning, related to the statistics of yielding within a representative volume element. This is the reason for the presence of 'kinematic hardening' in terms of the observed hysteresis.

4 Isotropic hardening: phenomenology and micromechanical origins

In the traditional approach of Critical State Soil Mechanics (CSSM, Roscoe & Schofield, 1963) the experimental normal compression curve is usually fitted to drive the modelled hardening process. This way of modelling neglects the underlying micromechanical processes that govern the hardening behaviour of the material. Most importantly, this aspect differs between geomaterials. As a consequence, if the conditions vary it is difficult to determine the associated parameters that are required for CSSM modelling without further experiments. A typical constitutive assumption is frequently introduced by proposing that the yield surface grows isotropically (Fig. 7a) at a certain rate that is determined by a curve fitting hardening parameter λ either in terms of growing void ratio, plastic volumetric strain, or total volumetric strain. For example, the original expression for the pre-consolidation pressure in those models is given by (see schematic representation in Fig. 7b):

$$p_c = p_c(v) = p_r \exp\left(\frac{v_\lambda - v}{\lambda}\right) \tag{26}$$

where $v = 1+e$ is the specific volume (with e being the void ratio); p_r is a reference quantity with the dimensions of stress that is normally having the value 1 kPa. In this model the void ratio is related to the logarithm of the pressure, by introducing the parameter, λ, the so-called normal consolidation compression modulus. This is an additional fitting parameter, which may not be necessary, as we later argue, if first principle of energy balance are considered adequately, at least when modelling brittle granular materials.

4.1 A brief review on breakage mechanics

Thus far we managed to provide an explanation to the micromechanics behind kinematic hardening, via the parallel Masing-Iwan model. In the previous subsection we have introduced the problem of observed isotropic hardening, and stated the common phenomenological model assumption in CSSM. In order to be able to explain isotropic hardening via first principles, it is necessary to employ a new theory, which

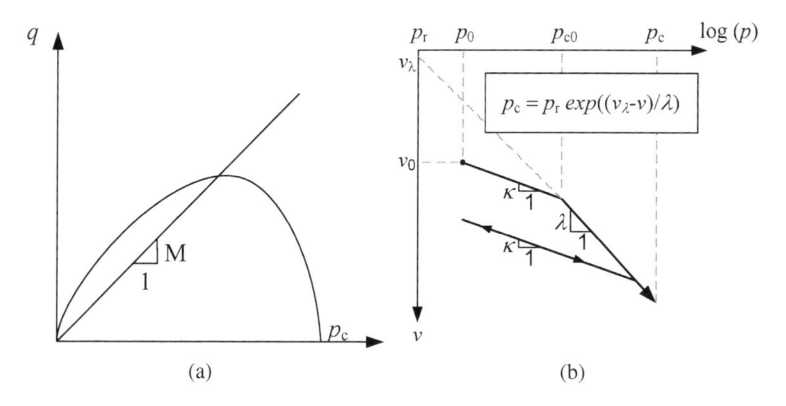

Fig. 7 (a) Typical yield surface in models of CSSM. (b) Schematics of isotropic hardening, representing the growths of the yield surface's largest pressure p_c.

we call Breakage Mechanics (Einav, 2007a & 2007b), along with the definition of a new internal (and measurable) variable, called 'breakage' B.

It is known that the shifting of the grain size distribution (gsd) due to comminution is the dominant governing mechanism of the isotropic hardening phenomenon observed on macroscopic scale in brittle granular materials. Therefore in micromechanics based constitutive modelling, this shifting gsd should be introduced to obtain the overall material behaviour. Einav (2007a) finds a way to express this requirement in a thermodynamically consistent way by relating the evolving gsd to the initial ($p_0(d)$) and ultimate ($p_u(d)$) ones via the internal variable of breakage B:

$$p(d) = p_0(d)(1-B) + p_u(d)B \tag{27}$$

This breakage B is measurable, given the corresponding gsd area definition in Fig. 8a. The current gsd $p(d)$, expressed in terms of Breakage (eq. (27)), is used for statistical homogenization. This is somewhat equivalent to the statistical averaging using the solid fraction $\phi(\eta)$ in section 3.1. However, while $\phi(\eta)$ is invariable in the previous definition, here the gsd $p(d)$ is evolving due to the crushing of the grains. The elastic strain energy of a volume element of granular material is given as the statistical average strain energy within all the grain size fractions within that volume:

$$\Psi = \int_{d_m}^{d_M} p(d)\psi(d,\boldsymbol{\varepsilon})\mathrm{d}d \tag{28}$$

where d_m and d_M are the minimum and maximum grain sizes, respectively. In this averaging process, the hypothesis of energy split (Einav, 2007a) is used, which suggests that the energy stored in a grain size fraction d is scaled by a non-dimensional factor $f(d) = (d/d_r)^n$:

$$\psi(d,\boldsymbol{\varepsilon}) = f(d)\psi_r(\boldsymbol{\varepsilon}) \quad . \tag{29}$$

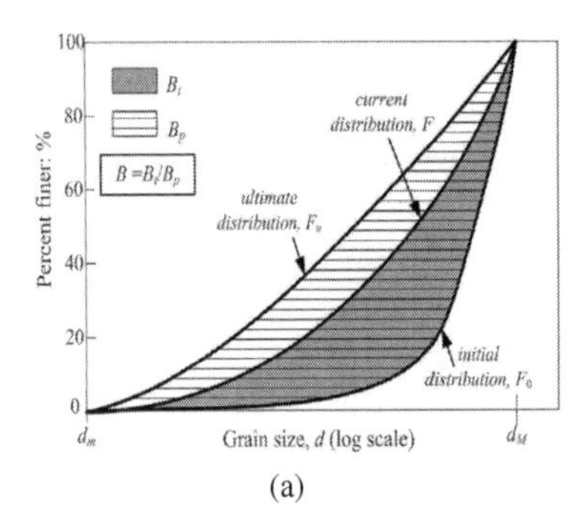

(a)

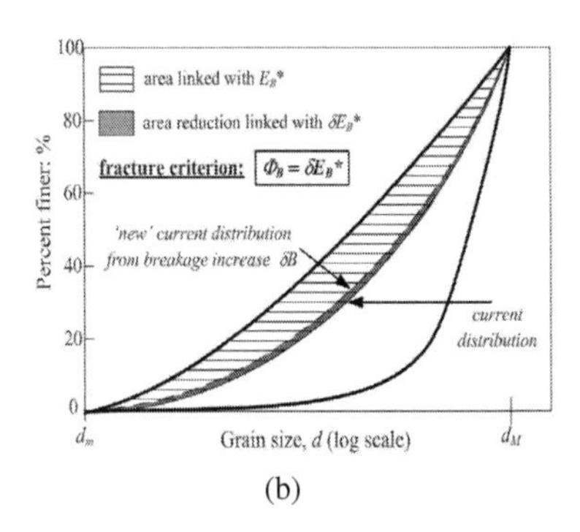

(b)

Fig. 8 The breakage measurement and evolution law (Einav, 2007a,c). The left figure (a) portrays the measurable definition of breakage. The right diagram (b) presents the breakage propagation criterion for granular materials. Φ_B is the breakage dissipation, denoting the energy consumption from incremental increase of breakage. δE_B^* is the incremental reduction in the residual breakage energy.

In the above $d_r = \sqrt{\int_{d_m}^{d_M} d^2 p_0(d) \mathrm{d}d} =< d^2 >_0^{1/2}$ is being used as the reference grain size; n represents the dimension of the system; $\psi_r(\boldsymbol{\varepsilon})$ is expressed in term of the total strain tensor $\boldsymbol{\varepsilon}$ and represents the elastic strain energy in the reference grain d_r. The energy scaling in eq. (29) has been verified by DEM analysis (Einav, 2007a). As a consequence of the energy split hypothesis, the Helmholtz energy potential is obtained in the form (Einav, 2007a):

$$\Psi = (1 - \vartheta B)\psi_r(\boldsymbol{\varepsilon}) \tag{30}$$

This expression, along with the equation of the proximity index property $\vartheta = 1 - < d^2 >_u / < d^2 >_0$, are the result of the same statistical homogenisation procedure. The proximity index property measures the "distance' between the initial and ultimate gsd's, with $< d^2 >_u$ and $< d^2 >_0$ being the second order moments of the ultimate and initial grain size distribution functions (by mass). From eq. (30), and consideration of thermodynamics, the energy conjugated to the breakage variable is defined as:

$$E_B = -\frac{\partial \Psi}{\partial B} = -\vartheta \psi_r(\boldsymbol{\varepsilon}) \quad . \tag{31}$$

The crushing of particles creates new surface areas which leads to energy release via the dissipation (termed Φ_B). It is then proposed that the breakage dissipation is driven by the loss of strain energy in the particles, as schematically portrayed in Fig. 8b and mathematically expressed by:

$$\Phi_B = \delta E_B^* \tag{32}$$

$$E_B^* = (1 - B)E_B \tag{33}$$

where E_B^*, the 'residual breakage energy', represents the available energy in the system for the crushing process (Einav, 2007a).

The yield criterion, which signifies the onset and then governs the evolution of breakage, can be worked out from (32) and (33) (Einav, 2007c):

$$y_B = (1-B)^2 E_B - E_c = 0 \quad , \tag{34}$$

where E_c is the critical breakage energy, a constant that arises directly from the derivation process. The introduction of the yield criterion requires the specification of the analysis to rate-independent processes, suggesting that during dissipation $E_B = \partial \Phi_B / \partial B$, and $\Phi_B = (\partial \Phi_B / \partial B)\delta B$.

4.2 General breakage models

We have briefly shown in the preceding section some of the fundamentals of Breakage Mechanics, based on first principles of energy balance and statistical homogenisation; this is a very convenient starting point for the formulation of micromechanics-based constitutive models of crushable granular materials. In this section we show how these principles could be applied to derive simple (and thermodynamically consistent) constitutive models. We also plan to highlight how the success of predicting the observed behaviour can be improved by considering the underlying elastic laws that govern the contact behaviour at the micro scales, not by adding more parameters.

The Helmholtz free energy potential Ψ and dissipation potential Φ, from which all other constitutive equations are derived, take the following general forms:

$$\Psi = (1-\vartheta B)\left[\psi_v(\varepsilon_v^e) + \psi_s(\varepsilon_v^e, \varepsilon_s^e)\right] \tag{35}$$

$$\Phi = \sqrt{\Phi_B^2 + \Phi_p^{v\,2} + \Phi_p^{s\,2}} \tag{36}$$

in which standard notations in soil mechanics are used: p and q denote mean effective stress and shear stress; ε_v^e and ε_s^e are elastic volumetric strain and shear strain, respectively. Functions ψ_v and ψ_s in (35) govern the elastic volumetric and shear behaviours of the model, and represent the 'unbroken' stored energy in a reference particle size d_r. The dissipation potential Φ comprises three parts corresponding to breakage dissipation Φ_B, plastic volumetric dissipation Φ_p^v and plastic shear dissipation Φ_p^s (Nguyen and Einav, 2009):

$$\Phi_B = \frac{\sqrt{2E_b E_c}}{(1-B)}\delta B \tag{37}$$

$$\Phi_p^v = \frac{p}{(1-B)}\sqrt{\frac{2E_c}{E_B}}\delta\varepsilon_v^p \tag{38}$$

$$\Phi_p^s = Mp|\delta\varepsilon_s^p| \tag{39}$$

In the above equations $M = q_u/p_u$ is the ratio between the ultimate shear stress q_u and ultimate volumetric stress p_u at failure (i.e., related to the ultimate mobilised friction angle); ε_v^p and ε_s^p are the plastic volumetric strain and shear strain, respectively. The constitutive relationships can be obtained from the free energy potential as follows:

$$p = \frac{\partial \Psi}{\partial \varepsilon_v^e} = (1-\vartheta B)\left(\frac{\partial \psi_v}{\partial \varepsilon_v^e} + \frac{\partial \psi_s}{\partial \varepsilon_v^e}\right) \tag{40}$$

$$q = \frac{\partial \Psi}{\partial \varepsilon_s^e} = (1-\vartheta B)\frac{\partial \psi_s}{\partial \varepsilon_s^e} \tag{41}$$

$$E_B = -\frac{\partial \Psi}{\partial B} = \vartheta\left[\psi_v(\varepsilon_v^e) + \psi_s(\varepsilon_v^e, \varepsilon_s^e)\right] \tag{42}$$

The yield function y^* in dissipative stress space is obtained as a result of the degenerate Legendre transformation of the dissipation potential (36) (see further details in Nguyen and Einav, 2009). Using (37-39), y^* can be replaced by the mixed breakage/yield function:

$$y_{\mathrm{mix}} = \frac{E_B(1-B)^2}{E_c} + \left(\frac{q}{Mp}\right)^2 - 1 \leq 0 \quad . \tag{43}$$

and the following flow rules apply ($\delta\lambda$ is a common non-negative multiplier): $\delta B = \delta\lambda\ \partial y^*/\partial E_B$, $\delta\varepsilon_v^p = \delta\lambda \partial y^*/\partial p$, $\delta\varepsilon_s^p = \delta\lambda \partial y^*/\partial q$. All of this is the outcome of careful consideration of thermodynamics (Nguyen and Einav, 2009). Examination of the yield criterion (43) highlights that the breakage energy E_B drives the isotropic hardening process and allows capturing this process in a natural way, without having to introduce any *ad hoc* parameters. The hardening of the material during comminution is directly driven by the competition between E_B and $(1-B)^2$: as B grows towards unity, E_B must increase to balance the equality in eq. (34). Since E_B is a function of the elastic stored energy, and governs the rate of dissipation during the crushing process, any improvement in the elastic behaviour of the model will automatically result in improvements in post yield behaviour. We believe that this aspect is a genuine step ahead, which allows us to be more economic in introducing model parameters, all by applying first principles.

The best example is our capability to eliminate parameters, such as λ in CSSM (see eq. (26) and Fig. 7b), which bears no direct physical meaning but is purely being a fitting parameter. This economy in model parameters is discussed at length in Einav (2007d), highlighting that the modelling of crushable granular materials using CSSM should require at least four additional ad-hoc parameters to achieve the same level of detail.

In the following we demonstrate how change in the elasticity law carries out changes to the post-yielding hardening behaviour. Nguyen and Einav (2009) further demonstrate the success of these changes in predicting experimental results.

4.3 Isotropic hardening via breakage model based on linear contact law

Linear elasticity is the simplest elastic law possible, which requires two constant parameters, e.g., the bulk and shear modulii K and G. For this purpose, ψ_v and ψ_s are defined as follows:

$$\psi_v = \frac{1}{2}K\varepsilon_v^{e2} \tag{44}$$

$$\psi_s = \frac{3}{2}G\varepsilon_s^{e2} \quad . \tag{45}$$

In this case, the breakage/yield surface (43) is rewritten in true triaxial stress space as:

$$y = \frac{\vartheta}{2E_c}\left(\frac{1-B}{1-\vartheta B}\right)^2\left(\frac{p^2}{K}+\frac{q^2}{3G}\right)+\left(\frac{q}{Mp}\right)^2 - 1 \leq 0 \quad . \tag{46}$$

Figure 9a shows the complete picture of the breakage/yield surface and its evolution in triaxial stress space.

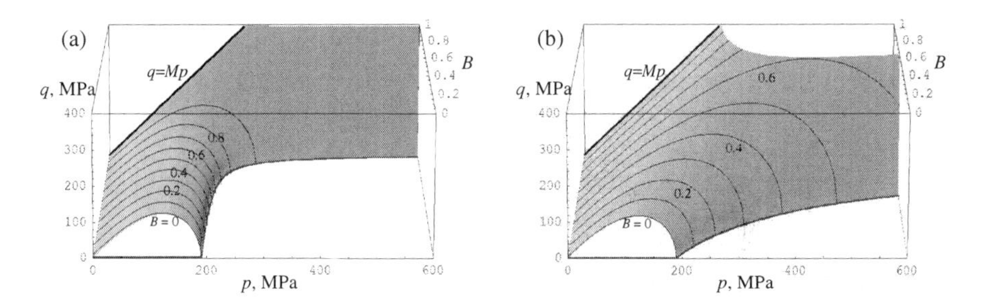

Fig. 9 Yield/breakage surface in p-q-B space, using: (a) linear elastic contact law (Einav, 2007d), and (b) Hertzian- like contact law (Nguyen and Einav, 2009).

Solving eq. (46) for $q = 0$ and $B = 0$, the initial crushing pressure p_{c0} in isotropic compression is given by:

$$p_{c0} = \sqrt{\frac{2KE_c}{\vartheta}} \quad . \tag{47}$$

As discussed by Einav (2000c), this relation bears striking relation to Griffith's (1921) criterion in fracture mechanics. As the breakage grows, we predict isotropic hardening of the yield pressure in terms of the breakage:

$$p_c = \left(\frac{1-\vartheta B}{1-B}\right) p_{c0} \tag{48}$$

since ϑ is between zero and one. This simple model was motivated by Einav (2007d) as a "student model" to illustrate some features of breakage mechanics models. In the next section we further advance the model, slightly increasing the complexity, but not the number of parameters.

4.4 *Isotropic hardening via breakage model based on non-linear contact law*

Enhancement to the model response can be made by developing appropriate nonlinear (hyper)elastic terms ψ_v and ψ_s, to account for pressure-dependent bulk and shear modulii. The following forms of ψ_v and ψ_s were derived for this purpose (Einav, 2007b; Nguyen and Einav, 2009):

$$\psi_v = p_r \frac{[\xi(\varepsilon_v^e)/p_r]^{2-m}}{\overline{K}(2-m)} \tag{49}$$

$$\psi_s = \frac{3}{2} p_r \overline{G} \left[\frac{\xi(\varepsilon_v^e)}{p_r} \right]^m \varepsilon_s^{e2} \tag{50}$$

where $\xi(\varepsilon_v^e) = p_r \sqrt[1-m]{\overline{K}(1-m)(\varepsilon_v^e - \varepsilon_{vr}) + 1}$ and p_r is defined as a reference pressure conveniently taken as 1 kPa; $m = 1/3$ for Hertzian contact, or $m = 1/2$ as per classical finding in soil mechanics, $\overline{K}$ and $\overline{G}$ are non-dimensional material constants, which replace the dimensional K and G in linear elastic contact model, and ε_{vr} the initial volumetric strain dependent on the maximum $e_{\max}$ and initial e_0 void ratios of the sample (Einav, 2007d): $\varepsilon_{vr} = -\log((1+e_0)/(1+e_{\max}))$.

The breakage criterion can be written in triaxial stress space as:

$$y = \frac{\vartheta(1-B)^2}{E_c} \left[\frac{p_r A^{\frac{2-m}{1-m}}}{\overline{K}(2-m)} + \frac{q^2 A^{\frac{-m}{1-m}}}{6(1-\vartheta B)^2 p_r \overline{G}} \right] + \left(\frac{q}{Mp} \right)^2 - 1 \leq 0 \tag{51}$$

where

$$A = \left(\frac{3p\overline{G} + \sqrt{9p^2\overline{G}^2 - 6\overline{G}m\overline{K}q^2}}{6(1-\vartheta B)p_r\overline{G}} \right)^{1-m} . \tag{52}$$

In this case, the crushing pressure p_{c0} in isotropic compression becomes:

$$p_{c0} = p_r \left[\frac{(2-m)\overline{K}E_c}{\vartheta p_r} \right]^{\frac{1}{2-m}} . \tag{53}$$

As breakage evolves, we predict isotropic hardening of the yield pressure in terms of the breakage:

$$p_c = \frac{(1-\vartheta B)}{(1-B)^{2/(2-m)}} p_{c0} . \tag{54}$$

Comparing eq. (54) and (48), it is clear that the elasticity law changes the hardening post yielding. The effects of this on the shape of the entire yield/breakage surface ($q \neq 0$) can be seen in Fig. 9. Nguyen and Einav (2009) highlight the success of this consideration in improving predictions, without adding fitting parameters. In particular, note the difference in the evolution of the surface with increasing breakage when the non-linear elasticity is applied (Fig. 9b) compared to the case of linear isotropic elasticity (Fig. 9a). Nguyen and Einav (2009) demonstrate the success of this consideration in improving predictions, without adding fitting parameters. Of particular interest are comparisons against the experiments by Wong et al. (1997), and thereafter the discussion concerning the energy balance during cataclastic shear.

5 Conclusions

We have illustrated the importance and the advantages of micromechanics-based constitutive models for geomaterials with particular attentions to issues related with isotropic and kinematic hardening. These phenomena are usually described by classical phenomenological models of plasticity at continuum scale, without resorting to the micro-structural changes at lower scales. Unlike phenomenological models, which require the introduction of *ad hoc* fitting parameters, micromechanics-based models have the advantage of facilitating explicit links between model parameters and the underlying microstructural properties. Moreover, the number of parameters can be kept minimal for even greater capability in model predictions.

In relation to kinematic hardening, we looked at two distinct spring-slider analog systems. We have argued that the use of the parallel spring-slider model is advantageous, compared to the series spring-slider model since it enables to describe random elasto-plastic media without introducing unphysical parameters. In relation to isotropic hardening, we looked at models based on the Breakage Mechanics theory. Thanks to the explicit links between the macroscopic model behaviour and the physics of grain crushing, the improvements of model predictions at macro scale can be achieved simply by enhancing the contact law at grain scale. No further parameters are required in this process.

Acknowledgements Itai Einav would like to acknowledge the Australian Research Council's for the Discovery Projects funding scheme (project number DP0774006). Giang Nguyen wishes to thank the University of Sydney for the financial support through the University of Sydney Postdoctoral Fellowship and the Bridging Support Grant.

References

1. Chaboche J.L. 2003. Thermodynamics of local state: overall aspects and micromechanics based constitutive relations. pp. 113–129 in Technische Mechanik, Manuskripteingang, 11. Juni 2003. Band 23, Heft 2-4.

2. Collins I.F., Einav I. 2005. On the validity of elastic/plastic decompositions in soil mechanics. *Proceeding of Symposium on Elastoplasticity*. Kyushu University, Japan, T. Tanaka and T. Okayasu (eds), 193–200.
3. Cundall P.A., Strack O.D.L. 1979. A discrete numerical model for granular assemblies. Géotechnique 29(1), 47–65.
4. Doghri I. 2000. Mechanics of deformable solids. Linear, nonlinear, analytical and computational aspects, Springer, Berlin.
5. Einav I., 2004. Thermomechanical relations between basic stress-space and strain-space models. Géotechnique. Vol. 54(5), 315–318.
6. Einav I., 2007a. Breakage mechanics–Part I: Theory, Journal of the mechanics and physics of solids 55(6), 1274–1297.
7. Einav I., 2007b. Breakage mechanics–Part II: Modelling granular materials, Journal of the mechanics and physics of solids 55(6), 1298–1320.
8. Einav I., 2007c. Fracture propagation in brittle granular matter, Proceedings of the Royal Society A: Mathematical, Physical and Engineering Sciences 463(2087), 3021–3035.
9. Einav I. 2007d. Soil mechanics: breaking ground, Philosophical Transactions of the Royal Society A: Mathematical, Physical and Engineering Sciences 365(1861), 2985–3002
10. Einav I., Collins I.F. 2008. A thermo-mechanical framework of plasticity based on probabilistic micro-mechanics. Journal of Mechanics of Materials and Structures, 3(5), 867–92.
11. Griffith A.A. 1921. The Phenomena of Rupture and Flow in Solids, Philosophical Transactions of the Royal Society of London. Series A, Containing Papers of a Math. or Phys. Character (1896-1934) 221(-1), 163–198.
12. Iwan W.D. 1966. A distributed element model for hysteresis and its steady-state dynamic response. J. App. Mech. Trans. ASME, 33(4), 893–900.
13. Iwan W.D. 1967. On a class of models for the yielding behaviour of continuous and composite systems. J. App. Mech., 34, 612–617.
14. Masing G. 1926. Eigenspannungen und verfestigung beim messing. Proc. 2nd Int. Congr. App. Mech.
15. Maugin G.A. 1992. The thermomechanics of plasticity and fracture. Cambridge University Press.
16. Mroz Z. 1967. On the description of anisotropic work hardening". J. Mech. Phys. Solids, 15, 163–175.
17. Nguyen G.D., Einav I., 2009. Cataclasis and permeability reduction: an energetic approach based on breakage mechanics, Pure and applied geophysics, 166 (2009) 1–32.
18. Schofield A.N., Wroth C.P. 1968. Critical state soil mechanics. London: McGraw-Hill.
19. Ostoja-Starzewski M., Ilies H. 1996. The Cauchy and characteristic boundary value problems of random rigid-perfectly plastic media. International Journal of Solids and Structures. 33(8), 1119–1136.
20. Weibull W. 1951. A statistical distribution function of wide applicability. J. Appl. Mech. (Trans. ASME) 18, 293–297.
21. Wong T., David C. and Zhu W. 1997. The transition from brittle faulting to cataclastic flow in porous sandstones: Mechanical deformation. Journal of Geophysical Research 102(B2), 3009–3025.

Mechanisms of localized deformation in geomaterials: an experimental insight using full-field measurement techniques

Gioacchino Viggiani

Abstract Deformation in geomaterials (soils, rocks, concrete, etc.) is often localized, e.g., in the form of shear bands or fractures. In experimental analysis of the mechanical behavior of such materials standard laboratory methods are insufficient as the majority of measurements are made at the sample scale and rarely at a local scale. X-ray tomography monitoring during loading allows high-resolution full-field observation of the development of deformation. However, such images only indicate clearly the deformation when there are significant changes in material density (i.e., volume changes) that produce a change in X-ray absorption. As such 3D Volumetric Digital Image Correlation (DIC) approaches have been developed that allow quantification of the full strain tensor field throughout the imaged volume. This paper presents results from triaxial compression tests on a clay rock (Callovo-Oxfordian argillite) and a granular material (Hostun sand) with in-situ synchrotron X-ray micro tomography imaging providing complete 3D images of the specimen at several stages throughout the test. These images have been analyzed using 3D Volumetric DIC to provide full-field displacement and strain measurements, which allowed the detection of the onset of strain localization and its timing relative to the load peak plus insight into the 3D structure of the localized zone.

1 Introduction

The importance of strain localization in the behavior of geomaterials, including rock and soil, has been known for a long time, and it has been thoroughly investigated in the laboratory. However, it should be kept in mind that in the presence of localized deformations, the meaning of stress and strain variables derived from boundary measurements of loads and displacements is only nominal, or conventional. Therefore, the most valuable experimental contributions to the understanding of localized

Gioacchino Viggiani
Laboratoire 3S-R, Grenoble Universities / CNRS, e-mail: cino.viggiani@hmg.inpg.fr

deformation are those measuring, in one way or another, the full field of deformation in the specimen – which is the only means by which test results can be appropriately interpreted. During the last two decades, so-called *full-field* measurement techniques have become more and more popular in laboratory experimental mechanics, not only in geomechanics (e.g., see the recent review paper [60]). Compared to more conventional measurement techniques, which are based on the use of transducers positioned at the specimen boundaries, full-field techniques provide what even multiple local measurements cannot, that is, a field record of a quantity (e.g., deformation, density, temperature, etc.) as opposed to point-wise data.

As far as experimental geomechanics is concerned, the use of full-field measurements in the laboratory is in fact not completely new. For instance, X-ray radiography was used already in the early 1960s in Cambridge as a non-invasive technique for measuring strain field in soil (e.g., [49, 50]). The technique was initially conceived with the purpose of following the positions of lead markers during a test on plane strain soil models. The underlying idea was that displacements of a grid of markers from one radiograph to the next can be interpreted as indications of a continuum displacement field and, through differentiation, a corresponding strain field. However, the radiographs revealed other, extremely interesting effects associated with local changes in the soil itself. In fact, local and relatively confined density changes, if sufficiently marked, produce corresponding contrasts in absorption of X-rays which can be seen on the radiographs. Such density variations allowed direct observation of narrow dilation bands forming in the soil models (e.g., Figure 1).

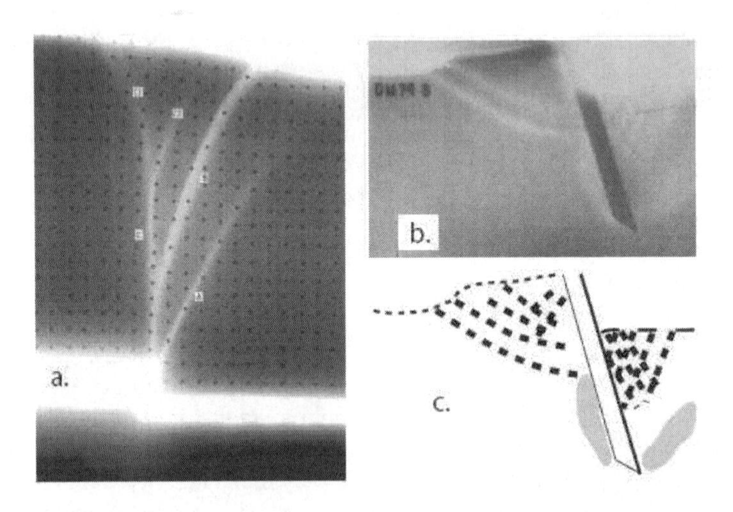

Fig. 1 (a) Dilation bands in dense sand over a displacing trapdoor (Stone 1985); (b), (c) dilation bands and zones of dilation around rotating blade in dense sand ((b) radiograph from Cambridge University archive; (c) interpretation of (b)). (From [41])

Other pioneering examples of the use of full-field measurement in experimental geomechanics include the method of *False Relief Stereophotogrammetry* (FRS), first used in soil mechanics by Butterfield and coworkers [16] and then extensively applied and developed in the 1980s to capture non-homogeneous deformation throughout tests under plane strain conditions (e.g., [18, 21, 22] for an overview). In essence, FRS is based on the analysis of successive pairs of photographs of the side of a specimen deforming under load and allows direct measurement of (incremental) deformations and determination of strain fields throughout a test. It is undoubted that FRS provided unprecedented insight for localization studies in geomaterials. However, due to recent, major technological development and rapid proliferation of alternative techniques making use of digital image analysis, FRS has been superseded by Digital Image Correlation (DIC) methods, which will be discussed later in this paper.

Outside of geomechanics, there have been many significant developments in the area of full-field measurements, which have opened new and attractive fields of investigation. While geomechanics was certainly pioneering some aspects of full-field measurements some forty years ago, it seems that more recently we are lagging behind our colleagues in solid mechanics – who use such techniques for "simple" materials such as metals and metallic alloys. However, materials such as sand, clay, sandstone, shale, granite, etc., (more generally: natural solids) are the quintessential heterogeneous, multi-scale materials. This paper aims to make the point that a wider use of full-field measurement in laboratory geomechanics not only is possible (and exciting), but also it is the only way to significantly advance our understanding of the mechanics of soils and rocks, especially in the hot topics of strain localization and fracture.

The structure of this paper is as follows. The next section gives the fundamentals of Digital Image Correlation, which is the most popular – and possibly obvious – full-field method available, and is essentially a mathematical tool for assessing the spatial transformation (including translations and distortions) between two digital (2D or 3D) images of a specimen deforming under applied load. In the subsequent section, we then present the method of X-ray Computed Tomography (CT), which can be used to study in three dimensions the internal structure and deformation processes in a specimen deforming under applied load. Finally, we show the power of combining the two methods of X-ray tomography and 3D DIC for quantitative analysis of the evolution of localized deformation. This is illustrated through two selected examples, both coming from recent experimental studies by our team – the former on a clay rock and the latter on a fine-grained sand, both tested in triaxial compression. I do not pretend that these are necessarily the best or the most advanced examples available, they are just those that I know the best. The paper concludes with a few general remarks concerning the lessons learned from these two examples and perspectives for current and future work.

2 Digital Image Correlation

Major developments in digital photography coupled with the (apparent) simplicity of digital image analysis, have recently and rapidly led to the proliferation of techniques alternative to the previously mentioned FRS that allow the same type of full-field measurement. These new methods, which may be grouped under the general name of Digital Image Correlation (DIC), make use of digital rather than analogue photographs, and therefore benefit from all the advantages of computer technology, including key aspects such as availability and affordability. In effect, any research group can easily acquire the necessary equipment and computer programs for performing DIC. This was not the case for FRS, which required special equipment in order to be used as a quantitative tool (including a *stereocomparator*, which is an advanced tool for 3D-photogrammetric measurements).

DIC techniques have been used increasingly over the last 20 years or so in a range of disciplines, including amongst others: solid and fluid mechanics (where the procedure is often referred to as Particle Image Velocimetry, PIV), medicine, animation and film special effects, and image registration. In recent years, the application of DIC to experimental mechanics has literally exploded. For example, Orteu [43] recently noted more than 350 journal papers dealing with DIC-based measurements in the field. An historical reference is provided by [54]. As far as geomaterials are concerned, the use of digital image analysis for characterizing soil fabric and its evolution under load is not new (e.g., [32, 40]). The application of digital image correlation is increasingly common for the monitoring of deformation in soil mechanics laboratory experiments (e.g., [13, 25, 26, 35, 48, 63]).

DIC is essentially a mathematical tool for assessing the spatial transformation (including translations and distortions) between two (2D or 3D) digital images. In practice, DIC is implemented as a computer program that allows regions of a photographed object to be tracked automatically from one digital image to the next, from which displacements can be deduced. This is achieved without specific markers and without manual intervention, provided there is some clear fabric or texture visible across the image. A number of commercial and academic codes exist to carry out DIC, although the implementation to develop in-house software is not too challenging. Whichever computer software is used, the basic idea of DIC remains the same, i.e., to determine the displacements, and perhaps deformations, to map one image onto another. A number of basic steps are generally involved: (i) definition of nodes distributed over the first image; (ii) definition of a region about each node (the so-called *correlation window*); (iii) calculation of a *correlation coefficient* for each displacement of the correlation window within an area (the *search window*) about the target node in the second image; (iv) definition of the discrete displacement (integer number of pixels (in 2D) or voxels (in 3D)) given by the displacement with the best correlation; (v) sub-pixel refinement (because the displacements are rarely integer numbers of pixels), which may also involve more complex transformations than simply rigid-body translation; (vi) calculation of the deformation based on the derived displacements. Each DIC implementation has its variations and additions to this general methodology. In particular, there are different approaches for the sub-pixel/voxel refinement after the initial discrete estimate of the nodal displacements has been made (based on the displacement in the search window of the correlation window that returned the maximum cross-correlation value). Perhaps the most straightforward approach is to use a functional description of the local correlation field (e.g., for 2D, describing the variation in correlation over the 9 positions within a 3×3, +/-1 pixel, search range about the best integer shift value) and some algorithm to find the maximum of this local cross-correlation surface. Alternatively the sub-pixel/voxel refinement can be carried out through iterative optimization of the transformation parameters with interpolation of the "gray-level" to reconstruct the image for each transformation and minimizing the misfit of the first and (transformed) second images. It should be noted that sub-pixel/voxel refinement, whichever method is used, is an essential step in DIC, especially when subsequently calculating strains from the displacements. Without sub-pixel/voxel refinement only integer numbers of pixel/voxel displacements can be resolved, which results in "stepped" displacement maps and strain images with very large strains corresponding to these steps and no strain between.

3 X-ray Computed Tomography

We recalled already in the introduction that the use of X-ray imaging in experimental geomechanics dates back to the 1960s. Figure 1 showed a few examples of radiographs obtained in Cambridge by Roscoe and coworkers, which allowed the detection of bands of localized dilation in granular soil. Similar observations were made by other authors, e.g., for triaxial compression [31], simple shear [52], directional shear [6, 7], plane strain compression [58], thick-walled hollow cylinders [4], as well as for flowing sand masses in hoppers [15, 39]. In all these examples, X-rays were used to obtain *radiographs*, i.e., 2D images produced by X-ray radiation on photographic plates. Radiographs represent maps of attenuation accumulated through the complete soil mass in the direction perpendicular to the image, which is related to soil density.

The above studies provided valuable qualitative information on localization patterning in sand specimens and sand box models. However, they all suffered from two major limitations: (i) lack of quantitative data on the observed density changes, and (ii) limitation to 2D images. Both limitations are overcome by X-ray *Computed Tomography* (CT). The principle of CT measurement consists of recording X-ray radiographs of a specimen at many different angular positions around the object. From these different projections, a three dimensional image of the object can be reconstructed with appropriate algorithms (usually based on a back projection principle); see for example [8] for a thorough description of the technique. X-ray CT is therefore a non-destructive imaging technique that allows quantification of internal features of an object in 3D. First developed for medical imaging, X-ray CT is now widely used in material sciences and proved its interest in various domains of geosciences, including geomechanics (e.g., [23, 38, 46]).

As far as applications of X-ray CT to soil mechanics experiments are concerned, a pioneering contribution is due to Arthur [5]. At that time X-ray CT was not truly available yet, but tomographic images were obtained by using a special moving source and photographic plate set up. The technique appeared extremely promising and strongly motivated subsequent studies by Desrues and coworkers in Grenoble, who started from the early 1980s to use X-ray CT as a quantitative tool for experimental investigation of strain localization in sand [17, 18, 20]; see [19] for a review. As an example, Figure 2 shows horizontal slices through a three dimensional CT image of a specimen of dry dense Hostun sand (at a stage towards the end of a triaxial compression test). Patterns of localized density variations are revealed as different intensities of the recorded X-ray radiation. The 3D mechanism that appears has some clear structure, which would be otherwise hidden (i.e., invisible from just looking at the specimen, which to external viewing just had a barrel shape at the end of the test). Similar structures are shown in Figures 3 and 4, obtained from another triaxial test on Hostun sand, [20]. The localization pattern in this case involves a cone and multiple sets of planes associated in pairs, each pair intersecting along a diameter at the top of the specimen. Figure 4 shows Desrues' interpretation of these patterns of localization, including a cross-section close to the top platen, and a section parallel to the axis (compare with Figure 3).

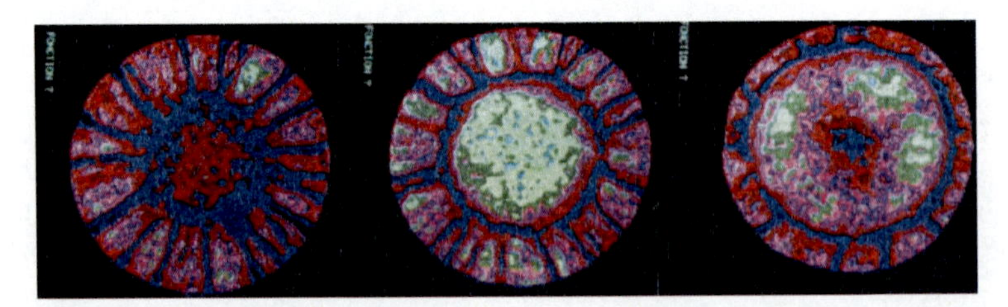

Fig. 2 Horizontal slices through a CT image volume of a triaxial compression specimen of dry dense Hostun sand showing complex patterns of density variations (image was acquired near the end the test). (Image courtesy of J. Desrues)

Similar applications of X-ray CT to *post-mortem* (i.e., after the test) observation of strain localization have been presented for a range of geomaterials. For example, Bésuelle and coworkers ([10, 11]) used post-mortem X-ray CT together with micro-structural analyses to investigate the features of localized damage (including shear and compaction bands) across the transitional regime from brittle faulting to cataclastic ductile flow in sandstone.

It appears that direct 3D observation of the internal structure of a specimen (be it a soil or a rock) can provide substantial advances in the understanding of strain localization in geomaterials. In particular, accessing the internal structure of a specimen during the test, i.e., *while it deforms* under applied load (as was done by Desrues [20]), allows the entire deformation process to be followed: prior to, at, and after the onset of strain localization. However, performing a test properly and X-ray scanning at the same time (this is typically called *in-situ* tomography) is extremely challeng-

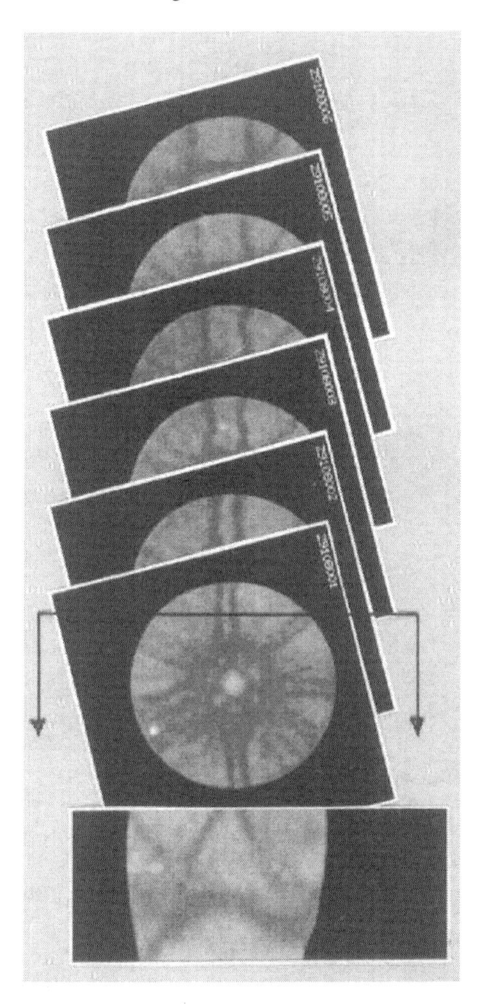

Fig. 3 Patterns of localized density variations revealed in slices through a CT image volume of a specimen of dry dense Hostun sand near the end of a triaxial compression test (note this is a different test to Figure 14). (From [20])

ing for geomaterials due to their specific testing requirements (confining pressure, pore pressure control, etc.). This has been achieved by a number of authors, including [2, 3, 30, 44, 45, 47, 62].

It should be noted that all the aforementioned in-situ studies have been performed using more or less "conventional" medical and industrial Computed Tomography (CT) systems. Another, much more powerful source of X-rays can be provided by *synchrotron radiation*, for which the X-ray beam is a thousand billion times "brighter" than the beam produced by a hospital X-ray machine (like the one used by Desrues, [20]). The higher energy and photon flux of synchrotron radiation allow for a much higher resolution, down to the micrometric scale. Such a resolution may

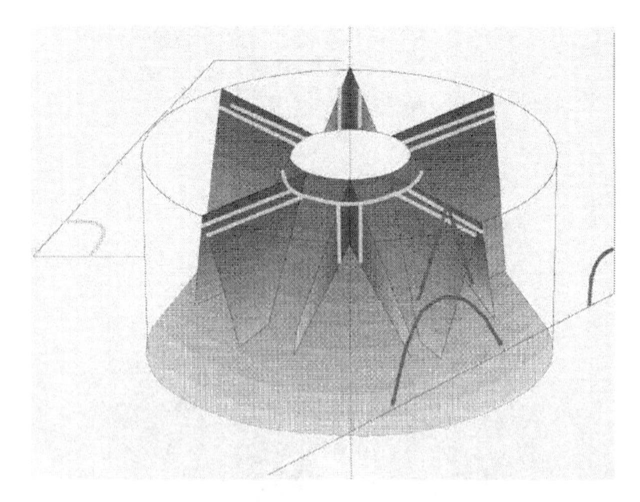

Fig. 4 Suggested 3D interpretation of the patterns of localization in Figure 3 (From [20])

be unnecessary for coarse-grained geomaterials such as sand, in which the width of a shear band is known to be roughly 10 to 20 times the mean grain diameter size (i.e., a few millimetres), unless one wishes to make observations at the scale of the grains (as we will discuss later in this paper). However, increased resolution is crucial for characterizing strain localization in fine-grained materials such as clays and clayey rocks, in which shear bands are much thinner, and in fact are often described as displacement discontinuities or slip surfaces, e.g., [60]. This is apparent from the relatively poor quality of the (few) available tomographic images of shear zones in fine-grained geomaterials obtained in the past [29, 44, 56].

4 Example 1: localized deformation in a clay rock

The first example, already presented in [12] and [34], comes from an experimental study on a clay rock called Callovo-Oxfordian argillite, issued from the ANDRA Underground Research Laboratory in France. The testing program was carried out at the European Synchrotron Radiation Facility (ESRF) in Grenoble, making use of X-ray micro tomography at beamline ID15A. The tests were conducted using a specifically built setup that could be placed in the X-ray beam, so that the specimens were scanned under load (in situ). Synchrotron X-ray radiation was selected because, as mentioned previously, it is the only one that can provide, thanks to its high photon flux, a combination of both fast scanning and high spatial resolution. The former is desired to minimize axial load relaxation, while keeping the specimen at constant axial strain during scanning (the acquisition of an entire specimen required four to six sections and in the most recent tests took 12 to 15 minutes).

High spatial resolution allows fine detail on the deformation process to be obtained (a voxel size of 14 μm was achieved for specimens 10 mm in diameter).

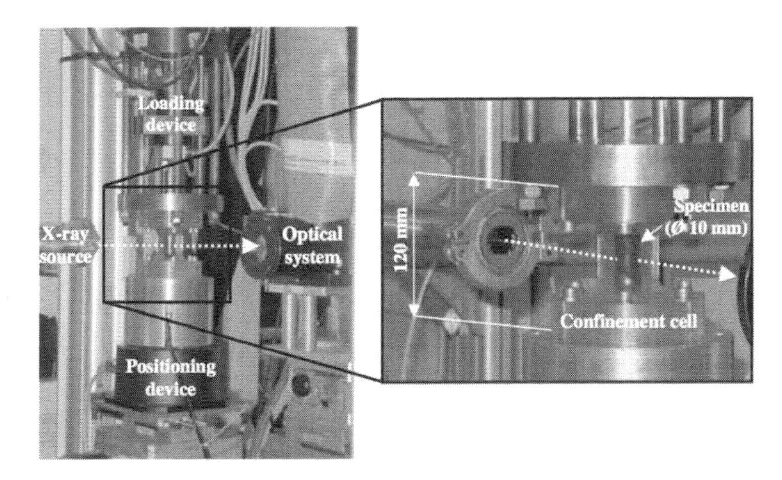

Fig. 5 Experimental setup, for in-situ triaxial X-ray tomography experiments: the complete set-up on the beamline (left) and zoom on the rock specimen inside the triaxial cell (right). (From [34])

Figure 5 shows the in-situ experimental setup, which included a triaxial apparatus and a loading system. The former is practically the same as a conventional triaxial testing system, except for its much smaller size and the shape of the confining cell. The cells were made from Plexiglas or polycarbonate to be as transparent to the X-rays as possible. Contrary to a conventional system, the tensile reaction force is carried by the cell walls and not by tie bars, which provides a clear path to the specimen for the X-ray beam, free of any obstacle (apart from the cell walls). The axial load and hence the deviator stress are applied using a motor-driven screw actuator. The loading system is placed in the X-ray beamline without interfering with the tomographic scans. See [33] for further details.

Callovo-Oxfordian argillite is a sedimentary rock composed of particles of calcite and quartz in a clay matrix, with a clay fraction of 40–45%. At the investigated depth (approximately 550 m below the ground surface), the material has an extremely low permeability (10^{-20}–10^{-22} m^2), a porosity of 15%, and a water content of 6%. The uniaxial compressive strength is about 20 MPa. Figure 6 shows the stress-strain response obtained from an undrained triaxial compression test on a 10 mm diameter specimen at 10 MPa confining pressure. The specimen was scanned at different steps: before and right after applying the confining pressure (steps 0 and 1, respectively), at different levels of axial strain during deviatoric loading (steps 2-7), and finally after removal of the confining pressure (step 8). Figure 7 shows four horizontal CT slices at different steps. The slices at steps 1, 3, 7 and 8 have been selected following a set of material points which were visible at step 1 and could be found on all subsequent images. At step 8, two open cracks can be seen at the specimen edges. Comparison of the images at steps 7 and 8 clearly indicates that crack opening in this test was essentially due to the removal of the confinement pressure. At the earlier step 3, which corresponds to the peak stress, no localized deformation is evident – even knowing where the cracks are eventually opening up later in the test. Furthermore, no visible difference can be seen between steps 1 and 3. In fact, localization becomes just visible in the CT images at step 4 (not shown in the figure – see [12]).

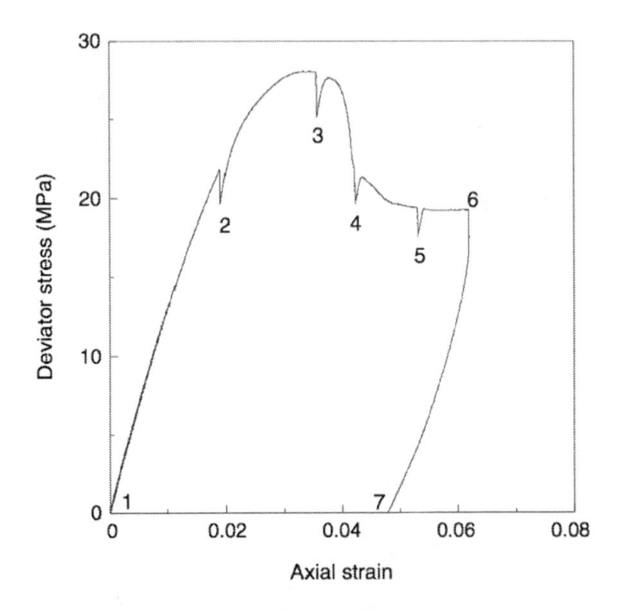

Fig. 6 Load curve (deviator stress versus axial strain response) for the discussed triaxial compression test on the Callovo-Oxfordian argilite at 10 MPa confining pressure. The numbers indicate the times at which CT images were acquired. (From [34])

Interestingly, the existence of distinct calcite inclusions in the argillite was particularly helpful to highlight relative displacement in the specimen otherwise invisible in the CT images. Figure 8a shows an example of one such inclusion, having an elongated vein shape and a length of a few millimeters. Figure 8b shows the same inclusion (in yellow) and the open fracture (in red) at step 8, i.e., at the end of the test. Figures 8c-h illustrate how, upon loading, this inclusion was severely strained by the localization. Shearing of the inclusion is apparent starting from step 5, which becomes more and more pronounced for increasing deformation (steps 6 through 8). However, this intense shearing is not accompanied by a variation of density measurable within the resolution of the X-ray CT. This indicates that the region of localized deformation in the central part of the specimen is undergoing shear without substantial volume changes, whereas crack opening can be observed towards the edge of the specimen.

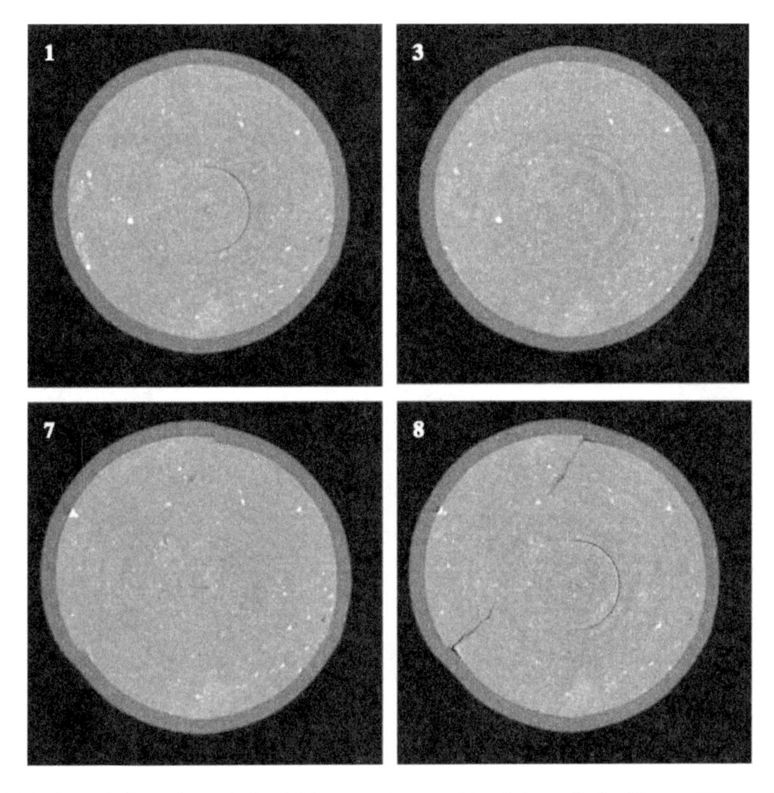

Fig. 7 Horizontal slices through the CT image volume of the Callovo-Oxfordian argilite specimen (∅ = 10 mm) at four different time steps (see Fig. 6). (From [34])

The above discussion illustrates a fundamental difficulty with experimental detection of strain localization through the use of X-ray CT, which is associated to the very nature of localized strain. The issue is that while localization can sometimes induce large volumetric deformation – either dilatancy (or crack opening) or compaction (compaction bands), depending on the material and loading conditions, in general volumetric strain in a shear band is small compared to the shear strain. Unfortunately, CT images only represent local mass density fields. If the material inside the region of localized deformation dilates (or contracts), then local mass density variations can be an effective means to track the regions of localized deformation. This was the case with the studies of shear banding in sand from Desrues presented in the previous section. However, for the experiments on the argillite, this was only the case with open cracks, where voids are created between the crack edges that are clearly visible on the tomographic images (e.g., step 8 in Figure 7). If the localized deformation is isochoric (no volume change), e.g., it consists of closed, shearing cracks, then it is invisible in CT images (except for the "lucky" cases where the shear cuts markers such as the inclusion in Figure 8).

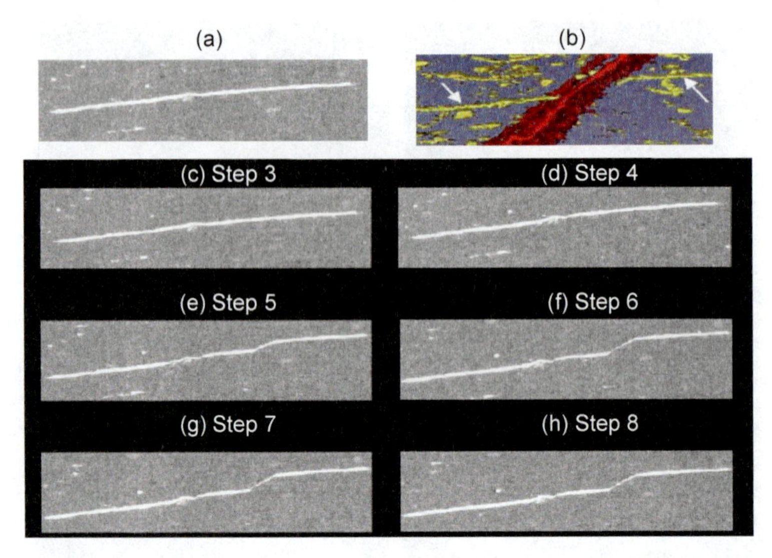

Fig. 8 (a) An inclusion in the Callovo-Oxfordian argilite at the start of the test (the image is a small vertical section extracted from the x-ray tomogram and the inclusion is a few millimeters long). (b) 3D view of the same inclusion (yellow) and cross-cutting open fracture (red) at step 8 (the end of the test) highlighted through thresholding and false-coloring of the gray-scale image. (c)-(h) the inclusion at different stages through the test showing increasing shearing. (From [12])

We will show hereafter how such a limitation can be in fact overcome by complementing X-ray CT with 3D digital image correlation (see Section 2). 2D DIC was previously applied to X-ray radiographs (e.g., [51, 55]). More recently, applications of DIC in three dimensions have been presented by e.g., [9, 14, 24, 53 and 59]. Our application of 3D DIC to the X-ray CT images of Callovo-Oxfordian argillite demonstrates that correlation of 3D digital images from X-ray tomography provides a means to effectively detect (within a given accuracy) localized deformation, independent of its nature (shear or volumetric strain). The details of the 3D DIC method used are thoroughly discussed elsewhere [34]. Herein it will be sufficient to bear in mind that each 3D image is decomposed into several cubical subsets, each of them containing 20^3 voxels. Due to the small deformation experienced by the argillite specimens, for this study the transformation between two images was assumed to be a rigid translation without any rotation or distortion.

Figures 9 and 10 show vertical and horizontal cuts through the DIC-derived (incremental) shear strain field volumes. In the pre-peak increment (2–3, see Figure 6), the onset of strain localization is visible at the bottom left corner of the vertical cut through the shear strain map (Figure 9). The horizontal cut in Figure 9 suggests that during this increment the localization developed with a conical shape. Note that the X-ray sections of the specimen at the end of the increment do not reveal any trace of localized deformation. In the post-peak increment (3–4, see Figure 6), a fully developed shear band through the specimen can be observed (Figure 10). As compared to the pre-peak increment, the zone of localized deformation is more planar both in the vertical and in the horizontal cuts. In fact, the post-peak localization band partly coincides with the pre-peak cone of localized deformation in the lower region, suggesting that the later shear zone evolved from the initial one. As for the previous increment, no clear sign of strain localization is visible in the X-ray sections of the specimen at the end of the increment. It is worth noting that the thickness of the shear bands in these strain volumes is exaggerated with respect to real size. This is due to the lower limit to resolvable thickness dictated by the dimensions of DIC subsets (280 μm side length), whereas CT images indicate that the thickness of the localization zones (be these dilating shear bands or open cracks) in the tested specimens was typically less than 70 μm.

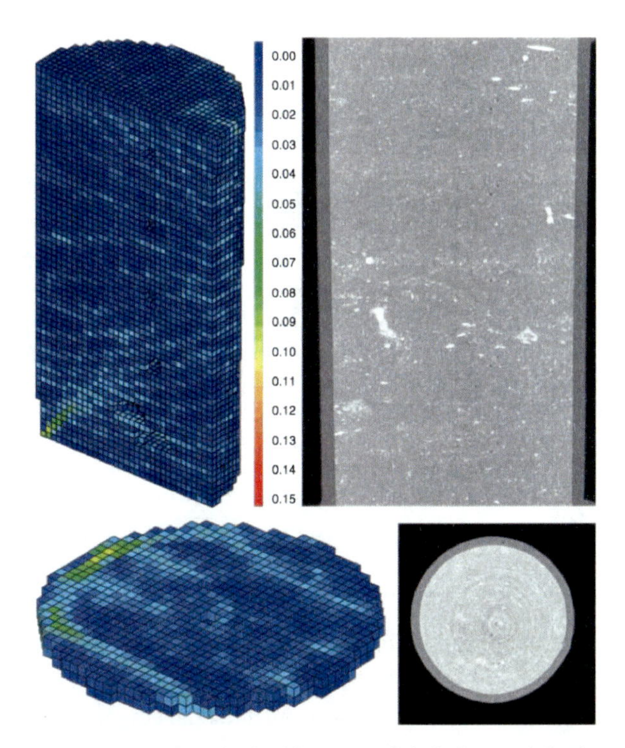

Fig. 9 DIC-derived deformation maps for load increment 2-3 (before peak load – see Fig. 6) of the triaxial test on the Callovo-Oxfordian argillite (∅ = 10 mm). Top and bottom right: vertical (along the specimen axis) and horizontal (close to the bottom of the specimen) sections through the CT volume image at step 3. Top and bottom left: incremental maximum shear strain maps for the equivalent sections (color scale is [0, 0.15]). (From [34])

One of the key conclusions from this study is that, through 3D DIC, it was shown that localized shear had initiated in the specimen already prior to the stress deviator peak. The important general message here is that the DIC analysis of 3D images from X-ray tomography can reveal patterns of deformation that could not be observed using only the gray scale images, in which they may remain hidden (if they

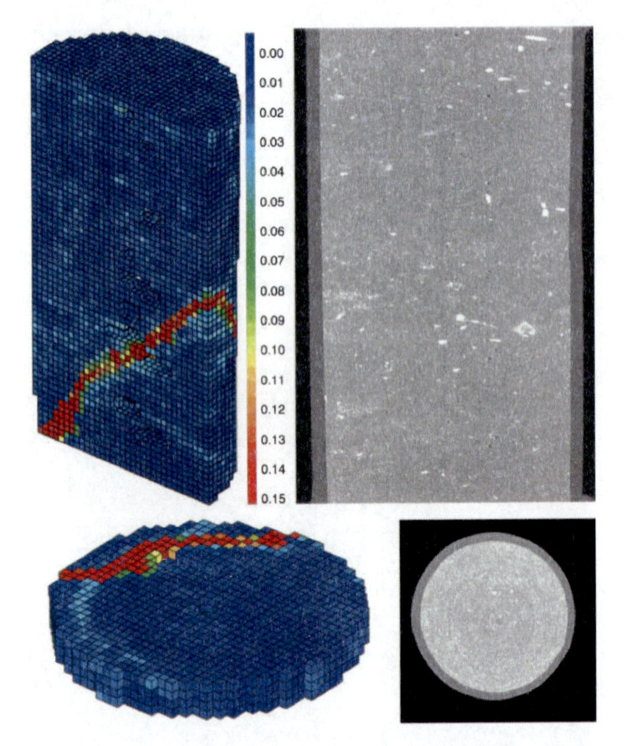

Fig. 10 As Figure 9 but load increment 3-4, i.e., after peak load. (From [34])

do not involve significant volume changes); this is even with high resolution synchrotron micro tomography.

5 Example 2: localized deformation in sand

As a second example, we present results from a triaxial compression test on a specimen of a natural granular material (Hostun sand) that was carried out with in-situ synchrotron X-ray micro tomography imaging, which provided complete high-resolution 3D images of the specimen at several stages throughout the test (see [28] for further details). X-ray images that capture the grain-scale detail of such materials are becoming more and more common (e.g., [36, 37, 42]) plus there are also a few examples of in-situ imaging where granular assemblies have been observed at different stages of deformation (e.g., [1, 36, 37]). Additionally, in the past, lower resolution X-ray tomography has been used to image the consequences of localized deformation in granular media (e.g., dilatant bands), but without being able to observe details at the grain-scale (e.g., the already cited [3] and [20]). Here the aim is to both observe the material evolution with grain-scale resolution and to image the deformation processes. To this end 3D-volumetric digital image correlation is employed to assess the 3D displacement and strain fields in the sample. To our knowledge this represents the first application of 3D-volumetric DIC to granular media; the granular nature of both the material (and thus images) and the kinematics represent new challenges. From such results the evolution of localized strain and its timing relative to the loading are investigated. It should be noted that here a natural sand, rather than an artificial granular medium such as glass beads, is investigated as this material is of interest in engineering applications but also as sand grains have (internal and surface) "character", which is important for DIC procedures.

The test described hereafter was carried out at the ESRF in Grenoble on beamline ID15A, using a setup very similar to the one described in the previous section for the experiments on the clay rock. Triaxial compression was performed on a dry specimen of S28 Hostun sand, under a confining pressure of 100 kPa. Deviatoric loading was strain controlled, with a screw driven piston descending at 60μm/min, which corresponds to quite a low strain rate (0.05 %/min for a 11 mm high specimen). The S28 (or RF) Hostun sand is a fine-grained, angular siliceous sand with a mean grain size (D_{50}) of about 300 μm. In this test, the sample was 11 mm in diameter and 22 mm high and had an initially dense packing. It should be noted that despite the small sample size (in comparison to standard triaxial tests on sands), the sample can be considered to be sufficiently large such that it remains mechanically pertinent (i.e., its response can be considered representative of that of a larger mass of the material); in fact the sample comprises roughly 50000 grains. These reduced dimensions were imposed by the X-ray imager width, which was just 14 μm (the sample needed to be smaller than this to not risk passing out of the field of view, although this does occur by the end of the test; see later).

X-ray tomography scans were carried out at key moments throughout the test, which are marked by (small) relaxations in the loading curve in Figure 11a. The sample stress-strain response shows a roughly linear initial trend followed by a curvature to the peak stress at around 11% nominal axial strain, after which the stress drops, to what is probably the beginning of a plateau, after which the test was stopped and the sample unloaded.

Figures 11b,c show a series of vertical slices through the X-ray tomography image volumes at different stages in the test (see Figure 11a). These slices are roughly perpendicular to the "plane" of localization that developed during the test. The voxel size of these images is $14\times14\times14$ μm^3, which represents a spatial resolution far greater than was previously possible with medical or laboratory scanners. As such it is possible to clearly identify the individual grains (recall that the mean grain size is around 300 μm or 21 voxels). These in-situ images show that the sample starts to lean to the right and there is a rotation of the upper platen in the latter part of the test, but there is no clear evidence of localized deformation. Porosities have been calculated from the images for each step based on overlapping cubic windows of side 61 voxels (854 μm) throughout the sample volume (see Figure 11d); these represent the accumulated porosity evolution from the start of the test. From these porosity fields an evolving inclined zone of localized dilation can be seen.

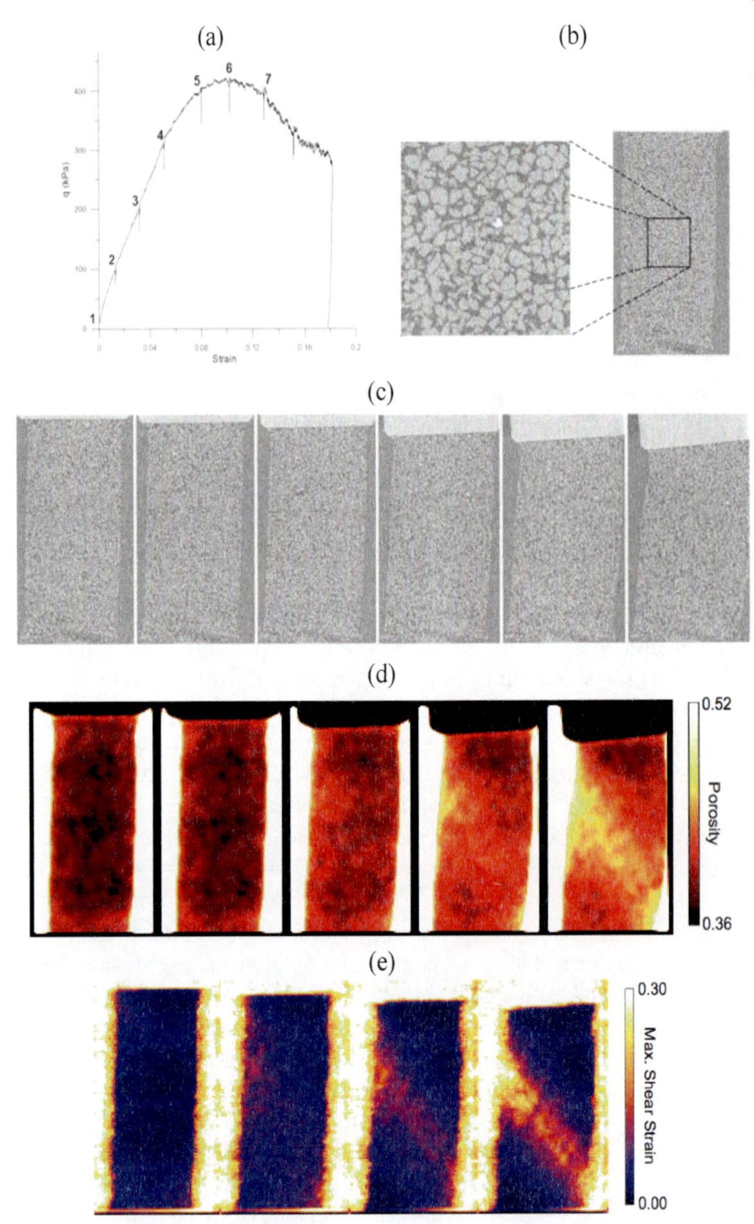

Fig. 11 (a) Deviatoric stress versus axial strain curve for the deviatoric loading part of the triaxial compression test. (b) Vertical slices through the 3D x-ray micro tomography image volumes of the sand specimen at stage 1 plus zoom and (c) for stages 2-7 of the triaxial compression. Median-projections, parallel to the evolved shear-band, of the 3D volumes of (d) the calculated porosity at stages 3-7 of the loading and (e) the DIC-derived incremental maximum shear strain for stages 3-4, 4-5, 5-6, 6-7. (Median values calculated for a projection across the sample parallel to the localisation feature). Note that the initial sample diameter was 11 mm. (From [28])

3D volumetric DIC has been carried out on consecutive pairs of 3D image volumes to provide the incremental displacement and strain fields (the results are thus averages over the given time interval). The key parameters in the DIC are the distance between the calculation nodes and the correlation window size; in this analysis these were, respectively, 20 voxels (or 280 μm) and a cube with sides of 21 voxels (or 294 μm) reduced to 11 voxels (or 154 μm) for the sub-pixel derivation.

Results from this DIC analysis indicate that, despite the granular nature of the material, smooth and relatively continuous displacement fields are measured. The evolved localization is evident as a band bounded by two "weak" discontinuities, i.e., strain are discontinuous, not the displacements. Figure 11e shows median projections of the 3D field of maximum shear strain $(\varepsilon_1 - \varepsilon_3)/2$ (where ε_1 and ε_3 are the major and minor principal strains) for increments 3-4, 4-5, 5-6 and 6-7. These strain images clearly show the evolution of a localized band that traverses the sample diagonally from left to right. It is important to note the zones of near zero strain away from the localized zone as these indicate that the DIC is clearly working well and the localization is real.

It is worth noting that the DIC strain analysis provides an incremental analysis, i.e., it indicates the deformation active in each strain increment. This is different from what can be seen with accumulated porosity changes shown in Figure 11d. As such it is seen from these incremental maps that the localization initiated in the increment 4-5, i.e., well before the peak load, and before this becomes clear in the porosity images. Furthermore, the localization is revealed to start as a broad zone that thinned with loading. In increment 6-7, the zone has a width of about 5 mm (i.e., about 17 D_{50}). It is also clear that the localized zone is not uniform, showing a degree of structure. The width of the band is in fact difficult to define. Recall that the strain images shown in Figure 11e are median projections across the volumes, which provides a global picture. However, one might also look at single slices through the 3D strain maps, which will reveal further details. For example, it appears that in places the shear band contains a narrower internal core of much higher strain. The DIC results show that the localized band is not uniform and, whilst it is well defined (and quite planar), there are aligned zones of either reduced or elevated strains at an angle "conjugate" to the main band. Note that Oda and coworkers [42] experimentally identified "columns" of aligned grains in a shear band in a sand; these columns were oriented similarly to the zones of low and elevated shear strains observed here. These aspects need to be further investigated, but they appear to be consistent with existing ideas that the underlying mechanism for shear banding in granular materials is by buckling of columns of grains (see for example [57]).

6 Conclusions

One of the key conclusions from both examples is that, through 3D DIC, it was shown that localized shear had initiated in the specimen already prior to the stress deviator peak. More generally, the important general message here is that complementing X-ray images with DIC significantly enhances their value. For the clay rock, the DIC analysis of 3D images from X-ray tomography revealed patterns of deformation that could not be observed using only the gray scale images. For the sand, the spatial resolution is such that the combination of these two methods can be used to study the kinematics of shear band formation *at the scale of the grains*. In fact, results from a "continuum" DIC analysis, like those presented in this paper, can only provide part of the story. It is clear that the deformation behavior of granular materials will involve significant relative displacements and rotations of the individual grains. In this respect, we have recently developed a completely original grain-scale "discrete" volumetric DIC method that permits the characterization of the full kinematics (i.e., 3D displacements and rotations) of all the individual sand grains in a specimen. This opens up new possibilities for understanding the mechanics of granular media (in three dimensions), as we can analyze individual grain kinematics and grain-grain interactions throughout the volume of a specimen and through the duration of a test [27]). The results thus far obtained indicate that a major step forward has been made, but the potential of the approach has still to be fully exploited.

Acknowledgements I would like to thank my close colleagues Steve Hall, Pierre Bésuelle, Nicolas Lenoir and Jacques Desrues, who are (at least) as much responsible as me of the results presented in this paper. I am also very grateful to Marco Di Michiel from the ESRF for his invaluable contribution to the X-ray CT experimental program, and to Michel Bornert at LMS (Paris) for his help with the DIC analysis for the clay rock. The experimental studies were conducted within the contexts of the French CNRS research network "GDR 2519 – Mesure de champs et identification en mécanique des solides", the ANR project μModEx (contract ANR-05-BLAN-0192), and the Ph.D. thesis by N. Lenoir funded by the French agency ANDRA.

References

1. Alshibli, K.A. and Alramhi, B.A. (2006) – Microscopic evaluation of strain distribution in granular materials during shear, *J. of Geotechnical and Geoenvironmental Engineering*, 132, 80-91.
2. Alshibli, K.A. and Hasan, A. (2008) – Spatial variation of void ratio and shear band thickness in sand using X-ray computed tomography, *Géotechnique*, Vol. 58, No. 4, 249–257.
3. Alshibli, K.A., Sture, S., Costes, N.C., Franck, M.L., Lankton, M.R., Batiste, S.N. and Swanson, R.A. (2000) – Assessment of localized deformation in sand using X-ray computed tomography, *Geotechnical Testing Journal*, Vol. 23, 274–299.
4. Alsini, A., Vardoulakis, I. and Drescher, A. (1992) – Deformation localization in cavity inflation experiments on dry sand, *Géotechnique*, Vol. 42, No. 1, 395–410.
5. Arthur, J.R.F. (1971) – New techniques to measure new parameters, *Proceedings of Roscoe Memorial Symposium on Stress–Strain Behaviour of Soils*, Cambridge, G.T. Foulis & Co., 340–346.
6. Arthur, J.R.F. and Dunstan, T. (1982) – Rupture layers in granular media, *Proceedings of IUTAM Conference on Defects and Failure in Granular Media*, Balkema, 453–459.
7. Arthur, J.R.F., Dunstan, T., Al-Ani, Q.A.J.L. and Assadi, A. (1977) – Plastic deformation and failure in granular media, *Géotechnique*, Vol. 27, No. 1, 53–74.
8. Baruchel, J. (2000) – *X-ray Tomography in Materials Science*, Paris, Editions Hermes.

9. Bay, B.K., Smith, T.S., Fyhrie, D.P. and Saad, M. (1999) – Digital volume correlation: three-dimensional strain mapping using X-ray tomography, *Experimental Mechanics*, Vol. 39, No. 3, 217–226.
10. Bésuelle, P., Desrues, J. and Raynaud, S. (2000) – Experimental characterisation of the localisation phenomenon inside a Vosges sandstone in a triaxial cell, *International Journal of Rock Mechanics & Mining Sciences*, Vol. 37, 1223–1237.
11. Bésuelle, P., Baud, P. and Wong, T. (2003) – Failure mode and spatial distribution of damage in Rothbach sandstone in the brittle-ductile transition, *Pure and Applied Geophysics*, Vol. 160, No. 5-6, 851–868.
12. Bésuelle, P., Viggiani, G., Lenoir, N., Desrues, J. and Bornert, M. (2006) – X-ray Micro CT for Studying Strain Localization in Clay Rocks under Triaxial Compression, in: *Advances in X-Ray Tomography for Geomaterials*, J. Desrues et al. Eds, ISTE, London, 35–52.
13. Bhandari, A.R. and Inoue, J. (2005) – Strain localization in soft rocks – a typical rate-dependent solid: experimental and numerical studies, *International Journal for Numerical and Analytical Methods in Geomechanics*, Vol. 29, 1087–1107.
14. Bornert, M., Doumalin, P., Maire, E. and Moulinec, H. (2004) – Full 3D investigation of the local strain field in particulate metal matrix composites, *Proc. 12th Int. Conf. on Experimental Mechanics ICEM12*, Bari, Italy, 1-8.
15. Bransby, P.J. and Blair-Fish, P.M. (1975) – Deformation near rupture surfaces in flowing sand, *Géotechnique*, Vol. 25, No. 2, 384–389.
16. Butterfield, R., Harkness, R.M., and Andrawes, K.Z. (1970) – A stereo-photogrammetric method for measuring displacements fields, *Géotechnique*, Vol. 20, No. 3, 308–314.
17. Colliat-Dangus, J.L., Desrues, J. and Foray, P. (1988) – Triaxial testing of granular soil under elevated cell pressure, in: *Advanced triaxial testing for soil and rocks*, R.T. Donaghe *et al.* Editors, STP977, ASTM, 290–310.
18. Desrues, J. (1984) – *La localisation de la déformation dans les matériaux granulaires*, Thèse de Doctorat es Science, USMG and INPG, Grenoble, France.
19. Desrues, J. (2004) – Tracking Strain Localization in Geomaterials Using Computerized Tomography, *Proc. of the International Workshop on X-ray CT for Geomaterials*, Kumamoto, Japan, November 6-7, 2003, Balkema, Lisse, The Netherlands, pp: 15-41.
20. Desrues, J., Chambon, R., Mokni, M. and Mazerolle, F. (1996) – Void ratio evolution inside shear bands in triaxial sand specimens studied by computed tomography, *Géotechnique*, Vol. 46, No. 3, 527–546.
21. Desrues, J. and Duthilleul, B. (1984) – Mesure du champ de déformation d'un objet plan par la méthode stéréophotogrammétrique de faux relief , *Journal de Mécanique Théorique et Appliquée*, Vol. 3, No. 1, 79–103.
22. Desrues, J. and Viggiani, G. (2004) – Strain localization in sand: an overview of the experimental results obtained in Grenoble using stereophotogrammetry, *International Journal for Numerical and Analytical Methods in Geomechanics*, Vol. 28, No. 4, 279 –321.
23. Desrues, J., Viggiani, G. and Bésuelle, P. (Editors) (2006) – *Advances in X-ray Tomography for Geomaterials*, ISTE, 452 pages.
24. Forsberg, F. and Sjödahl, M. (2004) – Tomographic 3D-DSP: measurement of internal deformations, *Proc. 12th Int. Conf. on Experimental Mechanics ICEM12*, Bari, Italy, 217–226.
25. Gudehus, G. and Nübel, K. (2004) – Evolution of shear bands in sand, *Géotechnique*, Vol. 54, 187–201.
26. Guler, M., Edil, T.B. and Bosscher, P.J. (1999) – Measurement of particle movement in granular soils using image analysis, *Journal of Computing in Civil Engineering*, ASCE, Vol. 13, No. 2, 116–122.
27. Hall, S.A., Bornert, M., Desrues, J., Pannier, Y., Lenoir, N., Viggiani, G. and Bésuelle, P. (2010) – Discrete and Continuum analysis of localised deformation in sand using X-ray micro CT and Volumetric Digital Image Correlation, *Géotechnique* (accepted for publication).
28. Hall, S.A., Lenoir, N., Viggiani, G., Desrues, J. and Bésuelle, P. (2009) – Strain localisation in sand under triaxial loading: characterization by X-ray micro tomography and 3D Digital Image Correlation, *Proceedings of the 1st Int. Symp. On Computational Geomechanics (ComGeo 1)*, IC^2E, 239-247.

29. Hicher, P.Y., Wahyudi, H. and Tessier, D. (1994) – Microstructural analysis of strain localisation in clay, *Computers and Geotechnics*, Vol. 16, 205–222.
30. Kawakata, H., Cho, A., Kiyama, T., Yanagidani, T., Kusunose, K. and Shimada H. (1999). – Three-dimensional observations of faulting process in Westerly granite under uniaxial and triaxial conditions by X-ray CT scan, *Tectonophysics*, 313, 293–305.
31. Kirkpatrick, W.M. and Belshaw, D. J. (1968) – On the interpretation of the triaxial test, *Géotechnique*, Vol. 18, No. 3, 336–350.
32. Kuo, C.-Y. and Frost, J.D. (1996) – Uniformity evaluation of cohesionless specimens using digital image analysis, *Journal of Geotechnical Engineering*, ASCE, Vol. 122, No. 5, 390–396.
33. Lenoir, N. (2006) – *Comportement mécanique et rupture dans les roches argileuses étudiés par micro tomographie à rayons X*, Ph.D. thesis, Grenoble (http://tel.ccsd.cnrs.fr/tel-00011996).
34. Lenoir, N., Bornert, M., Desrues, J., Bésuelle P. and Viggiani, G. (2007) – Volumetric digital image correlation applied to X-ray micro tomography images from triaxial compression tests on argillaceous rocks, *Strain - International Journal for Experimental Mechanics*, Vol. 43, No. 3, 193–205.
35. Liu, J. and Iskander, M. (2004) – Adaptive Cross Correlation for Imaging Displacements in Soils, *Journal of Computing in Civil Engineering*, ASCE, Vol. 18, No. 1, 46–57.
36. Matsushima, T., Katagiri, J., Uesugi, K., Nakano, T. and Tsuchiyama, A. (2007) – Micro X-ray CT at SPring-8 for Granular Mechanics, in: *Soil Stress-Strain Behavior: Measurement, Modeling and Analysis*, Ling et al. Eds., Springer, Netherlands, 225-234.
37. Matsushima, T., Uesugi, K., Nakano, T. and Tsuchiyama, A. (2006) – Visualization of Grain Motion inside a Triaxial Specimen by Micro X-ray CT at SPring-8, in: *Advances in X-Ray Tomography for Geomaterials*, J. Desrues et al. Eds, ISTE, London, 35–52.
38. Mees, F., Swennen, R., Van Geet, M., and Jacobs, P. (2003) – Applications of X-ray computed tomography in the geosciences, Geological Society, London, Special Publications; v. 215.
39. Michalowski, R.L. (1990) – Strain localization and periodic fluctuations in granular flow processes from hoppers, *Géotechnique*, 40, 389-403.
40. Muhunthan, B. and Chameau, J.L., (1997) – Void fabric tensor and ultimate state surface of soils", ASCE *J. Geotech. GeoEnviron. Engrg.* 123, 173–181.
41. Muir Wood, D. (2007) – The magic of sands, *Canadian Geotechnical Journal*, Vol. 44, No.11, 1329-1350.
42. Oda, M., Takemura, T. and Takahashi, M. (2004) – Microstructure in shear band observed by microfocus X-ray computed tomography, *Géotechnique*, Vol. 54, 539-542.
43. Orteu, J.J. (2009) – 3-D computer vision in experimental mechanics, *Optics and Lasers in Engineering*, Vol. 47, No. 3-4, 282-291.
44. Otani, J., Mukunoki, T. and Obara, Y. (2000) – Application of X-ray CT method for characterization of failure in soils, *Soils and Foundations*, Vol. 40, 111–118.
45. Otani, J., Mukunoki, T. and Obara, Y. (2002) – Characterization of failure in sand under triaxial compression using an industrial X-ray scanner, *International Journal of Physical Modelling in Geotechnics*, Vol. 1, 15–22.
46. Otani, J. and Obara, Y. (Editors) (2004) – *X-ray CT for Geomaterials – Soils, Concrete, Rocks*, Balkema.
47. Raynaud, S., Fabre, D., Mazerolle, F., Géraud, Y. and Latière, H.J. (1989) – Analysis of the internal structure of rocks and characterization of mechanical deformation by a non-destructive method: X-ray tomodensitometry, *Tectonophysics*, Vol. 159, 149–159.
48. Rechenmacher, A.L., and Finno, R.J. (2004) – Digital image correlation to evaluate shear banding in dilative sands, *Geotechnical Testing Journal*, Vol. 27, No. 1, 13–22.
49. Roscoe, K.H. (1970) – The influence of strains in soil mechanics, *Géotechnique*, Vol. 20, No. 2, 129–170.
50. Roscoe, K.H., Arthur, J.R.F. and James, R.G. (1963) – The determination of strains in soils by an x-ray method', *Civ. Eng. Public Works Rev.*, 58, 873–876 and 1009–1012.

51. Russel, S.S. and Sutton, M.A. (1989) – Strain-field analysis acquired through correlation of X-ray radiographs of a fiber-reinforced composite laminate, *Experimental Mechanics*, Vol. 29, No. 2, 237–240.
52. Scarpelli, G. and Muir Wood, D. (1982) – Experimental observations of shear band patterns in direct shear tests, *Proceedings of IUTAM Conference on Defects and Failure in Granular Media*, Balkema, 473–484.
53. Smith, T.S., Bay, B.K. and Rashid, M.M. (2002) – Digital volume correlation including rotational degrees of freedom during minimization, *Experimental Mechanics*, Vol. 42, No. 3, 272–278.
54. Sutton, M.A., Cheng, M., Peters, W.H., Chao, Y.J. and McNeill, S.R. (1986) – Application of an optimized digital correlation method to planar deformation analysis, *Image and Vision Computing*, 4,143–150.
55. Synnergren, P., Goldrein, H.T. and Proud, W.G. (1999) – Application of digital speckle photography to flash x-ray studies of internal deformation fields in impact experiments, *Applied Optics*, Vol. 38, No. 19, 4030–4036.
56. Tillard-Ngan, D., Desrues, J., Raynaud, S. and Mazerolle, F. (1992) – Strain localisation in the Beaucaire marl, in: *Geotechnical Engineering of Hard Soils–Soft Rocks*, Balkema, 1679–1686.
57. Tordesillas, A. and Muthuswamy, M. (2009) – On the modeling of confined buckling of force chains, *Journal of the Mechanics and Physics of Solids*, Vol. 57, No. 4, 706-727.
58. Vardoulakis, I. and Graf, B. (1982) – Imperfection sensitivity of the biaxial test on sand, *Proceedings of IUTAM Conference on Defects and Failure in Granular Media*, Balkema, 485–491.
59. Verhulp, E., van Rietbergen, B. and Huiskes R. (2004) – A three-dimensional digital image correlation technique for strain measurements in microstructures, *Journal of Biomechanics*, Vol. 37, 1313–1320.
60. Viggiani, G. and Hall, S.A. (2008) – Full-field measurements, a new tool for laboratory experimental geomechanics. Keynote paper, in: *Proceedings of the 4th International Symposium on Deformation Characteristics of Geomaterials* (Burns S.E., Mayne P.W and Santamarina J.C. (Edts.)), IOS Press, Atlanta, 2008, 3-26.
61. Viggiani, G., Lenoir, N., Bésuelle, P., Di Michiel, M., Marello, S., Desrues, J. and Kretzschmer, M. (2004) – X-ray micro tomography for studying localized deformation in fine-grained geomaterials under triaxial compression, *Comptes rendus Mécanique*, Vol. 332, 819–826.
62. Vinegard, H.J., de Waal, J.A. and Wellington, S.L. (1991) – CT studies of brittle failure in Castlegate sandstone, *Int. J. Rock Mech. Min.*, Vol. 28, 441–448.
63. White, D.J., Take, W.A. and Bolton, M.D. (2003) – Soil deformation measurement using particle image velocimetry (PIV) and photogrammetry, *Géotechnique* Vol. 53, No. 7, 619–631.

Two-dimensional Distinct Element Method (DEM) modeling of tectonic fault growth in mechanically layered sequences

Martin P.J. Schöpfer, Conrad Childs and John J. Walsh

Abstract The 2D Distinct Element Method (DEM), in which rock is represented as an assemblage of cylindrical particles that can be cemented together, is used for modeling the growth of normal faults in layered sequences. The outcrop-scale models are comprised of strong and weak layers, modeled as bonded and non-bonded particles, respectively, which are deformed at various confining pressures using a pre-defined fault at the base of the periodically layered sequence. The models successfully reproduce a wide range of fault geometries seen in outcrop, such as lithologically controlled fault dip variations, fault bifurcation and segmentation, and also illustrate the progressive removal of asperities and the linkage of fault segments, which leads to the formation of fault bound lenses. A sensitivity study reveals that fault zone complexity decreases with decreasing strength of the cohesive beds and increasing confining pressure. A suite of models comprised of sequences with different proportions of strong and weak beds suggests that the overall fault zone dip increases with decreasing fraction of weak material in the sequence.

1 Introduction

Tectonic faulting is the predominant deformation mechanism in the upper crust. Following Anderson's theory of faulting (Anderson 1905, Anderson 1951) three end-member fault types can be distinguished. These are normal, reverse and strike-slip faults which develop if the vertical stress component is the maximum, minimum or

Martin P.J. Schöpfer
Fault Analysis Group, UCD School of Geological Sciences, University College Dublin, Belfield, Dublin 4, Ireland, e-mail: martin@fag.ucd.ie

Conrad Childs and John J. Walsh
Fault Analysis Group, UCD School of Geological Sciences, University College Dublin, Belfield, Dublin 4, Ireland

intermediate principal stress component respectively (throughout this paper compressive stresses are positive and $\sigma_1 > \sigma_2 > \sigma_3$.

The Coulomb-Mohr failure angle hypothesis (Mohr 1882) states that the acute angle between two conjugate faults is $90° - \phi$ (where ϕ is the friction angle) and that the intersection of conjugate pairs of faults is parallel to the intermediate principal stress axis. Using this hypothesis and assuming friction angles of 30°–40°, the dip of the three end-members of faults may be predicted. Normal faults should exhibit dips of 60°–70°, reverse faults are expected to dip 20°–30°, and strike slip faults should be vertical.

The Andersonian theory of faulting, which is based on the Coulomb-Mohr failure angle hypothesis and the assumption that one of the principal stresses is vertical has been applied successfully to faults on all scales from large crustal scale faults (e.g., Mandl 1988) to small faults in isotropic soils (e.g., Parry 1995). Generally speaking Andersonian theory is in agreement with outcrop and experimental observations of fault orientations in homogeneous materials. However the dips of faults offsetting mechanically heterogeneous rocks are not as simple as Andersonian theory predicts. Faults are generally not simple planar discontinuities as many textbook diagrams suggest but often exhibit complex dip and/or strike changes on various scales and display complex patterns of segmentation (Fig. 1). Providing a mechanical basis for this complexity using Distinct Element Method (DEM) modeling is the main aim of this study.

2 Geometry of faults in layered sequences

When a fault surface propagates through a rock volume it rarely does so as a single continuous surface but as an irregular and, to a greater or lesser extent, segmented array. Segmentation is due to local retardation or acceleration of a propagating fault tip-line controlled by the mechanical heterogeneity of the rock volume; this process has been referred to as 'tip-line bifurcation' (Childs et al. 1996b). Following bifurcation the different segments within a fault array remain kinematically and mechanically linked components of a single structure.

The lithological control on fault segmentation and the formation of segment boundaries is often clearly observed in cross-sections of normal faults offsetting sub-horizontal sequences with high competence contrast between interbedded lithologies (Peacock and Zhang 1993, Childs et. al. 1996a; Fig. 1). Overstepping segments in cross-section, may, with increasing displacement, link together to give a staircase fault geometry where fault dip changes across bedding interfaces i.e. fault plane refraction (as in Fig. 1a). Fault refraction can also from differences in the angle of internal friction of interbedded lithologies or can be caused by different types of faulting (tensile, hybrid or shear failure, e.g., Sibson 1998) in different layers, depending on the effective stress and the strength of the interbedded layers (Fig. 2; see Section 3). For example, steeply dipping fault segments within the stronger layers

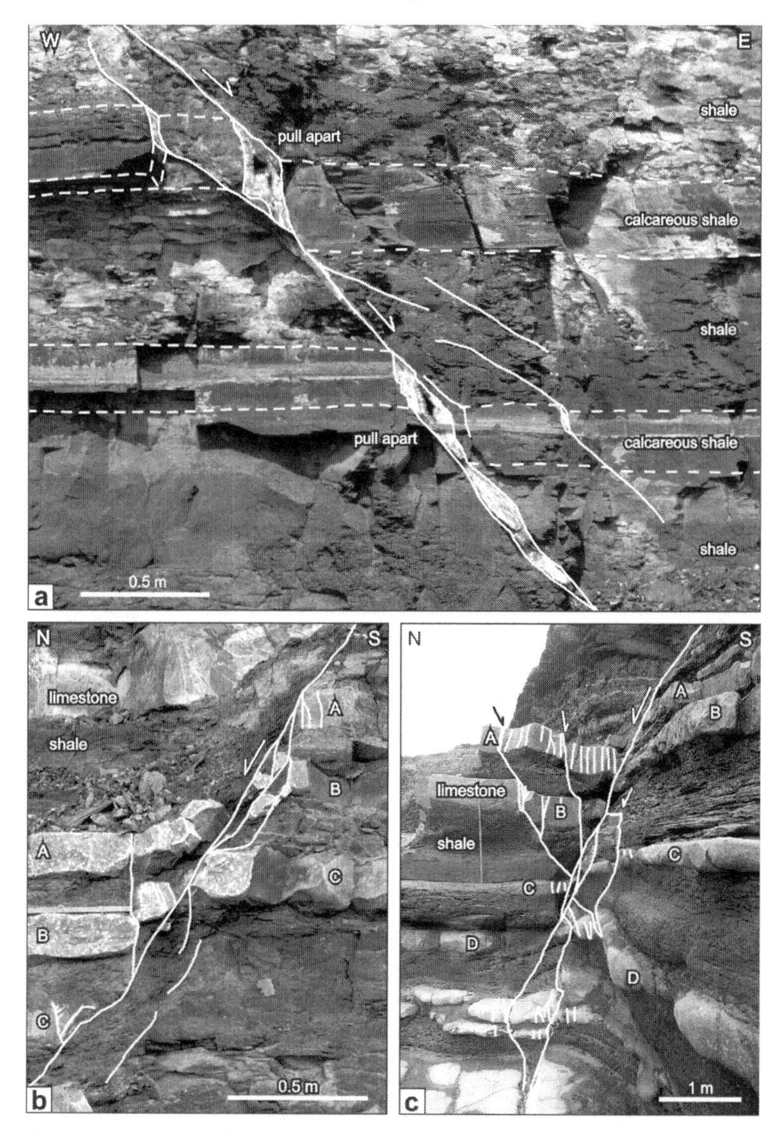

Fig. 1 Field examples of small normal faults in mechanically layered sequences. **a** Fault exhibiting lithologically controlled dip variations, with vertical fault traces in the strong, calcareous shale beds, and $\approx 50^{\circ}$ dipping faults in the weaker shale beds. Fault displacement along this refracting fault led to the development of pull-aparts, which are partially infilled with calcite. Kimmeridge Clay Formation, Dorset, UK. **b** and **c** Complex fault zones in limestone shale sequences, Kilve foreshore, Somerset, UK. The fault in **b** contains fault bound lenses, comprised of fractured and rotated limestone blocks, and a synthetic hangingwall splay. The fault in **c** also contains fault bound lenses at various scales. Two antithetic fault splays exhibit lithologically controlled fault dip variations, with steep fault traces in the strong limestone beds.

of a bedded sequence are often attributed to initial Mode I fracturing (Fig. 1a), i.e. movement of the walls normal to the fracture plane.

Another class of fault zone complexity is splays. A splay is a minor fault that branches from a larger main fault (Figs. 1b and 1c). Splays can develop during the propagation process at the tip of a fault (as horsetails), to accommodate space problems resulting from slip around fault bends or due to linkage of initially overlapping faults (as breaching faults). Splays along dip slip faults, especially in layered sequences, are very common structural features. Typically splays have the same slip direction as the master fault and are referred to as synthetic splays (Fig. 1b). Splays which have a dip direction and sense of offset which is opposed to the main fault, termed antithetic splays (Fig. 1c), typically form to accommodate space problems caused by irregularities or dip variations of the master fault (e.g., Mandl 1988).

The bifurcation, overstep formation and breaching processes described above ultimately result in the incorporation of fault bound lenses within fault zones. Irregularities on fault surfaces, for example due to fault refraction, are zones of concentrated strain which will be bypassed by the formation of a later, and presumably straighter fault surface; this process has been referred to as asperity bifurcation (Childs et al. 1996b). Once such a lens has been incorporated into a fault zone it will continue to be deformed and transected by secondary shears and eventually will become fault rock (Childs et al. 2009). These processes, together with the formation of splays result in the formation of extremely complex internal fault zone structure. While conceptual models of the development of fault zone complexity have been developed, the detailed mechanical, quantitative and predictive constraints are not available. Development of predictive approaches to fault zone structure would be of significant economic importance (e.g., Mandl 1988, 2000) in, for example, predicting the location and character of branch lines which may act as conduits for mineralizing brines or risking the likelihood of segmentation of faults bounding potential hydrocarbon traps. A simple approach for predicting fault dip variations in layered sequences (Fig. 1a) is given in the next Section.

3 Origin of fault dip refraction

Fault dip changes (refraction; Fig. 1a) can be attributed to a variety of mechanisms (Ferrill and Morris 2003, and references therein): (i) post-faulting differential compaction, (ii) active faulting, with slip along layers or intersecting faults, (iii) linkage of an originally vertically-segmented fault and (iv) fault initiation with dip controlled by rock properties and effective stresses. Two of these mechanisms (iii and iv), which are not mutually exclusive, underpin the most popular models for the growth of faults within layered sequences.

A quantitative insight into fault dip variations in mechanically layered sequences can be obtained by means of a Mohr diagram (Fig. 2a). The strong layers are represented using a Mohr envelope given by the 2D Griffith criterion that merges with a Coulomb criterion where the slope is identical (Sibson 1998). The weaker lay-

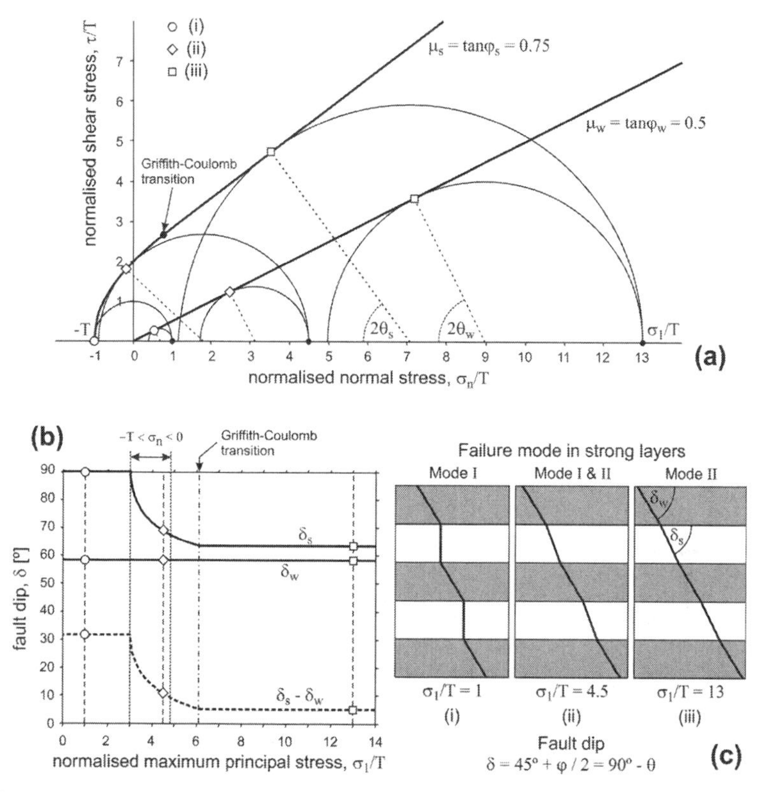

Fig. 2 Mechanical explanation for lithologically controlled dip variations. **a** Mohr diagram in which failure envelopes for a strong cohesive, and a weak incohesive, material are plotted. Three solutions (i, ii, iii) for synchronous normal faulting in those materials are shown with different symbols (○, ◇ and □). The normal and shear stress acting on planes along which failure occurs is plotted as open symbols according to the normalized maximum principal stress, σ_1/T (○ = 1, ◇ = 4.5 and □ = 13). **b** Plot of fault dip vs. σ_1/T for the two failure envelopes shown in **a**. The fault dip difference between the strong and weak material is also plotted as dashed line. **c** Diagrams illustrating the fault traces for the three different examples for which Mohr's circle solutions are shown in **a**.

ers are represented with a Coulomb criterion without cohesion with a friction angle which is lower than the friction angle of the strong beds. The abscissa and ordinate in the Mohr diagram shown in Fig. 2a are normal and shear stress, σ_n and τ respectively, normalized by the tensile strength T of the strong beds. Mechanical equilibrium demands that the normal stress acting on the interface between the two different materials is identical. In the example shown the maximum principal stress σ_1 is identical, both in magnitude and direction, in the two materials and normal to the interface. This corresponds to an Andersonian stress state of normal faulting and horizontal layering. In order to obtain the fault dip in both materials it is necessary to assume that the onset of faulting occurs simultaneously, that is, the two Mohr circles describing the stress states in both layers touch their corresponding failure envelopes (Fig. 2a).

Three solutions that satisfy the above mentioned criteria are shown in Fig. 2a (examples i, ii and iii). Because the weak material has no cohesion and is characterized by a straight failure envelope the fault dip in those beds is always $45° + \phi/2$ (i.e. 58°; Fig. 2b). Example (i) is at very low effective vertical, or overburden, stress ($\sigma_1/T = 1$; circles) and the solution predicts that the strong layers fail via vertical extension fractures (Mode I), resulting in significant fault dip variations (dip difference $> 30°$; Figs. 2b and c). Example (ii) is for a higher maximum principal stress ($\sigma_1/T = 4.5$; diamonds) and the Mohr circle of the stronger beds touches the envelope where the normal stress on the failure plane is greater than the minimum possible stress ($-T$) but still negative (i.e. tensile). Hence the fault dip in the strong bed is smaller than in the previous example and the failure type is mixed (Mode I & II), resulting in fault dip difference between the beds of $\approx$10° (Figs. 2b and c). In example (iii) the interface normal stress is so high ($\sigma_1/T = 13$; squares) that the strong beds fail in shear (Mode II), with fault dips of $45° + \phi/2$ (i.e. 63°; Fig. 2b). Hence the dip difference of faults within the strong and weak beds is half the friction angle difference (i.e. 5°; Figs. 2b and c). If the friction angles of the strong and weak beds were identical then no dip difference would be expected.

Although this analysis can give a somewhat quantitative insight into the mechanical origin of fault dip variations in layered sequences it is associated with a number of assumptions which are not necessarily correct in a natural system: (i) Laboratory data of rocks indicate that the angle θ between failure planes and the maximum principal stress increases with increasing confinement, but that the θ-angle is typically smaller than predicted by the slope of the failure envelope ($45° - \phi/2$). (e.g., Paterson and Wong 2005) (ii) Laboratory data also show that neither the 2D nor the 3D Griffith criterion accurately predict either the failure angle or the stress state during failure (Ramsey and Chester 2004) (iii) The assumption of rigidity and that failure occurs simultaneously in the strong and weak beds may not be easily justified, because field examples demonstrate that failure occurs first in the stiffer and more brittle beds, while the weaker beds deform permanently in a ductile manner (e.g., Peacock and Zhang 1993) (iv) Faulting in mechanically layered sequences is often characterized by the development of precursory monoclines (the amplitude prior to failure depends on a number of factors, e.g. rock rheology, pressure/temperature and strain rate). Folding of beds leads to interfacial shear stress which can rotate the orientations of the principal stress within the beds and hence modify the fault dips (Mandl 1988).

The main limitation of this approach is, however, that it is impossible to predict the complexity of faults (e.g. segmentation; see Section 2) and how it changes as a function of strength contrast, confining pressure and the proportion of weak and strong beds comprising the sequence. In the next Section the numerical method utilized in this study to address these and other questions is presented.

4 Distinct Element Method

The Discrete Element Method (DEM) is a broad class of methods that models finite displacements and rotations of discrete bodies (Cundall and Hart 1992). The DEM is capable of modeling the growth of discontinuities, such as faults, without the limitations of continuum mechanics. The elements interact with each other via a force displacement law and can be of arbitrary shape, rectangular blocks and spheres being the most common ones. The DEM has been used to model a wide range of tectonic processes such as the formation of shear zones and deformation bands (Morgan and Boettcher 1999, Kock and Huhn 2007), displacement transfer and linkage of pre-existing faults (Walsh et al. 2001, Imber et al. 2004), normal faulting in layered sequences (Schöpfer et al. 2006, 2007a, 2007b, Egholm et al. 2008), fault bend folding (Strayer et al. 2004, Benesh et al. 2007), fault-propagation folding (Cardozo et al. 2005, Hardy and Finch 2007), fold and thrust belts (Burbridge and Braun 2002, Strayer and Suppe 2002) and collisional orogens (Vietor and Oncken 2005).

In this study a 2D approach with cylindrical particles as implemented in commercially available PFC2D software (Itasca Consulting Group 1999) is used. A comprehensive review of PFC is provided in Potyondy and Cundall (2004). A more technical discussion of the DEM with emphasis on granular mechanics and programming is given in Pöschel and Schwager (2005). A general overview of various (mainly engineering) applications of the DEM (blocks and spherical elements) can be found in the monograph by Jing and Stephansson (2007).

4.1 Particle Flow Code (PFC)

The commercially available Particle Flow Code in two dimensions (PFC2D, Itasca Consulting Group 1999) models the movement and interaction of circular particles using the DEM. The particles are treated as rigid discs and are allowed to overlap at particle-particle and particle-wall contacts. Walls are rigid boundaries that allow the user to define boundary conditions, e.g. constant velocity or stress, but are not accelerated due to interaction with particles. The amount of overlap, U^n, is small compared to particle size and is proportional to the contact force (Fig. 3a). Both normal F^n and shear forces F^s arise at contacts; the contact plane orientation is normal to the vector joining the centers of two particles and the contact point is located at the centre of the overlap area (Fig. 3a). The shear force F^s is computed in an incremental fashion. When the contact is formed, the shear force is initialized to zero. Each subsequent relative shear-displacement increment ΔU^s produces an increment of elastic shear force ΔF^s that is added to the current shear force. The contact slip model is an intrinsic property of two entities (either particle-particle or particle-wall) and is always active. The slip model is defined by a particle (or wall) contact friction coefficient μ_c (by default the minimum friction of the two entities forming the contact is taken), which limits the maximum shear force at the contact (Fig. 3a).

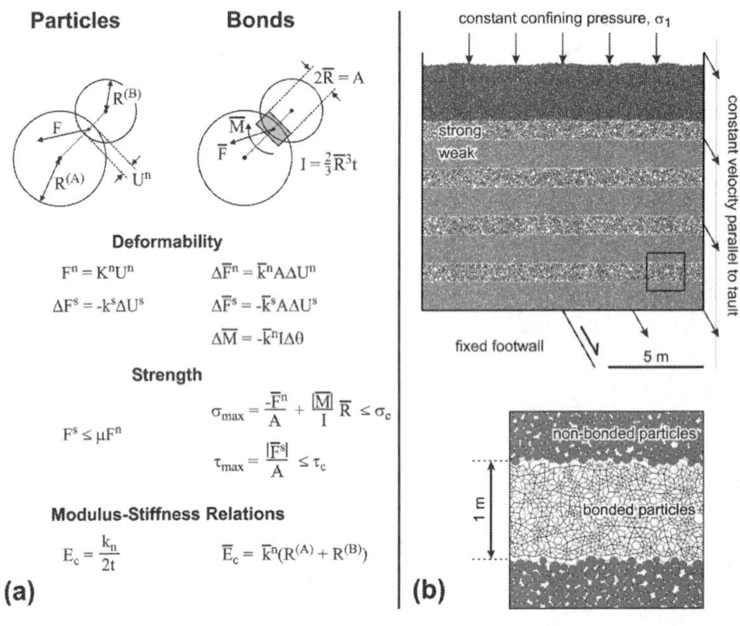

Fig. 3 Principles of PFC2D and model boundary conditions. **a** Force-displacement behavior of particle-bond system (after Potyondy and Cundall, 2004). Normal and shear forces, F^n and F^s, develop at particle contacts due to normal and tangential displacements, U^n and U^s. At unbonded contacts the shear force is limited by a contact friction coefficient, μ. At bonded contacts additional normal and shear forces and a moment $\overline{M}$ develop, where A and I are the area and moment of inertia of the bond cross-section, respectively. The bond breaks if either the maximum normal or shear stress acting on the bond periphery, σ_{max} and τ_{max}, exceeds the normal or shear strength, σ_c and τ_c, of the bond, respectively. The equations given are for cylindrical particles, with thickness t (in the present study t = 1). **b** PFC2D model consisting of > 23,000 bonded (white) and non-bonded (grey) cylindrical particles. The blow-up of the model shows particles joined by bonds (drawn for clarity as lines joining particle centers) and illustrates the resolution of the models.

Bonds can exist between particles, but not between particles and walls. If either the tensile σ_c or shear strength τ_c is exceeded the bond will break and is removed from the system (Fig. 3a). In contrast to the often-used contact bond (e.g., Strayer and Suppe 2002, Benesh et al. 2007), which does not have stiffness and width and can only transmit force, a parallel bond represents a cemented contact which can transmit both force $\overline{F^n}$ and $\overline{F^s}$) and moment $\overline{M}$ (Potyondy and Cundall 2004; Fig. 3a).

The calculation cycle in PFC is a time stepping algorithm that requires the repeated application of the law of motion to each particle (force = mass times acceleration), a force displacement law (Fig. 3a) to each contact, and a constant updating of wall (rigid boundaries) positions. The equations of motion are integrated in PFC using a centered finite-difference scheme, which requires a timestep. PFC calculates automatically the critical timestep (which depends on mass and stiffness of the system) that is necessary to ensure that the solution produced is stable and uses an actual timestep which is a fraction of that critical timestep.

4.2 Faulting model materials and boundary conditions

In contrast with continuum methods, where the rheology of the model material is defined using constitutive laws, the macroscopic response of the (bonded) particle assemblage in DEM models has to be calibrated using a numerical laboratory. The microproperties (particle size and size distribution, particle and bond modulus, contact friction, bond strength) are adjusted, mainly by trial and error, to obtain the desired model macroscopic response calibrated to laboratory rock deformation data.

The particles in this study have a uniform size distribution with radii ranging from 31.25 to 62.50 mm; hence particles represent volumes of rock, rather than individual grains. The particles have contact Young's modulus E_c of 50 GPa and a normal to shear stiffness ratio k_n/k_s of 3.0 (the normal stiffness is given by the modulus-stiffness relation provided in Fig. 3a). The contact friction coefficient μ_c is 1.0 for particles comprising the strong layers, and 0.5 for particles within the weaker interbeds and the top layer (Fig. 3b). The bonds between particles comprising the strong layers have a contact Young's modulus $\overline{E_c}$ of 50 GPa and a normal to shear stiffness ratio $\overline{k_n}/\overline{k_s}$ of 3.0 (the normal stiffness is given by the modulus-stiffness relation given in Fig. 3a). The width of each bond $\overline{R}$ is equal to the radius of the smaller particle (Fig. 3a). The bond strength was varied in this study, with average bond normal strengths (σ_c) of 300, 250, 200 and 150 MPa and average bond shear strengths (τ_c) equal to half the normal strength. The bond normal and shear strengths are randomly selected from a normal distribution with coefficients of variation of 1/12 and 1/6, respectively.

The strength of bonded materials is sample size dependent (Potyondy and Cundall 2004, Schöpfer et al. 2007a,), thus proper calibration requires tests on samples at a scale appropriate to the model. In the multilayer models presented in this paper, the basic mechanical unit is one bed; hence the width of the calibration samples is equal to the thickness of the strong layers in the multilayer models, i.e. 1 m (see below). The mechanical properties of the bonded material were investigated using unconfined compression tests. These tests were used for calculating the bulk elastic properties (Young's modulus E, Poisson's ratio ν) and provided the unconfined compressive strength (UCS). Additionally direct tension tests on dog-bone shaped samples with a central thickness of 1 m were performed at various confining pressures in order to define the failure envelope in the tensile stress field ($\sigma_3 < 0$).

The mechanical properties of the non-bonded material were investigated using confined compression tests. Since the material is cohesionless and exhibits no (bulk) elasticity the only bulk property that was calculated for each test is the friction coefficient.

The four different bonded materials (σ_c of 300, 250, 200 and 150 MPa) have similar elastic properties, with a Young's modulus range of 21–22 GPa and Poisson's ratios in the range of 0.24–0.31. The average UCS values systematically decrease with decreasing bond strength and are 128, 106, 83 and 64 MPa. The tensile strengths T of these four materials are about one third of their UCS, i.e. lower than for natural rock where UCS/T is > 10, because circular discs with a narrow particle size range and no rolling resistance are used (Schöpfer et al. 2009). Coulomb failure envelopes with a tension cut-off were fitted to the peak stress data of the four different bonded materials. A Coulomb criterion without cohesion provides a good fit to the non-bonded material. The four bonded materials exhibit friction angles ϕ ranging from 27–29°, the non-bonded of 25°. Again, these friction angles are relatively low compared to natural rock (with the exception of phyllosilcate-rich rock) because circular particles with no rolling resistance are used (Schöpfer et al. 2009). Nevertheless, apart form UCS/T and to some degree ϕ, the bonded materials match the behavior of strong, brittle, sedimentary rocks (limestones, sandstones), whereas the non-bonded material is similar to weak, ductile rock (e.g. shale).

Multilayer models are created using the specimen generation procedure described in Appendix A in Potyondy and Cundall (2004); models are 15 m wide, 13 m high and consist of $> 23{,}000$ particles (Fig. 3b). Layering is introduced by assigning particles to three different groups, strong layers, weak layers or the top layer. The top layer is 3 m thick and its primary function is to act as a buffer between the multilayer sequence and the topmost layer of particles to which the confining pressure is applied. The model is confined by applying a force equal to a particle diameter times the desired stress to particles at the surface of the model. After model confinement, bonds are installed between particles comprising the strong layers, which in this study are always 1 m thick and interbedded with weak, i.e. non-bonded, layers of variable thickness. Localization of a single through-going fault is achieved by introducing a pre-cut 'fault' at the base of the multilayer sequence. The dip of the basement fault is 60° and the L-shaped wall on the hangingwall side of the pre-cut fault moves downward with constant velocity (Fig. 3b).

5 Results

In this Section the main findings of the numerical modeling are summarized. Section 5.1 is a summary of a study in which the early stages of fault growth (i.e. localization), within one model are investigated (Schöpfer et al. 2006). Section 5.2 is a brief description of one particular model that illustrates how fault bound lenses, similar to those shown in Figs. 1b and c, develop (Schöpfer et al. 2007a). In Section 5.3 the impact of layer strength and confining pressure on fault zone structure is illustrated using a suite of 16 models (Schöpfer et al. 2007b). In Section 5.4 multilayer model results for various proportions of weak material in the periodically layered sequences are shown. A geometric model that can explain the observed fault dip variations is then described.

5.1 Fault localization

Normal faults propagating through mechanically layered sequences at low effective vertical stress are expected to exhibit fault dip variations, with steeply dipping faults in the strong beds, and shallowly dipping faults in the weaker interbeds (Fig. 2). The Mohr failure envelope analysis presented in Section 3 assumes that the onset of faulting is synchronous in the strong and weak beds, an assumption which might not always be valid. A multilayer model in which the strong beds have an unconfined compressive strength of 106 MPa was faulted at a confining pressure of 23 MPa, which, assuming lithostatic conditions and an overburden density of 2500 $\mathrm{kg\,m^{-3}}$, corresponds to a depth of faulting of approximately 1 km. The model was analyzed in 1 cm throw increments up to a final throw of 10 cm, at which a through-going fault developed. Because the displacement of the fault is small in comparison to particle size, contours of incremental maximum shear strain are plotted to visualize fault localization. Stress and strain paths for selected regions that straddle the (future) fault are also analyzed to gain an insight into how the modeled fault localizes (e.g. failure mode).

Four selected stages of fault localization are shown in Fig. 4a. Each of these diagrams represents a stage where one of the four strong beds (plotted in white) failed in the previous 1 cm throw increment. The fault within the strong beds localizes first as steeply dipping fractures which propagate from the top of each bed towards the bottom. This is interpreted to be due to outer-arc extension within a precursory monocline (elastic flexure of beds). Only after all four strong beds have failed do inclined shear bands, which link the steeply dipping fractures in the strong beds, develop within the weaker interbeds. Although the fault generally propagates from bottom to top, the topmost bed failed before the underlying bed (compare t = 8 cm with t = 9 cm).

This interpretation is supported by stress-strain paths of two selected localities, A and B, as shown in Fig. 4a at t = 9 cm. The volumetric strain in region A, which is located in the topmost strong bed, increases due to layer parallel extension; this extension is due to the horizontal component of displacement along the pre-defined fault. The elastic flexure of the bed is supported by a plot of the rotational component of strain (Fig. 4b). Horizontal extension leads to a decrease in minimum principal stress, which is sub-horizontal. At a throw between 7 and 8 cm, the topmost strong bed fails with an associated dramatic increase in volumetric strain, and an increase in the rotation rate and stress relaxation (reflected in increases in minimum principal stress). The stress-strain behavior within the strong bed is therefore consistent with the geometric interpretation, that vertical fractures within the strong layers nucleate as Mode I fractures.

The weaker interbeds exhibit, not surprisingly, a quite different stress-strain history. Elastic flexure of the adjacent strong beds leads to compaction and rotation (transpression) and an increase in maximum principal stress (which is subvertical). The last strong bed fractures at a throw of 8–9 cm, after which shear bands in the weak layers develop. Localization in the weak material leads to a volume increase, an increase in the rotation rate and a decrease in mean stress (both σ_1 and σ_3 de-

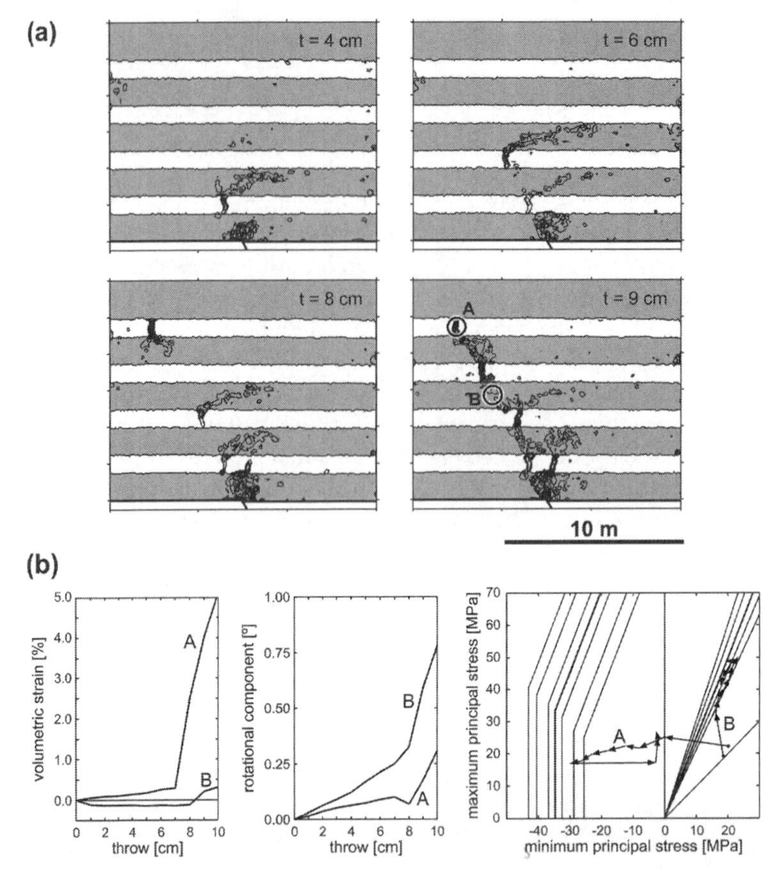

Fig. 4 Localization of a normal fault in a PFC2D model. Unconfined compressive strength of strong beds is 106 MPa and confining pressure is 23 MPa. **a** Incremental maximum shear strain contour plots (contour interval is 0.005) at four different vertical displacements (t = throw). White and grey layers consist of bonded and non-bonded particles, respectively. **b** Strain and stress paths for two selected locations, A and B, shown in **a** at t = 9 cm. Positive volumetric strain is dilation and all rotations shown are clockwise. In the principal stress plot each arrow corresponds to the change of stress in a 1cm throw increment. The lines are best-fit failure envelopes obtained from numerical lab tests. For the strong material the different curves represent the 0.01, 0.5, 0.25, 0.50, 0.75, 0.95 and 0.99 percentile of the probability distribution. For the weak material the average ± 1 and ± 2 standard deviations are shown.

crease along a path limited by the failure envelope obtained for this material). The failure planes within the weak layers are basically dilational shear bands; tensile failure is impossible because the material is cohesionless.

Two important conclusions can be drawn from this model: (i) Large dip variations, and related fault refraction, are due to different types of failure of layers (extension vs. shear). (ii) Normal faults in brittle/ductile sequences at low confining pressure localize first in strong layers as steeply dipping Mode I fractures and are later linked via shallow dipping faults in weak layers.

It is clear that a fault surface with significant dip variations as illustrated by this model cannot accommodate a significant amount of displacement without strain incompatibility problems (e.g. volume increase or decrease). As summarized in Section 2, initial fault zone complexities are often bypassed by new faults to form fault bound lenses; a model that illustrates that the DEM is not only capable of modeling early stages fault growth, but also large displacements is presented next.

5.2 Fault growth

As a fault propagates through a rock volume it rarely does so as a perfect plane. The modeled normal fault shown in Fig. 5, for example, happened to be convex upwards at a throw of 0.5 m. Further displacement along this non-planar fault trace would lead to serious space problems, i.e. overlap and associated volume loss in the upper part and dilation in the lower part. Significant amounts of volume loss in nature can be accommodated by, for example, pressure solution, which is not implemented in the current model, and might not always keep pace with the displacement rate of a natural fault. A volume increase in the form of a pull-apart (Fig. 1a) is often observed at very low confining pressure, but in this particular model the confining pressure is too high (46 MPa) to permit the formation of a large cavity. Hence, in order to accommodate further displacement a new fault strand develops in the footwall of the former to form a fault bound lens (t = 1 m in Fig. 5). From this stage onwards displacement is accommodated by slip along both fault strands and rotation of the area bound by these two fault strands (t = 1.5 m). However, the newly developed fault strand is itself not perfectly planar and the upper portion of the fault trace is convex upwards. The same process described above is repeated and a new fault strand develops further into the footwall to form a second fault bound lens (t = 2–3 m).

Another interesting feature of this model is the formation of antithetic fault splays. These develop were the trace of an active fault strand is locally concave up, or exhibits a 'kink'. One of these localities is where the first fault strand (at t = 0.5 m) merges into the pre-defined fault at the base of the model. The fault developed within the multilayer sequence happened to be slightly steeper than the predefined fault. At this 'kink', an antithetic adjustment fault develops that accommodates the difference in the horizontal (heave) and vertical (throw) displacement components on either side of the kink (fig. I.2-43 in Mandl 1988). Although this example illustrates the origin of antithetic faults, they are clearly due to the boundary conditions used. However, a similar process can be seen on a smaller scale in this model, namely within the topmost layer, where an antithetic fault splay develops at a throw of 1.0–1.5 m (Fig. 5).

The final model fault geometry is complex, and in fact very similar to that seen in outcrop (e.g. Fig. 1c). In both models and outcrop, fault bound lenses are comprised of rock volumes which are rotated and fractured to variable degrees, and antithetic adjustment faults develop to accommodate local fault dip variations. This model

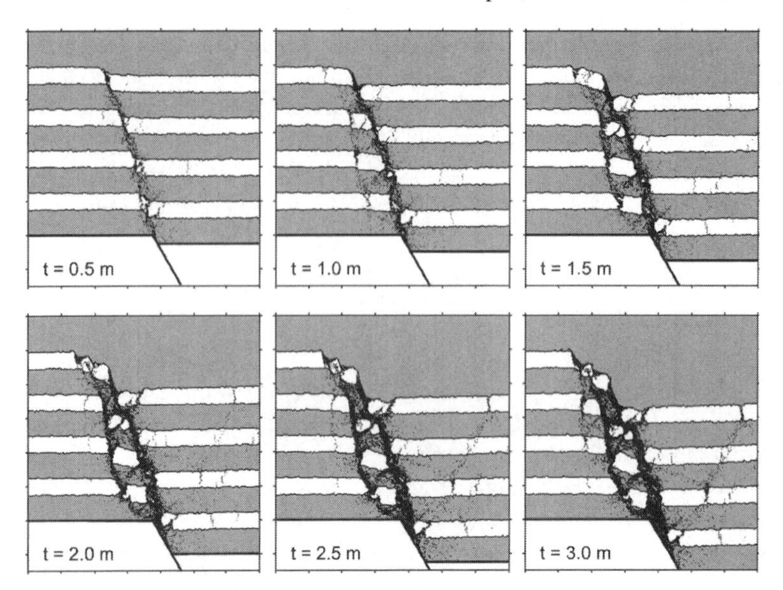

Fig. 5 Formation of fault bound lenses in a PFC2D model. Unconfined compressive strength of strong beds is 128 MPa and confining pressure is 46 MPa. White and grey layers consist of bonded and non-bonded particles, respectively, and t = throw. Black lines are particle separations > 1 cm, i.e. lines joining particles that where initially neighbors and are now separated by a distance greater than 1 cm.

illustrates that the DEM is a powerful tool for modeling the growth of faults up to large displacements and produces results which are very similar to structures seen in outcrop. A common difficulty when studying natural faults, however, is that it is often difficult to infer the mechanical properties of the sequence during faulting and the effective stresses at which faulting occurred. The next section provides a short summary of a suite of models that was run in order to investigate the control of the mechanical properties of the layers and confining pressure on fault zone structure.

5.3 *Impact of strength and confining pressure on fault zone structure*

A suite of 16 models was run to investigate the impact of strength contrast and confining pressure on fault zone geometry. The unconfined compressive strength of the strong beds is 128, 106, 83 or 64 MPa, and the confining pressure is 23, 46, 74 or 100 MPa, which, assuming lithostatic conditions and an overburden density of 2500 kg m^{-3}, correspond to depths of faulting of approximately 1, 2, 3 and 4 km. These 16 models, at a throw of 2 m, are arranged in Fig. 6 so that the layer strength decreases to the right and confining pressure increases downwards. The models highlight a variety of important features of both the geometry and growth of fault zones.

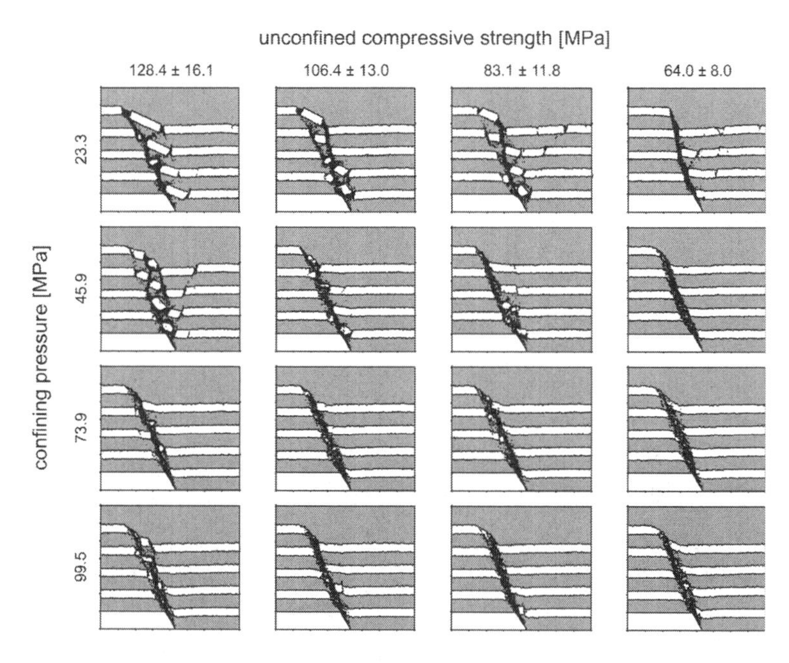

Fig. 6 PFC2D models illustrating the impact of layer strength and confining pressure on fault zone geometry. The unconfined compressive strength plus/minus one standard deviation of the cohesive beds (white) and the applied confining pressure are given. Plotting method is identical to that used in Fig. 5 (particle separations > 10 cm are shown) and the throw is 2 m.

At low confining pressure and high strength contrast, fault zones are initially highlysegmented giving rise at higher displacements to paired and, sometimes,multiple slip surfaces between which displacement is partitioned (upper left corner in Fig. 6). These wide faultzones are generated by steeply dipping faults localizing first within the strong layersand then by later linkage via shallow dipping faults in the weak layers (see also Fig. 4). Asperitiesarising from associated fault refraction and segmentation are progressively removed,with resulting fault-bounded blocks rotating within the fault zone as displacementaccumulates. The complexities associated with low confining pressures and highstrength contrasts appear to be also accompanied by an increased prevalence ofantithetic faults associated with fault segmentation and block rotation.

At high confining pressure and low layer strength the same mode of failure (shear) does occur in the strong and weak layers and this has clear implications for fault zone structure. The fault geometries in the weakest and deepest models are similar to ductile shear zones (lower right corner in Fig. 6) and fracturing is, in contrast to the high strength/low pressure models, pervasive within a monocline, rather than discrete.

Whilst this summary is somewhat qualitative; nevertheless, the results shown in Fig. 6 indicate that the mechanical properties of the sequence and the stress state during faulting both exert an important control on fault zone geometry and growth.

5.4 Impact of layering on fault dip

All multilayer faulting models presented so far are comprised of strong layers with a thickness of 1 m and interbedded weak layers with a thickness of 1.5 m, with only one model shown for each bed strength and confining pressure. Natural rocks and also DEM model materials are inherently heterogeneous and therefore exhibit a significant variability in terms of elasticity, strength and fracture/fault geometry. The suite of 35 models presented in Fig. 7 shows the impact of the relative proportions of strong and weak materials in the sequence on fault geometry and fault dip, and also illustrates how fault geometry varies between different realizations of statistically identical models (Fig. 7). Different realizations can be obtained in PFC by varying the seed number of a random number generator. Each different seed number provides a model in which the populations of particle and bond properties are identical, but in which the locations of those particles and bonds are different. Because fracturing is a path dependent process, slight variations in local heterogeneities can give rise to quite different fracture geometries. In that respect, a DEM model does not provide a unique solution to a problem. However, it does provide unique insights into the variability of structures that develop in materials which are, from a statistical point of view, identical.

The strong beds in the models shown in Fig. 7 have an unconfined compressive strength of 128 MPa and are 1 m thick. The thickness of the weak, cohesionless beds is varied (2.0, 1.75, 1.5, 1.25, 1.0, 0.75 and 0.5 m) and layering is expressed as a fraction of weak material in the sequence. A constant thickness of the strong beds was used because bonded particle models exhibit a scale-dependency of strength, whereas the strength of non-bonded materials is almost scale independent. The applied confining pressure is 46 MPa.

Approximately 70% of the models shown in Fig. 7 have paired or multiple slip surfaces and it appears that the most complicated fault zones develop where the fraction of weak material is 0.5 and 0.43. Although fault geometry is highly variable, one general trend can be observed: The overall, or bulk, dip of the fault zones increases with decreasing fraction of weak material in the sequence, f_w. The dips of the fault zones δ_f were measured by hand, a somewhat subjective procedure, and plotted vs. f_w in Fig. 8a. Despite the significant data scatter, a negative correlation can be seen between δ_f and f_w. Although a wide range of functions could be fitted to these data, the function chosen expresses the bulk fault dip δ_f as a function of the fault dips in the strong and weak layers, δ_s and δ_w respectively, and f_w, and assumes that the fault is a single continuous surface (Fig. 8b). The two variables, δ_s and δ_w, were determined using a least-square method (the function is almost linear for $\delta_f >$ 45°). Despite the scatter and the simplified assumption that the fault is a single con-

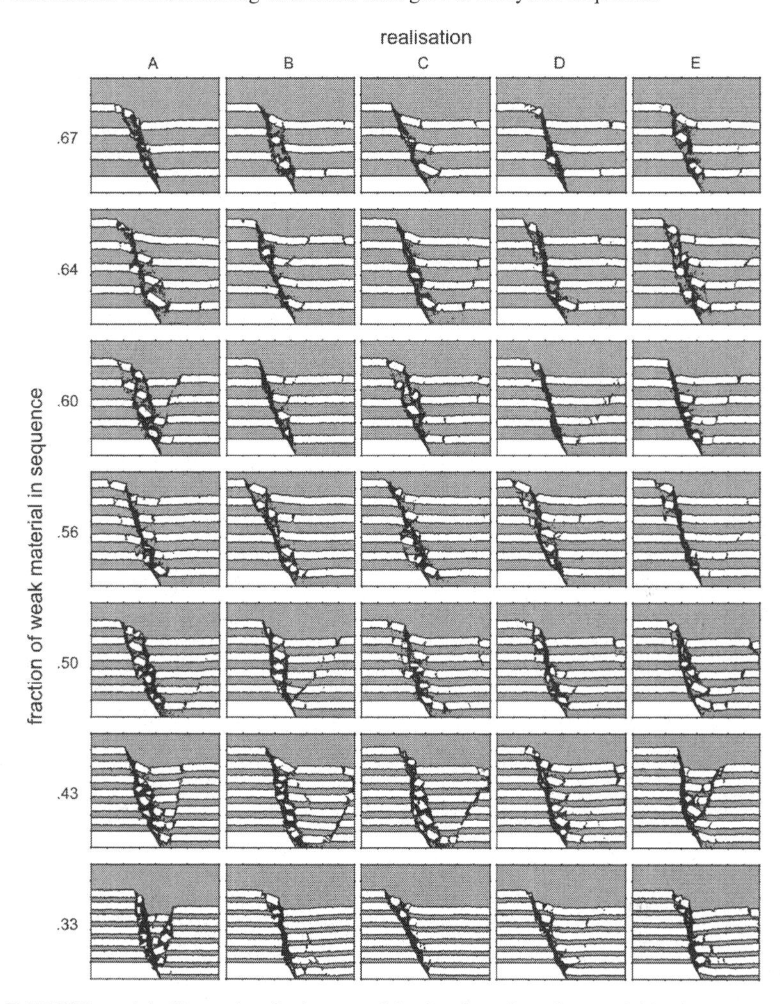

Fig. 7 PFC2D models illustrating the impact of the fraction of weak material in the sequence, f_w, on fault zone geometry. Unconfined compressive strength of strong beds is 128 MPa and confining pressure is 46 MPa. The fraction of weak material decreases towards the bottom and for each fraction five realizations (A–E) are shown. Plotting method is identical to that used in Fig. 6 and the throw is 2 m.

tinuous surface, the best-fit parameters are meaningful and somehow expected for the given layer strength and confining pressure, i.e. faults in the strong and weak beds have dips of 83° and 65°, respectively. Faults in the strong beds should in theory be vertical at the applied confining pressure, whereas faults in the weak layers are expected to have dips of 57.5° (45° plus half the friction angle of the weak material, which is 25°). In light of the high variability of the data and the absence of data for very low and very high fractions of weak material it is not clear whether the difference between best-fit fault dips and theory is significant.

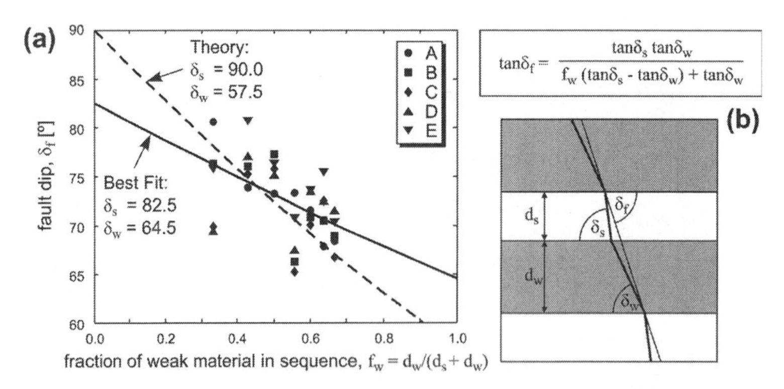

Fig. 8 Results of fault zone dip measurements from the suite of models shown in Fig. 7. **a** Plot of overall fault dip δ_f vs. fraction of weak material in the sequence, f_w. For each realization (A–E) a different symbol is used. The plotted solid curve is the best-fit solution obtained from a least-square regression and the dashed curve is the theoretical prediction using the equation given in **b**.

However, much more work, both in terms of modeling and field measurements, is required to arrive at a firm conclusion regarding the possible relation between fault dip and the relative proportions of strong and weak layers.

6 Summary and discussion

The Distinct Element Method, as implemented in PFC2D, is a powerful tool for modeling the growth of faults in layered sequences. The models reproduce a wide range of fault geometries seen in outcrop and offer unique insights into the growth history of those structures. The models also provide quantitative constrains on the mechanics of faulting and could potentially be used to assess the likelihood of particular structures and their frequency in the subsurface, below the resolution of geophysical imaging methods (e.g. seismic reflection data).

The mechanical behavior of the model materials used in this study is relatively simple and excludes time-dependent behavior (e.g. creep, sub-critical fracture growth) and the effect of fluids (e.g. pressure-solution, pore-pressure). Some of these effects can be relatively easily implemented in the present modeling scheme (e.g., Potyondy 2007, Al-Busaidi et al. 2005), and might provide even more realistic results.

Acknowledgements The staff of Itasca is acknowledged for past and present support. MPJS's PhD thesis project was funded by Enterprise Ireland (PhD Project CodeSC/00/041) and a Research Demonstratorship at University College Dublin. This research was also partly funded by an IRCSET (Irish Research Council for Science, Engineering and Technology) Embark Initiative Postdoctoral Fellowship and a Science Foundation Ireland (SFI) Research Frontiers Programme (RFP) grant.

References

1. Al-Busaidi A, Hazzard JF, Young RP (2005) Distinct element modeling of hydraulically fractured Lac du Bonnet granite. J Geophys Res 110:B06302.
2. Anderson EM (1905) The dynamics of faulting. Trans Edinb Geol Soc 8:387-402.
3. Anderson EM (1951) The dynamics of faulting and dyke formation with applications to Britain. Oliver and Boyd, Edinburgh.
4. Benesh NP, Plesch A, Shaw JH, Frost EK (2007) Investigation of growth fault bend folding using discrete element modeling: Implications for signatures of active folding above blind thrust faults. J Geophys Res 112:B03S04.
5. Burbidge DR, Braun J (2002) Numerical models of the evolution of accretionary wedges and fold-and-thrust belts using the distinct-element method. Geophys J Int 148:542-561.
6. Cardozo N, Allmendinger RW, Morgan JK (2005) Influence of mechanical stratigraphy and initial stress state on the formation of two fault propagation folds. J Struct Geol 27:1954-1972.
7. Childs C, Manzocchi T, Walsh JJ, Bonson CG, Nicol A, Schöpfer MPJ (2009) A geometric model of fault zone and fault rock thickness variations. J Struct Geol 31:117-127.
8. Childs C, Nicol A, Walsh JJ, Watterson J (1996a) Growth of vertically segmented normal faults. J Struct Geol 18:1389-1397.
9. Childs C, Watterson J, Walsh JJ (1996b) A model for the structure and development of fault zones. J Geol Soc Lond 153:337-340.
10. Cundall PA, Hart R (1992) Numerical modeling of discontinua. Eng Comp 9:101-113.
11. Egholm DL, Clausen OR, Sandiford M, Kristensen MB, Korstgård JA (2008) The mechanics of clay smearing along faults. Geology 36:787-790.
12. Ferrill DA, Morris AP (2003) Dilational normal faults. J Struct Geol 25:183-196.
13. Hardy S, Finch E (2007) Mechanical stratigraphy and the transition from trishear to kink-band fault-propagation fold forms above blind basement thrust faults: A discrete-element study. Mar Petrol Geol 24:75-90.
14. Imber J, Tuckwell GW, Childs C, Walsh JJ, Manzocchi T, Heath AE, Bonson CG, Strand J (2004) Three-dimensional distinct element modelling of relay growth and breaching along normal faults. J Struct Geol 26:1897-1911.
15. Itasca Consulting Group (1999) Particle Flow Code in Two Dimensions. Minneapolis, MN, USA.
16. Jing L, Stephansson O (2007) Fundamentals of Discrete Element Methods for Rock Engineering - Theory and Application. Developments in Geotechnical Engineering 85. Elsevier, Amsterdam.
17. Kock I, Huhn K (2007) Numerical investigation of localization and micromechanics in a stratified soil specimen. J Struct Geol 29:1679-1694.
18. Mandl G (1988) Mechanics of tectonic faulting. Models and Basic Concepts. Elsevier, Amsterdam.
19. Mandl G (2000) Faulting in brittle rocks. Springer, Heidelberg.
20. Mohr O (1882) Über die Darstellung des Spannungszustandes und des Deformationszustandes eines Körperelements und über die Anwendung derselben in der Festigkeitslehre. Civilingenieur 28:113-115.
21. Morgan JK, Boettcher MS (1999) Numerical simulations of granular shear zones using distinct element method 1. Shear zone kinematics and the micromechanics of localization. J Geophys Res 104:2703-2719.
22. Parry RHG (1995) Mohr circles, stress paths and geotechnics. Chapman & Hall, London.
23. Paterson MS, Wong T-F (2005) Experimental rock deformation - the brittle field. Springer, Heidelberg.
24. Peacock DCP, Zhang X (1993) Field examples and numerical modelling of oversteps and bends along normal faults in cross-section. Tectonophysics 234:147-167.
25. Pöschel T, Schwager T (2005) Computational Granular Dynamics, Springer, Heidelberg.
26. Potyondy DO (2007) Simulating stress corrosion with a bonded-particle model for rock. Int J Rock Mech Min Sci 44:677-691.

27. Potyondy DO, Cundall PA (2004) A bonded-particle model for rock. Int J Rock Mech Min Sci 41:1329-1364.
28. Ramsey JM, Chester FM (2004) Hybrid fracture and the transition from extension fracture to shear fracture. Nature 428:63-66.
29. Schöpfer MPJ, Abe S, Childs C, Walsh JJ (2009) The impact of porosity and crack density on the elasticity, strength and friction of cohesive granular materials: Insights from DEM modelling. Int J Rock Mech Min Sci 46:250-261.
30. Schöpfer MPJ, Childs C, Walsh JJ (2006) Localisation of normal faults in multilayer sequences. J Struct Geol 28:816-833.
31. Schöpfer MPJ, Childs C, Walsh JJ (2007a) 2D Distinct Element modeling of the structure and growth of normal faults in multilayer sequences. Part 1: Model calibration, boundary conditions and selected results. J Geophys Res 112:B10401.
32. Schöpfer MPJ, Childs C, Walsh JJ (2007b) 2D Distinct Element modeling of the structure and growth of normal faults in multilayer sequences. Part 2: Impact of confining pressure and strength contrast on fault zone geometry and growth. J Geophys Res 112:B10404.
33. Sibson RH (1998) Brittle failure mode plots for compressional and extensional tectonic regimes. J Struct Geol 20:655-660.
34. Strayer LM, Suppe J (2002) Out-of-plane motion of a thrust sheet during along-strike propagation of a thrust ramp: a distinct-element approach. J Struct Geol 24:637-650.
35. Strayer LM, Erickson, AG, Suppe J (2004) Influence of growth strata on the evolution of fault-related folds - Distinct-element models. In: McClay KR (ed) Thrust tectonics and hydrocarbon systems. AAPG Memoir 82:413-437.
36. Vietor T, Oncken O (2005) Controls on the shape and kinematics of the Central Andean plateau flanks: Insights from numerical modeling. Earth Planet Sci Lett 236:814-827.
37. Walsh JJ, Childs C, Meyer V, Manzocchi T, Imber J, Nicol A, Tuckwell G, Bailey WR, Bonson CG, Watterson J, Nell PAR, Strand JA (2001) Geometrical controls on the evolution of normal fault systems. In: Holdsworth RE, Strachan RA, Magloughlin JF, Knipe RJ (eds) The nature of the tectonic significance of fault zone weakening. Geological Society of London, Special Publication 186:157-170.

When geophysics met geomechanics: Imaging of geomechanical properties and processes using elastic waves

Stephen A. Hall

Abstract Geophysics measurement techniques can be considered as non-destructive, remote methods to analyse the physical properties of geologic materials hidden at depth in the earth, within geotechnical structures or within laboratory test specimens. The objective of this paper is to highlight at least some of the potential to use geophysics methods to aid geomechanical investigations, both at the laboratory and the real-world scales, by characterising property distributions and their changes plus, eventually, the mechanisms by which they change. The focus is primarily on geophysical imaging using elastic waves, whose propagation is controlled by a material's elastic properties and density. The former can be thought of as the summation of contributions over a range of length scales: grains, discontinuities (including cemented or uncemented grain contacts), inter- or intra-granular cracks, fractures and layers, which can all be anisotropic or can produce an anisotropic aggregate material. Examples are provided, at the laboratory and hydrocarbon reservoir scales, of geophysical characterisation of these different geomechanical attributes through anisotropy analyses, time-lapse measures of deformation and "full-field" velocity imaging.

1 Introduction

The aim of this paper is to outline how geophysics, and in particular elastic wave (ultrasonic, acoustic, seismic) methods, can be used to gain insight into geomechanical properties and processes occurring at depth in the subsurface or within laboratory test specimens. Geomechanics is the study of the mechanical behaviour of geologic materials (i.e., soils and rocks) and structures, from the laboratory scale to the mountain-massif scale with application to tunnel design, slope stability, petroleum engineering and underground waste disposal, to mention but a few. Geophysics, in

Stephen A. Hall
Laboratoire 3S-R CNRS / Grenoble Universities, e-mail: stephen.hall@hmg.inpg.fr

the current context, can be considered to be the use of non-destructive and remote methods to analyse the physical properties of geologic materials hidden at depth in the earth, within geotechnical structures or within laboratory test specimens. To achieve such imaging various different physical phenomena are exploited, with the principal methods involving elastic (ultrasonic, acoustic or seismic) or electromagnetic wave propagation and gravity.

As geophysics aims to provide information on rock/soil property distributions inside the geologic masses of interest, and geomechanics requires information on these properties to model/understand the mechanical behaviour of the materials and structures, there should clearly be a close link between these two disciplines. However, this is not generally the case. The objective of this paper is to highlight at least some of the potential to use geophysics to aid geomechanical investigations, both at the laboratory and at the real-world scales, by characterising property distributions and their changes plus, eventually, the mechanisms by which they change. The discussion paper of this paper focuses on geophysical imaging using elastic waves. By far the biggest use of elastic wave imaging in geophysics is in the hydrocarbon industry where seismic methods are used to image the subsurface. In fact, the primary data for characterisation of the 3D geometry of subsurface geologic structures are from seismic (reflection) surveys. A second usage of such data is to provide insight into the rock and fluid property distributions, e.g., the identification of potential oil and gas deposits. More recently it has been found that repeated imaging of producing subsurface reservoirs can provide a means of monitoring changes in both the rock and fluid properties and the 3D geometries, e.g., due to subsidence and compaction; so-called time-lapse or 4D seismic monitoring (e.g., [35, 2]). Aside from the hydrocarbon industry applications, seismic and ultrasonic methods are also used, but to very much a lesser extent, to image around underground mining tunnels (e.g., [52]).

At the laboratory scale ultrasonic wave measurements have been used for a relatively long time as a non-destructive method in laboratory soil and rock mechanics to study of changes in elastic properties due to deformation associated with loading. Additionally, recording of acoustic emissions, which constitutes an extension of earthquake monitoring carried out routinely at the real-world scale, has, since around the 1960s, become a popular means to track the evolution of damage in deforming rocks.

The paper starts by a discussion on the sensitivities of elastic wave propagation, in geomaterials, to the physical properties and changes in them due to geomechanical processes. Examples of a few applications at the laboratory and real-world scales are subsequently presented. The discussion and examples are drawn mainly from the author's own direct experience, a consequence of which is that the focus is primarily rocks although some discussion of application to soils is provided (for more details on elastic waves in soils the reader is referred to [66]). The aim is not to provide an exhaustive review, rather it is to present some of the potential for the use of advanced elastic wave imaging methods in geomechanics. Furthermore this overview is aimed at the mechanics audience of the Symposium on Mechanics of Natural Solids in Horto, Greece, in September, 2009.

2 Elastic wave propagation in deforming geomaterials

In this section a brief overview of wave propagation in geomaterials is presented, starting from the underlying wave equation and progressing to a discussion on the sensitivities of wave propagation to geomechanical properties and thus their changes. In this context, propagation of ultrasonic, acoustic and seismic waves can be considered as elastic in that as the waves pass through a medium they cause no permanent changes to the medium. Clearly if the waves are of sufficient magnitude they can cause damage, but here the consideration is of the non-damaging, elastic case. It should also be noted that even if no permanent damage is caused by the passage of the wave, a geophysicist might consider the wave to be inelastic if energy is lost by the wave as it propagates.

Different types of wave propagation exist and can be exploited in characterising materials or a site. Firstly there are body waves and surface waves, which, as their names suggest, refer to waves travelling through the body of a medium and those travelling along the surface. Here the former is considered, as the aim is to investigate how such elastic body waves might be used to characterise the internal structure and properties of geomaterials. Secondly there are longitudinal and transverse (or shear) waves (also known as P- and S-waves); the former involves particle motions in line with the direction of energy propagation and the latter particle motions perpendicular to the energy propagation, S-waves can thus be polarised. The majority of the following discussions will consider P-waves, but are, for the most part, applicable to S-waves also.

2.1 *Wave equation*

From a theoretical standpoint, propagation of a mechanical wave through a medium without any resultant permanent damage can be described by the elastic wave equation, as can be found in any text book on geophysics or mechanics (e.g., [1]). In its most general form the wave equation is,

$$c_{ijkl}\frac{\partial^2 u_l}{\partial x_j \partial x_k} = \rho \frac{\partial^2 u_i}{\partial t^2}, \tag{1}$$

where u is the displacement, t the time, x the spatial coordinate, ρ the material density and c_{ijkl} the elastic stiffness tensor. Clearly from this equation, the key material properties influencing the wave propagation are the density and elastic properties. This can be viewed conversely such that the density and elastic properties are in fact the information that might be gained from measurements of waves propagating through a medium.

The primary measure of wave propagation that can be made is the travel-time ("time-of-flight") between two points, which, if the distance between the two points is known, can be interpreted to give the velocity of propagation in the medium (velocity=distance/time). This velocity is, in terms of elasticity theory, a function of the elastic properties and the density. For a longitudinal wave (P-wave) propagating along the 1-axis the velocity can be defined as $v_p = \sqrt{c_{1111}/\rho}$, or, for an isotropic medium, $v_p = \sqrt{(\lambda + 2\mu)/\rho}$ (λ and μ are the Lamé coefficients, the latter being the shear modulus). For a shear wave propagating in the same direction with particle motion in the 2-direction $v_s = \sqrt{c_{1212}/\rho}$, or, for an isotropic medium, $v_s = \sqrt{\mu/\rho}$. At this point it is worth noting that the wave equation (1) is given in its anisotropic form, i.e., in terms of the elasticity tensor, c_{ijkl}, this is because geomaterials are, more often than not, anisotropic. In fact isotropy is more the exception rather than the rule, despite the common neglect of anisotropy. An implication of this is that the velocities at a point in a material will depend on direction and furthermore, for shear waves, on the direction of particle motion, which implies that for the same direction of propagation two shear waves can exist with different velocities (this gives rise to the phenomenon of shear-wave splitting or birefringence). Because anisotropy in general exists in geomaterials, much of the following discussion on the link between geomechanics and geophysics concerns the anisotropic case.

The wave equation (1) has a solution of the form $u(t,x) = A\exp[i(kx \pm \omega t)]$, with A amplitude, k wavenumber and ω angular frequency. This identifies the other key components of a wave, in addition to travel-time, that might be measured: the amplitude and frequency, plus also the wavenumber (inverse of the wavelength, but this is less commonly considered). Amplitude measures can provide insight into the attenuation of the wave (i.e., the loss of energy from the wave to other forms of energy such as heat, through friction) and scattering effects, either by small scatterers causing diffraction type effects or by discrete interfaces that reflect or refract the propagating wave. For the latter case, the wave amplitude might be used to characterise the properties of the interface.

In summary the main measurements of wave propagation through a medium that might be made are the travel-times and the wave amplitudes plus the frequencies and wavelengths. Whilst the main material property that might be characterised is the velocity, which is linked to the elastic properties and density.

2.2 *Sensitivity of elastic wave propagation to geomechanical properties and processes*

There are a number of factors that will influence the propagation of elastic waves in a given geomaterial. The principal structural characteristics include the lithology (i.e., the rock/soil type, including the mineralogy, cementation etc.), the porosity and the fluid content of the pores. In addition there are a number of environmental factors, with the most important being the effective stress (function of the confinement and the interstitial pore fluid pressure) and the saturation state (i.e., the type of fluid in the pores and the volume fractions in the case of multiple fluids). In the following the key mechanical aspects that contribute to spatial and temporal changes in the elastic wave propagation through a general geomaterial are discussed (a summary is also provided in Fig. 1). The effects of fluids are neglected here, but this is obviously a major factor since a change in the saturating fluid (e.g., replacing air by water) can imply a significant change in the average elastic moduli. In addition, both the lubricating effect and mobility of a fluid in the pore space can lead to interesting frequency dependent effects.

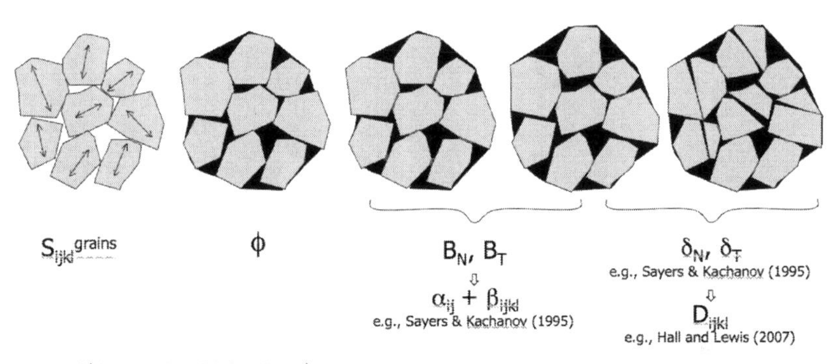

$c_{ijkl} = f(\phi + LPO + SPO + D_{ijkl})$

- ϕ - porosity:
  - Assumed isotropic (non-isotropic components appear in SPO and D_{ijkl})
  - Can change with deformation (dilation/compaction)
- LPO - Lattice preferred orientation:
  - Leads to anisotropic elastic properties if the crystallographic axes are aligned
  - Can vary with deformation through grain rotation, but variations are expected to be small
- SPO - Shape preferred orientation:
  - Grain, grain boundary and contact organisation
  - Can vary with grain reorganisation
- D_{ijkl} - Damage – at all scales below that of observation (i.e., of dimensions much less than the propagation wavelength):
  - grain boundary cracks (cement breakage)
  - grain breakage
  - coalescence to fractures

Fig. 1 Summary of the different geomechanical/structural contributions from the grain-scale upwards, to the elastic properties of a geomaterial (particularly for cemented granular rocks such as sandstones).

As discussed above, the key factors controlling elastic wave propagation, from a theoretical standpoint, are the elastic properties, c_{ijkl}, and the density, ρ. The latter is considered to be isotropic and can be thought of as the volume average of the constituent mineral and fluid phases, e.g., $\rho_{aggregate} = (1-\phi)\rho_{solid} + \phi\rho_{pore}$, where ϕ is the porosity. The elastic properties of an aggregate material are, however much more complicated to describe, the following discussion addresses the key aspects (good overviews can also be found in [67, 66], for soils, and [51], for rocks).

The elastic properties of a rock can be described in terms of the contributions of a number of factors/mechanisms over a range of length scales: grains, discontinuities (including cemented or uncemented grain contacts), inter- or intra-granular cracks, fractures and layers. Each of these can be anisotropic or can act to produce an anisotropic aggregate material. The summation of these different contributions can be most easily described as the sum of their compliances ($s_{ijkl} = c^{-1}_{ijkl}$), based on an assumption of non-interaction (after, for example [71]),

$$s_{ijkl} = s^{grains}_{ijkl} + s^{contacts}_{ijkl} + s^{damage}_{ijkl}. \tag{2}$$

At the smallest scale, the grain elasticity, s^{grains}_{ijkl}, determined for each grain by its crystal lattice structure, is described by some average over all the constituent grains of different mineralogy (this averaging is in general carried out using a "Voigt-Reuss-Hill average", see [38] for example). These crystal elasticities are anisotropic and therefore, with a preferred orientation of crystals (lattice preferred orientation, LPO), can produce an elastic anisotropy in the aggregate. The effect of LPO fabrics on elastic properties of rocks has primarily been investigated in relation to studies of the deep Earth (e.g. [50]), for example, to map olivine alignment and thus mantle flow (e.g. [4]). Fewer studies exist on the characterisation of LPO effects in sedimentary rocks (examples include [7, 82, 38]). Whilst this effect is possibly small for sandstones for example, rocks and soils rich in clay minerals can exhibit very significant LPO-induced anisotropies (see the example given later where anisotropy increases with mica content). The crystallographic contribution to the elasticity will not be very sensitive to the loading unless there are significant grain rotations, thus the LPO contribution can generally be considered to be a static parameter (except in the case of mantle flow cited previously).

At the next scale up, additional contributions to the elasticity and particularly the anisotropy are grain-scale structure effects that might be generally described as shape preferred orientation (SPO) effects. This involves primarily microcracks or thin pores and grain-boundary contacts ($s^{contacts}_{ijkl}$ in (2)) as well as small-scale layering or non-random distributions of mineral phases associated with the formation of the material. Microcracks, thin pores and grain-boundary contacts all represent highly compliant features and therefore can have a significant impact on the overall elastic properties of a material giving both significant anisotropies and considerable variations under loading. There has been extensive work on the characterisation of such crack-induced anisotropy and its load-dependence through experimental observations (e.g. [54, 40, 48, 70, 68, 15, 21]), as well as through theoretical models of the structural (e.g. [5, 30, 32, 33, 9, 69, 70, 71, 34, 68]) and load-dependent aspects (e.g. [53, 81]).

A significant amount of work has also been presented on the experimental characterisation of the *evolution* of (micro-crack) damage using elastic wave velocities, examples include [64, 75, 14] (see later sections for a discussion on the issues associated with such analyses). Such damage can occur through different mechanisms, depending on the rock and the loading conditions: at low pressures the main damage mechanism in rocks and soils would be cement breakage or destructuration, whilst at higher confining pressures grain breakage is likely to be important. As the damage mechanisms progress, microcracks can coalesce to form fractures, which has implications for elastic wave imaging as, at the laboratory scale with ultrasonic wave frequencies, these fractures might no longer be considered within the type of effective medium context described in the following, although at the large scale with longer wavelengths effective medium assumption will still hold (until the fractures become faults that is). Below, a framework for describing cracked or fractured materials (without evolution) is outlined, based on the developments of [69, 70, 71]; these representations might be used to forward model the likely effect of cracks or fractures on the elasticity of a medium or within an inversion scheme to determine the crack or fracture properties from elastic wave data.

From a theoretical point of view, the contribution $s_{ijkl}^{contacts}$ due to microcracks, thin pores and grain-boundary contacts can be described, following [69, 70, 68], in terms of second and fourth order "*crack density tensors*", α_{ij} and β_{ijkl},

$$s_{ijkl}^{contacts/cracks} = \frac{1}{4}\left(\delta_{ik}\alpha_{jl} + \delta_{il}\alpha_{jk} + \delta_{jk}\alpha_{il} + \delta_{jl}\alpha_{ik}\right) + \beta_{ijkl}, \tag{3}$$

where δ_{ij} is the Kronnecker delta. For a distribution of r planar discontinuities with surface areas $A^{(r)}$ and surface-normal vectors $n_i^{(r)}$ in a volume V, these crack density tensors can be defined in terms of two crack compliance terms that describe the additional normal and tangential compliances, B_N and B_T, due to each discontinuity,

$$\alpha_{ij} = \frac{1}{V}\sum_r B_T^{(r)} n_i^{(r)} n_j^{(r)} A^{(r)}, \tag{4}$$

$$\beta_{ijkl} = \frac{1}{V}\sum_r (B_N^{(r)} - B_T^{(r)}) n_i^{(r)} n_j^{(r)} n_k^{(r)} n_l^{(r)} A^{(r)}, \tag{5}$$

([69, 70, 68]). B_N and B_T characterise the normal and tangential displacement discontinuities due to normal and tangential tractions acting upon the cracks faces. Therefore, α_{ij} and β_{ijkl} can be thought of as characterising the cumulative magnitude and orientation distribution function of the additional normal and shear compliances due to all the compliant discontinuities present in a sample plus the relative ease to undergo shear or opening along the different orientations. These will be controlling factors not just for the elasticity of a rock but also in defining the failure. For the special case where for all the cracks $B_N = B_T$ it follows that $\beta_{ijkl} = 0$. In this case any set of cracks can be represented by three orthogonal effective *scalar* crack sets that are fully described by the second order crack density tensor, α_{ij}, which has principal axes defined by the three scalar crack densities, α_{11}, α_{22} and α_{33} ([70]). Δs_{ijkl} will thus be orthotropic and aligned with α_{ij}. Deviations from such a scalar crack model, i.e., for $B_N \neq B_T$, will give rise to non-zero values of β_{ijkl}, however [70] and [68] argue that in most cases the magnitude of β_{ijkl} will be small.

Finally, at the field-scale and for long wavelength elastic waves, there is an additional contribution to the elasticity associated with, often preferentially oriented, large-scale discontinuities such as fractures and layers. For the latter, a classic paper describing the anisotropy due to layering is [80] and the effect can also be due to a periodic layering of the medium, e.g., due to seasonal variations in deposition. For aligned fracture sets a similar description to that used for micro-cracks, described above, can be used as presented, for example, by [71],

$$s_{ijkl}^{fractures} = \frac{1}{4}\sum_m (Z_{ik}^{(m)} n_l^{(m)} n_j^{(m)} + Z_{jk}^{(m)} n_l^{(m)} n_i^{(m)} + Z_{il}^{(m)} n_k^{(m)} n_j^{(m)} + Z_{jl}^{(m)} n_k^{(m)} n_i^{(m)}), \quad (6)$$

where Z_{ij} is the 3x3 fracture system compliance tensor describing the excess compliance due to the most general form of low aspect ratio fracture sets. If the fractures are rotationally invariant about their normal axis and have no significant preferential slip direction, Z_{ij} may be simplified to three non-zero elements given by just two independent normal and tangential fracture compliances, Z_N and Z_T,

$$Z_{ij} = Z_N n_i n_j + Z_T(\delta_{ij} - n_i n_j). \quad (7)$$

For modelling, various different descriptions for Z_N and Z_T exist; condensed summaries can be found, for example, in [20, 22].

In the following sections examples of how the different geological / geomechanical characteristics can be perhaps determined, in laboratory samples or at the large scale, e.g., hydrocarbon reservoir scale, are presented.

3 Laboratory Scale

Ultrasonic measurement for the study of changes in elastic properties with loading has a relatively long history in laboratory soil and rock mechanics. In laboratory soil mechanics the main interest in using elastic wave measurements has been to define the small strain shear modulus, "'G_0", after it was found that the shear modulus measured using standard mechanical loading tests varied with the applied loading amplitude (strain); i.e., the response was non-linear. Using shear-wave measurements in a sample under load allowed a more representative shear-modulus (G_0) to be measured under any loading condition, thus the evolution of G_0 could also be followed (e.g., [83]). In laboratory rock physics/mechanics and geophysics, the main interest has been measurement of permenant and non-permenant non-linear elastic changes under loading, these aspects are discussed in more detail in the following.

3.1 *Stress sensitivity*

Measurement of the dependence on the confining pressure of ultrasonic wave propagation velocities in rocks is the bread-and-butter of laboratory rock physics. Such analysis is usually carried out by confining a sample of rock under increasing hydrostatic pressures and measuring the change in velocity along the core axis. This provides curves of velocity versus pressure, such as the sketch in Fig. 2, where the velocity increases with confining pressure until a plateau is reached. This can be interpreted as a closure of the compliant cracks and grain contacts until a stable configuration is reached, but no new permanent changes, i.e., damage, have been induced (e.g, [54, 40, 48, 15, 21], amongst many others). One objective of such measurements is to define the pressure sensitivity of rocks, with the holy grail, for some, of using such data to interpret seismic velocity variations in-situ, e.g., in a subsurface hydrocarbon reservoir under production, as an indicator of the in-situ pressure changes. With correct calibration this may be possible, however it is often difficult to separate the changes due to pressure variations from changes due to associated variations in saturation (e.g., oil being replaced by water).

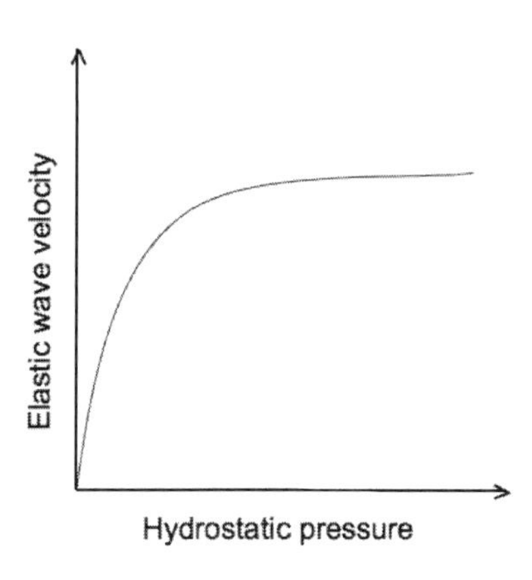

Fig. 2 Schematic of the typical laboratory observation of the variation of rock properties, such as P-wave velocity, with confining pressure under the assumption of elastic deformation; similar trends are commonly observed for S-wave velocities, bulk moduli and permeability for example.

3.2 *Anisotropy, crack density tensors and damage*

Another application of the "rock physics" type measurements described above is the quantification of anisotropy and subsequent analyses of rock fabrics. However, characterisation of velocity anisotropy requires measurements along many directions. In fact full characterisation of anisotropy really requires the impractical acquisition of a large number of measurements along many different axes, e.g. using spherical samples ([63]). More practical analyses can be carried out using fewer measurements based on simplifying hypotheses of higher order symmetries, such as hexagonal or orthorhombic. The simplest approach is through desktop (unconfined) measurements, which represent a relatively simple means to achieve good characterisation of velocity anisotropy, e.g., with circumferential velocity analysis (CVA); e.g. [28] and Fig. 3a [23]. However, desktop measurements do not allow changes with loading to be assessed nor more realistic in-situ stress states state to be analysed. Acquiring multi-axis velocity measurements under pressure is more challenging.

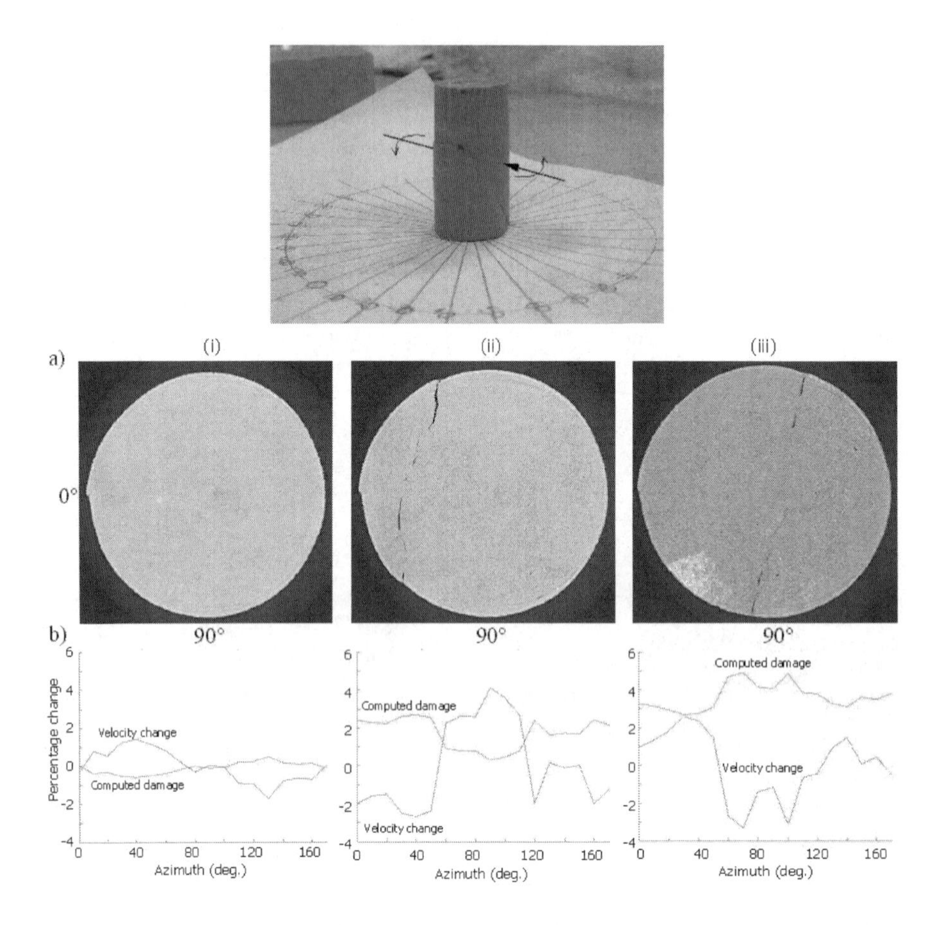

Fig. 3 Results of desktop circumferential velocity analysis (CVA) measurements for a sample of Beaucaire Marl, a sitff clay, after [23]. Ultrasonic velocity measurements were made across the diameter of the sample at three different levels around the circumference as indicated in the upper image (the three different measurement levels can be seen as traces on the sample surface), (a) then shows slices through x-ray tomography image volumes at the corresponding positions and (b) shows the measured velocities and interpreted "damage" magnitudes, based on the deviation from an assumed undamaged velocity value (however for the top level (i) the variations are perhaps more due to an anisotropy of lithologic origin). (For scale: the sample was 20 mm in diameter and 40 mm in height.)

Standard rock-mechanics apparatus with velocity measurement capabilities are usually designed with transducers mounted in the piston allowing ultrasonic wave propagation along the specimen axis. Even with both P- and S-wave transducers, this does not provide sufficient data to characterise anisotropy. A solution to this problem is to install sensors on the surface of the sample, through the sample-jacket (e.g. [76]), but this requires a more complex set-up. A number of issues are overcome by using a poly-axial loading cell for cubic samples, in which both P- and S-wave ultrasonic transducers could be placed on all six faces ([40]); this is however a specialised apparatus. An alternative approach (e.g., [29, 21]) is to make multi-axial measurements, using a standard pressure-cell with P- and S-wave transducers mounted only in the piston, by using specimens cut along different axes relative to the rock fabric. Such a methodology and associated results (from [21]) are discussed further in the following.

In the following an example of how laboratory ultrasonic data might be used to characterise the grain-contact and micro-crack structure in a rock is discussed, which is based on the work of [21]. By cutting three orthogonal core plugs from a single specimen of rock, "sister" samples that can be considered to be approximately the same can be tested separately, but the analysis of the results may be carried out as if all the data were acquired from the same sample. To this end [21] used samples cut as shown in Fig. 4: 1 parallel to the well core axis, 2 normal to the well core axis and 1 at 45 to the well core axis. Each sample was loaded under isotropic compression from 0 to 45 MPa with P-wave and 2 orthogonal S-wave velocity measurements made throughout. This provided multi-axial velocity data that could be inverted for anisotropy, and more precisely the crack density tensors of (3), using an inversion strategy to assess the crack structure and the changes under loading building on previous work by [68] and theoretical developments of [69, 70]. The approach of [21] also included an analysis of the petrofabric to provide an *a priori* measure of the intrinsic LPO anisotropy to be included in the inversion (the LPO contribution was determined through electron back-scattered diffraction - EBSD - and x-ray texture goniometry - XRTG - methods). This allowed the separation of the intrinsic and extrinsic components of the anisotropy, and thus the more accurate characterisation of the grain-boundary/crack structure variation under loading. The resultant crack description includes the contributions due to all the compliant crack-like features present at the grain-scale (microcracks, grain-boundary contacts, etc.) and accommodates the influence of their filling materials (clays, etc.) through their overall effect on the crack compliances. Another aspect of the approach was to, where appropriate data were available, assess 2nd and 4th order crack-density tensors, which provided insight into the relative sizes of the normal and tangential compliances of the cracks/contacts (see equations (3)-(5)). Figure 4 shows the different measured velocities as a function of confining pressure for four example samples as well as the inverted 2nd order crack densities plus the back-calculated velocities based on these, which act as a check on the modelled crack density tensors. Figure 5 presents a summary of the results for all the samples studied by [21] showing the separated contributions to the anisotropy from the LPO and the cracks/grain-contacts, which interestingly (for this example) both show a correlation with the mica mineral content (micas are highly anisotropic, flat minerals).

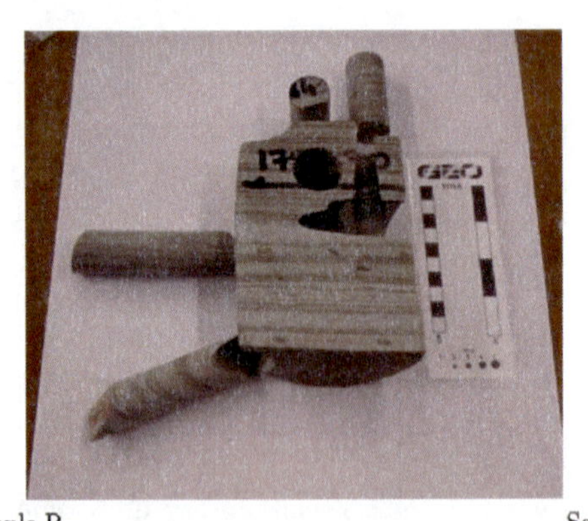

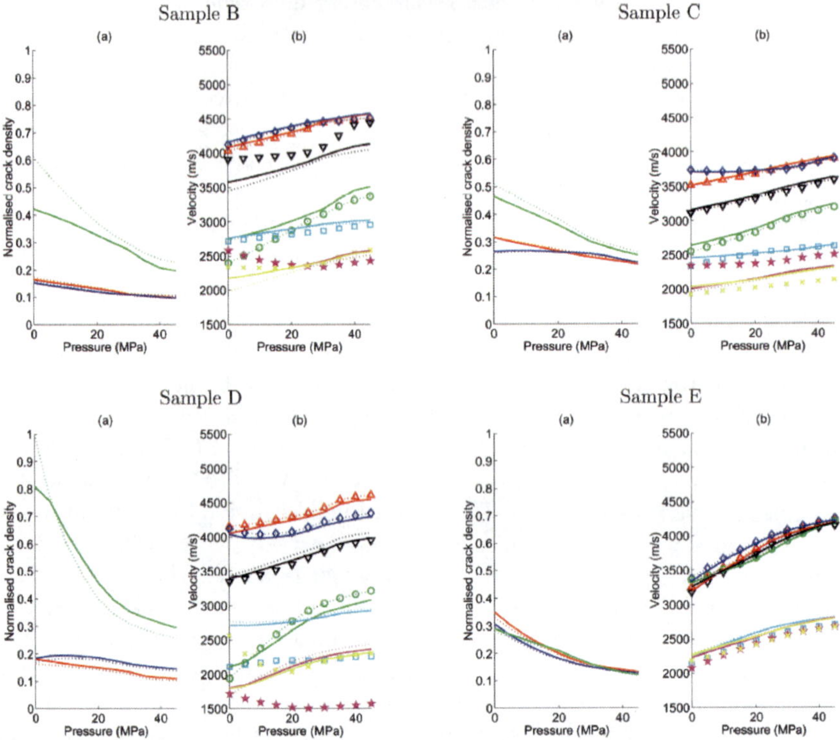

Fig. 4 Upper plot: Illustration of the cutting of the four "sister" core-plugs from the well core to carry out the multi-axis measurements. Lower plots: Four examples of inverted scalar crack densities and back-calculated velocities for a few different samples of a reservoir sandstone, from [21]: (a) α_{ij} as a function of pressure calculated using just the P-wave velocity data for the three principal directions (red - α_{11}, blue - α_{22}, green - α_{33}). (b) Velocity data (symbols) and back calculated velocities (lines) corresponding to the calculated α_{ij}s. (Green - $v_{p_{33}}$, red - $v_{p_{11}}$, blue - $v_{p_{22}}$, black - $v_{p_{13}}^{45^o}$, cyan - $v_{s_{12}}$, magenta - $v_{s_{13}}$, yellow - $v_{s_{23}}$.)

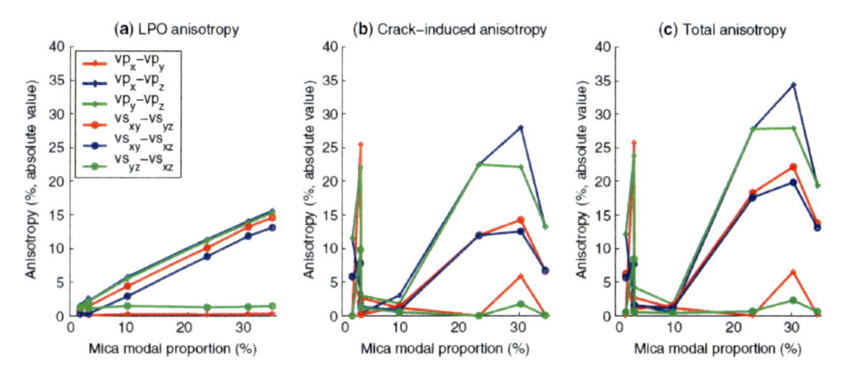

Fig. 5 Anisotropy versus mica (mineral) content, at 45 MPa confining pressure, for all the samples of reservoir sandstone analysed by [21]. (a) and (b) show the separated contributions to the velocity anisotropy due to, respectively, the LPO-induced velocity anisotropies (calculated based on known crystal elastic properties and EBSD/XRTG measurements) and the cracks (based on the inverted crack density tensors), which together produce the total (observed) anisotropies shown in (c).

In addition to the "active" ultrasonic measurements described above, "passive" listening for noise associated with damage formation (e.g., cracking), i.e., acoustic emission (AE) measurements, are very useful for tracking active deformation during a loading test (e.g., [44, 45, 12, 76, 64, 75], amongst many others). Location of the AE events in space is possible if a number of transducers are used, and details of the rupture process (e.g., shear or opening) might also be inferred (e.g., [86]), however the latter remains quite a challenge.

3.3 Full-field measurement of elastic properties: ultrasonic tomography

Problems arise with the approaches described above in the presence of heterogeneities, either pre-exisiting or evolving in the form of deformation structures (deformation in geomaterials is rarely uniformly distributed and is often localised, e.g., in to shear-bands or fractures). With heterogeneity, measurement of wave propagation along just a few directions in a sample does not allow a good characterisation of the elastic properties nor of the change in them associated with deformation; only averages along the (limited number of) propagation paths can be described. These issues are well illustrated in Fig. 3, where the anisotropy at the top of the sample is well characterised by CVA measurements, but the localised damage is clearly detected though not well characterised (Figures 3b,c, which also highlight how sensitive elastic waves are to the damage as the observed effect on the velocities is very significant). Ultrasonic tomography might overcome some of the challenges as it represents a more "full-field" measurement method (see [84], for a discussion on the need for, and use of, full-field measurements in geomechanics).

Ultrasonic tomography involves inverse analysis of data from measurements along many crossing propagation paths that pass through the studied medium. At the real-world scale this might be between two boreholes, whereas in the laboratory the measurements would involve ultrasonic transducers placed on the surface of the sample. In general two types of elastic-wave tomography can be considered, travel-time and attenuation. The theory is similar for both, involving attribution of the measured data, e.g., the total travel-time along a ray-path between pairs of emitting and receiving transducers (often referred to as the time-of-flight or travel-time), to the summation over the path-lengths within all the grid cells of the inversion model traversed by each ray. The combination of many such measurements, for a set of intersecting ray-paths between emitting and receiving transducers placed at different locations over the sample, allows, by inversion, a map of the variations in propagation velocity (or attenuation) in the sample to be determined (in a least-squares sense). Therefore ultrasonic tomography provides access to the spatial variations of the elastic properties in a test sample. The reader is referred to the many texts in the geophysical literature for a presentation of tomography methods, an example oriented towards geomechanics is [65]. A small number of previous ultrasonic tomography studies for application to geomaterials have been presented, including for imaging of large inclusion heterogeneities in soils [10] and of localised deformation in rocks [12, 11, 73, 77] and soils [42, 25, 23].

In the following ultrasonic tomography results are presented for two examples, the first for a stiff clay and the second for a sandstone, both exhibiting localised deformation. The procedure mimics the type of analysis often carried out at the large-scale between two boreholes known as cross-well tomography. In the first example, for the stiff clay, a simple acquisition procedure was performed involving two transducers placed accurately at a number of different positions along the side-walls of the sample to yield a number of intersecting raypaths (see Fig. 6a). In the second example, for the sandstone, the acquisition was made using two arrays of 64 transducer elements yielding 64x64 intersecting raypaths in a single acquisition (see Fig. 7), which provides an unprecedented (for geomaterials) data coverage for the ultrasonic tomography.

Figure 6 shows an example of the application of ultrasonic tomography to a dry, tabular sample of stiff clay (Beaucaire marl), which was previously deformed under plane strain compression. In this case, the travel-time data used in the tomographic inversion were acquired using just two transducers placed on opposite sides of the sample at different offset positions so as to achieve measurements along a series of crossing ray-paths (as indicated in Fig. 6a). The resultant map of ultrasonic velocities is shown in Fig. 6c and the localised shear band shows up as a zone of increased velocities, which is consistent with the expected compaction associated with localisation for this test sample. Some variability is also seen in the band, which may represent varying degrees of compaction. However, these results are potentially affected by artefacts of the tomography technique, for example the band offset from the real position in space (marked by the dashed line) and the high velocity zones in the corners at the ends of the band.

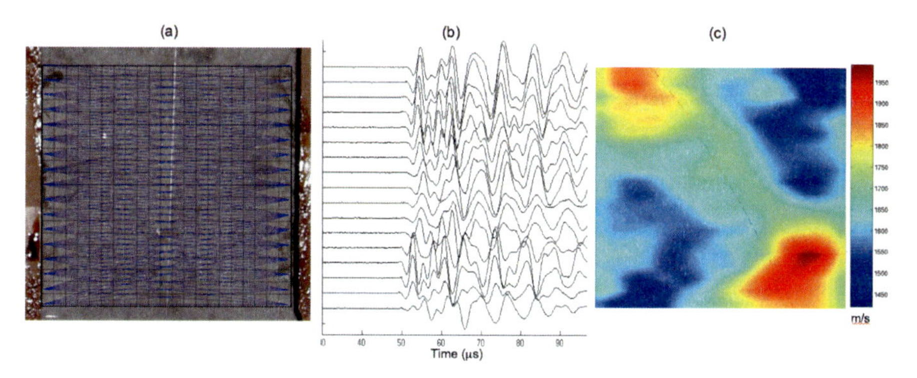

Fig. 6 Ultrasonic tomography analysis on a parallelpipedic sample of Beaucaire Marl (stiff clay) that had previously been deformed under plane-strain conditions such that a localised (shear) deformation band developed diagonally across the sample (from top left to bottom right in (a). (a) Photograph of the front face of the sample with the tomography inversion grid and (straight) raypaths between each source and receiver over which travel-times were determined. (b) Example of the data acquired for each source-receiver pair, in this case just for the raypaths for transducer positions directly opposite each other across the sample. (c) Final ultrasonic velocity map from the 2D tomography analysis, which shows a band of higher (increased) velocities that roughly coincides with the trace of the localised deformation zone indicated by the dotted line.

As a second example of the laboratory ultrasonic tomography, results are presented for a post-mortem analysis of a sandstone specimen deformed under triaxial compression. The specimens were cylinders with two opposite flattened faces to provide contact surfaces for the ultrasonic transducer arrays (see Fig. 7). Notches were made in these flattened faces to enforce the expected shear bands to develop in the middle of the sample and in the direction perpendicular to the velocity measurements. Figure 7 also shows example waveforms for propagation between transducer elements directly opposite each other across the sample, which might be compared to the data presented in Fig. 6b; in this case, the data acquisition is much denser than in the previous example. If the sample were homogeneous this "plane wavefront" would be expected to show equal arrival times and similar waveforms for each recording. However, some variability can be seen in the wavefront, including in the arrival times e.g., with, in particular, later arrivals around the two positions that correspond to the notches indicating reduced velocities.

Figure 8 shows an example of an inverted velocity field for a notched cylinder sample of Vosges sandstone deformed under axisymmetric triaxial compression at 20 MPa confining pressure, which resulted in an inclined shear-band between the two notches. This velocity field, as with others for similar tests, indicates the resolution of the localised shear-band as a zone of reduced velocity, despite the fact that these zones are more dense than the surrounding less-deformed rock (as observed through x-ray tomography). These apparently contradictory observations indicate that the damage phenomena involved the creation of a high density of new cracks/contacts, likely associated with micro-cracking through destructuration (cement breakage) - grain crushing is unlikely at this level of confining pressure -, and the porosity reduction (density change) effect is small. Whilst the low-velocity zone in Fig. 8, which corresponds to the localised deformation band, appears to be well resolved, it is in fact much wider than the width identified on the corresponding x-ray tomography images. On the one hand this could indicate the greater sensitivity of the elastic waves to the damage processes (only quite strong density changes can be resolved in the x-ray images) and so damage around the concentrated deformation zone in the core of the shear band might be detected. However there is a spatial resolution limit to the ultrasonic tomography related to the wavelength of the waves. In this case the dominant wavelength was around 800 kHz, which for a velocity of 1800 m/s corresponds to about 2 (or so) mm ($v = f\lambda$: velocity = frequency x wavelength). In some texts it is stated that the resolution limited of elastic wave tomography is about 3λ (so about 9 mm in this case, which is larger than the "expected" shear-band width of around 5 mm), although features smaller than this will be detected, but likely with a smeared image and increased misfit of the inverted model to the data. To achieve greater spatial resolution will require advances in the acquisition to have smaller wavelengths and perhaps the use of different inversion strategies such as full-waveform tomography (e.g., [60]); which are both under investigation.

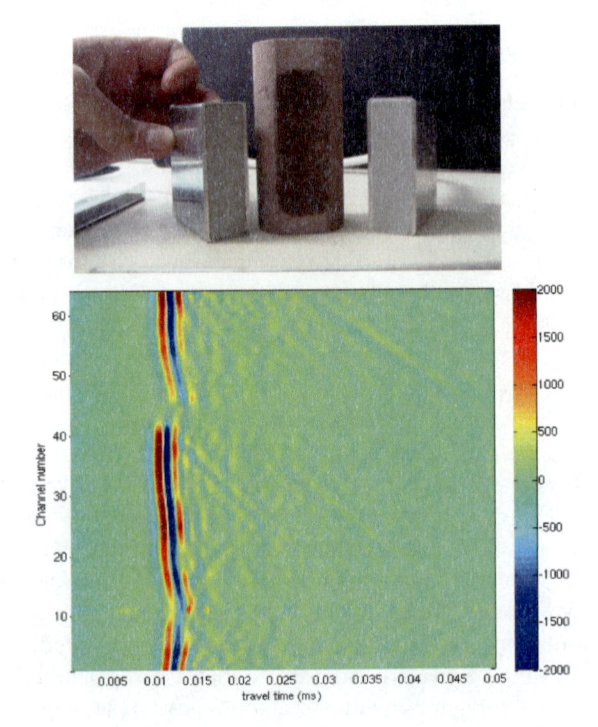

Fig. 7 Upper image: Example of the sandstone specimens used in the study of ultrasonic imaging of localised deformation (the specimen is about 80 mm high) and the ultrasonic transducer arrays ("barettes") used for the measurements (each one has 64 transducer elements distributed over the height at 0.75 mm vertical spacing). Lower image: Example of the data acquired for each source-receiver pair (for the raypaths for transducer positions directly opposite each other across the sample). These data are equivalent to those in Fig. 6b, but with waveforms plotted in colour corresponding to the amplitude. The vertical positions of the two "notches" (see main text) correspond to the lower amplitude zones and variations in velocity can be seen in the arrival time differences.

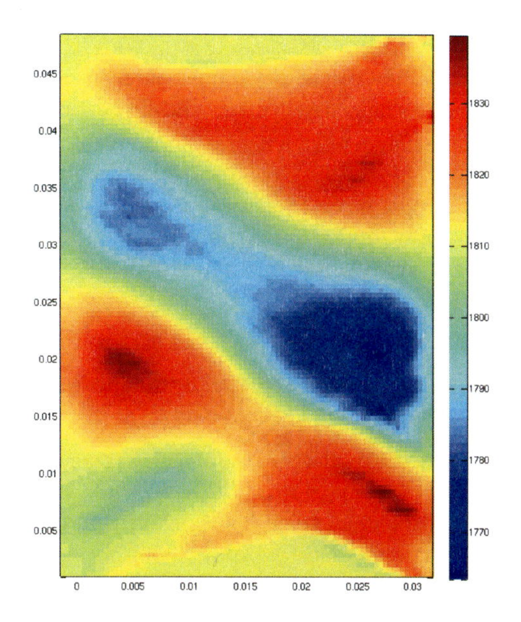

Fig. 8 Result of the 2D ultrasonic velocity tomography over the central zone of a sample of Vosges sandstone deformed under 20 MPa confining pressure (colourbar indicates velocities in m/s). A band of low velocities is seen crossing diagonally across the sample corresponding to the location of the localised shear deformation feature that developed during loading between the notches (the notches are at the edges of the specimen near the upper and lower extremities of the low velocity band). This low velocity zone indicates a zone of damage (i.e., concentrated micro-crack development) associated with the localised deformation.

4 Reservoir Scale

The laboratory scale analyses described above are very useful in gaining an understanding of the fundamental processes of the mechanics of geomaterials and how deformation alters elastic properties or, conversely, how elastic properties might allow characterisation of deformation effects. Such analyses might also help understand observations of spatial and temporal changes in elastic properties of the subsurface detected in seismic data. In the following a few examples of such observations are presented to show how geophysics can illuminate geomechanical properties and processes remotely in the subsurface. It is not the aim to outline fully the methods just to present some the important aspects and examples of the results that can be attained; an overview of some of the methods is given by [22], for example.

Standard seismic methods can be used, as described in the introduction, to determine the structural elements of the subsurface (e.g., lithologic layers, faults and folds), which is essential in the prospection for hydrocarbons. Furthermore such structural data can be used to define a model for geomechanical simulations of subsurface operations. In addition to structural information, there is also significant insight to be gained about the subsurface rocks and fluids from seismic attributes such as travel-times and amplitudes. So-called seismic inversion approaches permit, after calibration by direct borehole measurement data, the transformation of a seismic image of the subsurface into an approximate image of lithologies and perhaps fluid saturations that might be used to populate a model of the subsurface (e.g., [58]). However extraction of absolute properties from such data is difficult.

In the following, two examples of how seismic data can provide valuable insight into geomechanical properties and processes at the reservoir scale are discussed. The first is an application of seismic anisotropy analysis to map the directionality of seismic properties (velocities) spatially in the subsurface. The second is an example of how repeated seismic images of a producing hydrocarbon reservoir can be compared to quantify compaction and subsidence due to the pressure reduction associated with fluid extraction from the subsurface.

4.1 Seismic anisotropy detection and mapping

Delineation and characterisation of fracturing is important in the successful exploitation of many hydrocarbon reservoirs, as fractures can significantly affect permeability and thus production, e.g. during horizontal drilling or water-flooding. Extensively reported surface observations suggest that fractures often exist in semi-ordered and semi-parallel fracture sets and that they can be open under certain deformation histories and in-situ stress states. However, it is difficult to identify, quantify or characterise fracturing in the subsurface, by observation or by inference. Some quantification, e.g. of fracture orientation, can be acquired from borehole data, but these data are isolated one dimensional measurements, representing either single points or a series of points in potentially highly variable arrays of fractures. These observations are therefore very susceptible to the sampling, e.g. if there are widely spaced fractures the borehole might not intersect them, thus not providing a correct impression, plus spatial variations are not well captured. Thus, as with many reservoir characterisation challenges, there is a need to use some form of remote sensing technique, such as seismic imaging, to provide information on the in-situ properties or conditions.

With the appropriate seismic data acquisition it is possible to assess the directionality of the seismic properties and to investigate the anisotropic nature of the rocks. This will not necessarily give absolute measures, but could provide important relative data on the elastic anisotropies and eventually their evolution. The techniques employed involve mapping variations with azimuth (of observation) of seismic reflection travel times or amplitudes (e.g. [46, 47, 20, 26, 17]) or shear wave splitting (e.g. [39, 59]). From these measurements, estimates of the magnitude and principal orientations of the in-situ seismic anisotropy can be derived. Such "seismic anisotropy" characterisation has been identified as having significant potential for the characterisation of subsurface reservoir fracturing. This is based on the hypothesis that preferentially aligned fracture sets present in the subsurface at sub-seismic scale, can produce seismic anisotropy, as discussed in Section 2.2. Thus, measurements of anisotropy from seismic data might be used to delineate fracturing and investigate their properties. As a result observations of azimuthal anisotropy in seismic reflection data are increasingly being used to characterise subsurface fracturing.

Figures 9 and 10 show examples of the results of an analysis of azimuthal variations in seismic reflection amplitudes (after [27, 26]) for the top of a North sea hydrocarbon reservoir unit at around 2500 m depth below the sea floor. In both cases seismic anisotropy magnitudes and principal directions are plotted that show an overall tendency to parallel faults and there are spatial distributions that, in the second example, seem to be to some extent fault-bounded. These observations suggest an association between faulting and anisotropy development.

A common hypothesis used in interpreting results such as those in Figs. 9 and 10 is that the observed anisotropies reflect the in-situ fracturing, i.e., higher magnitude anisotropy indicates a greater degree of fracturing (which could either be a larger number of fractures or increased size of the fractures). The correspondence of the anisotropy variations and larger-scale faulting in Figs. 9 and 10 does seem to support such a hypothesis, potentially indicating damage caused by the faulting process. Alternatively this correspondence might arise due to local stress reorientations, because of the structural heterogeneities represented by the faults, causing spatially varying preferential closure of cracks and fractures. In fact , the most likely the origin of the measured anisotropy is a combination of these two factors (see [22]). There may well also be other contributions such as from LPO. Therefore to really asses potential fracture distributions from such seismic anisotropy data requires that other possible causes of anisotropy are accounted for, e.g., by first determining the LPO anisotropy through petrofabric measurements, then assessing the stress-dependent and stress-independent SPO anisotropy using core-scale ultrasonic measurements, as discussed earlier. Therefore there is still some way to go before these analysis yield fully quantitative results on the underlying geomechanics. But the potential is there.

One of the most intriguing potential applications of these seismic anisotropy mapping approaches (as with many other aspects of seismic imaging) is the potential to monitor changes over time, e.g., during the production of hydrocarbons from or the injection of CO_2 into subsurface reservoirs. This could allow changes in stress fields, fracture patterns or the development of new damage to be tracked with the aim to see how the production/injection process is changing (geomechanically) the reservoir rocks and thus their hydraulic properties. A few results of such analyses have been published, but thus far the full potential of this has yet to be realised (in part due to lack of appropriate data). Analysis of such results will also require advances in the understanding and modelling of the geomechanical processes and their impact on the seismic data.

4.2 Time-lapse seismic and monitoring of fluid-extraction induced reservoir deformation

Seismic imaging can provide high-resolution 3D images of the subsurface and, if several images are acquired of a reservoir during its production, i.e., time-lapse seismic, temporal changes often can be identified and perhaps characterised, e.g.,

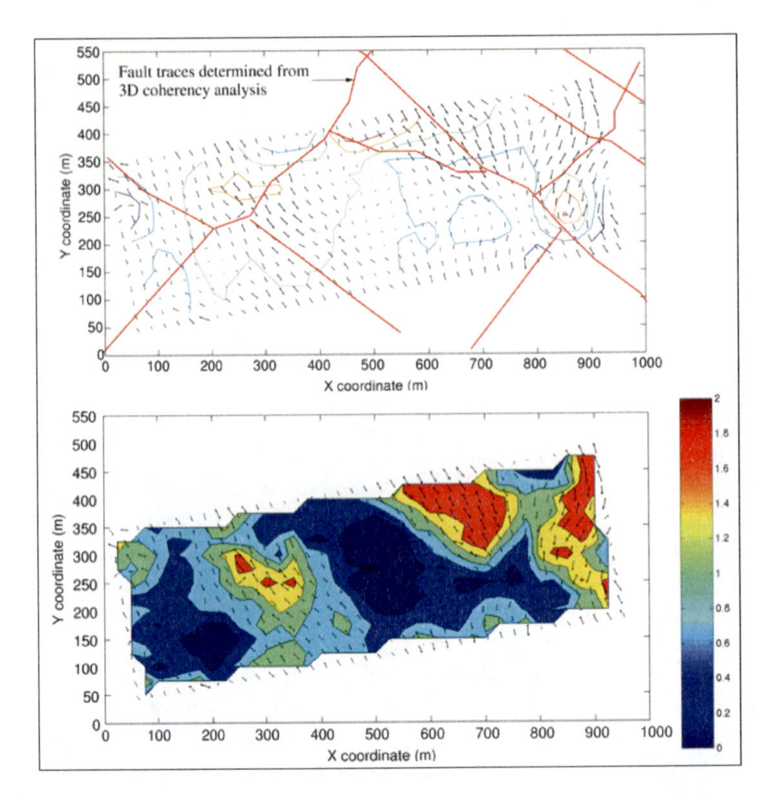

Fig. 9 Map of seismic reflection amplitude anisotropy magnitude and orientations for part of a North Sea hydrocarbon field, after [26]. On the upper plots, traces of larger-scale geologic faults, determined through other analysis, are plotted with dashed lines and the contours indicate the topography of the analysed geologic horizon. The arrows in both plots indicate the direction of the principal direction of anisotropy, that might correspond to an underlying fracture orientation (in the absence of other contributions to the elastic anisotropy). It is seen that these orientations show an overall tendency to parallel faults, being most marked for the NW–SE-oriented faults and in the far eastern section. This correlation supports an association between faulting and anisotropy development, potentially including damage caused by the faulting process. Alternatively this might indicate local stress reorientations because of the faulting causing preferential closure of cracks and fractures (and most probably a combination of the two factors, see [22])

.

[35, 56]. The primary aim of the time-lapse seismic method is to monitor changes in reservoir properties, such as saturation and pressure, mainly based on assessment of changes in seismic reflection amplitudes. However, it is also expected, and it has been observed, that there will be geomechanical changes associated with a change of pressure conditions in subsurface reservoirs. In a more general sense this issue is associated with the extraction of any fluids from the subsurface, i.e., also the extraction of water from aquifers. With regards to hydrocarbon production, such effects were observed first at Goose Creek, Texas, in the early part of last century ([61]). The effect has since been observed in many other locations worldwide such

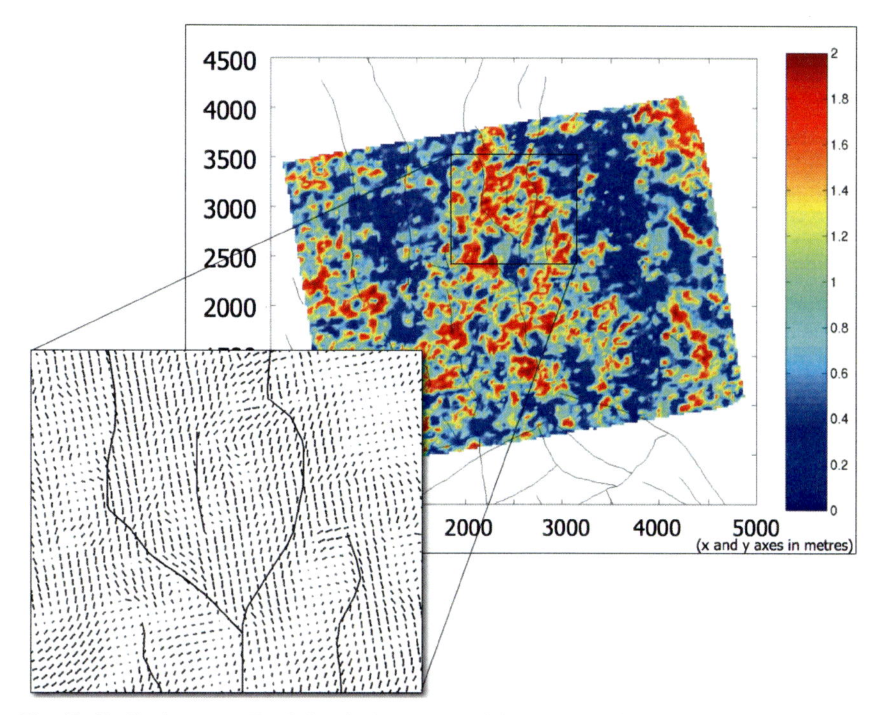

Fig. 10 Similar image to Fig. 9, but for larger area of the same North Sea hydrocarbon field (after [27]), showing the reflection amplitude anisotropy magnitude and fault traces plus the anisotropy orientations in the zoom. Again a good correspondence of the anisotropy magnitude and orientation variations with the larger-scale geologic faults is observed. Note that the location of this analysis is roughly the same as that for the compaction and subsidence analysis in Fig. 11.

as Ekofisk, e.g., [79], and Valhall., e.g., [57, 2], in the North Sea; Groningen in The Netherlands, e.g., [72]; Wilmington, Lost Hills and Belridge in California, e.g., [13], and reference therein; Lake Maracaibo in Venezuela; and Lacq in France, e.g., [41]. In each case, the effects of the depletion-induced deformation have been significant both in terms of hydrocarbon production with changes in permeability and formation of reservoir-bounding faults and socio-economic consequences such as flooding and surface structure damage. Summaries of the processes and consequences of deformation resulting from subsurface fluid extraction can be found in [6] and [36]. Until recently, the only tools to investigate the problem were observations of seafloor/land subsidence such as from bathymetry or synthetic aperture radar SAR measurements, e.g., [13], backed up by minimal subsurface observations, e.g., measured reservoir pressure changes or marker displacements. To understand the 3D system, geomechanical modelling must be used. However, such models can be constrained only by the minimal surface or down-hole observations. Additionally, model resolution is normally limited; thus, geologic under/overburden is largely neglected. Although some studies have been relatively successful at matching the observations, in a general sense, and have allowed predictions about future displacements, e.g., [57, 74], they cannot capture the full heterogeneity of the system.

[16] observed differences in the travel-times from the surface to the reservoir level between two seismic images from different times during the production of hydrocarbons from a subsurface reservoir; such analysis has also been carried out more recently for other fields, e..g., [24, 2, 18], and increasingly since. Such an approach sheds new light on subsurface processes in terms of the combined effects of vertical displacements and strain-induced velocity changes, but the resolution of true 3D effects (shear and lateral displacements/strains, e.g., [79, 74]) cannot be assessed properly as the analysis is just carried out in the vertical direction.

[19] presented a methodology for vector analysis of the image displacements between successive 3D seismic image volumes that provides 7D analysis (3-component displacement vectors defined in 3D over the given time increment) of in situ subsurface deformation around hydrocarbon reservoirs. The proposed method is essentially a form of 3D digital image correlation, a method that is increasingly used in experimental mechanics (e.g., [85, 84]). Figure 11 shows an example (from [24, 19]) of the results from such an analysis applied to a compacting North sea chalk reservoir, in which the downward displacement of the top of the reservoir unit around different producing wells can be seen in the vertical component of the displacement vectors. The horizontal components, shown in the zoom, also indicate some structure associated with the compaction/subsidence bowls, although these are less well defined due to the lower level of variation along geologic horizons in the seismic images compared to in the vertical direction (roughly perpendicular to the geologic layering).

The results discussed in the preceding paragraph, and those of other workers in the field, illustrate that the correlation of repeated 3D seismic images of a producing reservoir can yield interesting images of displacement fields between the images. However, there is a significant challenge in resolving these "image" displacements into real displacements. This challenge arises because the seismic images have a vertical axis of *travel-time*, which can be interpreted in terms of depth only if the velocity of all the rocks through which the waves pass is known. This is a processing procedure routinely carried out in the seismic imaging industry for single data sets. However, in a time-lapse context there are a greater number of ambiguities and so the image displacement fields from the correlation of images contain contributions from the changes in travel-times due to true spatial displacements of the subsurface (change in the travel-path length) and also contributions from the change in the velocities of the rocks due to strain effects; this was observed in the original paper by [16]. Generally an increase in travel-time for the same distance travelled is observed due to stretching of the overburden, although compaction of the reservoir will cause decreased travel-times across the reservoir. A suggestion on how to separate these components was given by [19], in which the image correlation approach was extended to separate the accumulated travel-time effects from local displacement effects through an iterative procedure involving assumptions on how the velocities change with strain. With this approach it was possible, in part to separate the accu-

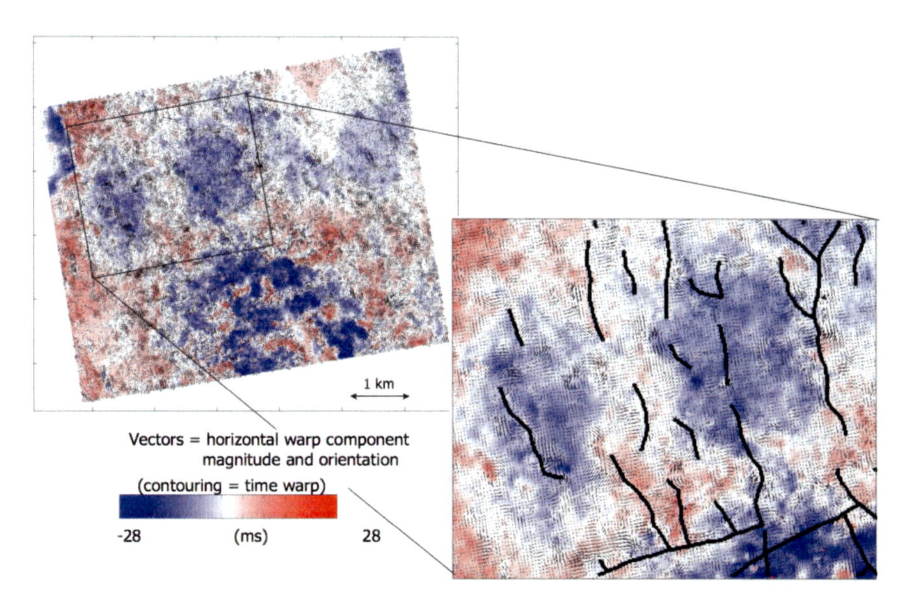

Fig. 11 Time-lapse analysis of subsidence and compaction in a North Sea hydrocarbon reservoir analysed through derivation of 3D image displacements between two seismic images of the reservoir; after [26] and [19]. These data were acquired 7 years apart, during which time hydrocarbons were extracted from the reservoir through a number of wells, whose positions coincide with the locations of the main downshift zones (blue areas). Note that the displacements shown in colour are the vertical shifts in "two-way travel-time" (which will have a contribution from the real displacement and the overburden velocity change - see main text). The zoomed section shows two producing areas in more detail with the traces of geologic faults overlain; clear features/trends in the horizontal displacement vector field (overlain) are seen that relate to the main vertical displacement features and the fault traces. Some correspondence is also seen with the anisotropy field for the same area (in Figure 10),which was analysed in the second data set used in this time-lapse analysis and thus could be expected to have an origin in the compaction/subsidence.

mulated overburden effects to reveal a local compaction signature in the reservoir, but realistic, quantitative values of strain have not yet been realised.

The analysis of [24, 19] discussed above was for the same North Sea reservoir analysed for seismic anisotropy presented in the preceding section and the regions shown in Figs. 10 and 11 are approximately the same. Interestingly some correspondence between the two measured fields (the subsidence/compaction and the anisotropy) can be seen, indicating that there is an origin to, at least part, of the observed anisotropy from the compaction/subsidence processes. This contribution could be due to the development of damage around the subsidence zones or perhaps a non-linear stress/strain effect associated with preferential opening of existing cracks or grain contacts, in any case it indicates the sensitivity of the anisotropy measures to an underlying geomechanical process (although the story is inevitably much more complex). In the same context it is worth mentioning the results of [55], in which shear-wave splitting results indicate anisotropy in the overburden of this field with orientations of anisotropy axes that follow the outline of the main subsidence bowl.

As with the laboratory examples, the focus here has been the use of controlled or active source analyses of geologic structures, but another important data type that can be used to assess both active geomechanical processes and image geomechanical properties or changes in them are passive seismic methods (so called as, like for acoustic emission analyses, it involves simply listening for seismic waves from natural, or at least non-intentional, sources). The obvious example of passive seismic methods is earthquake monitoring, which has been carried out at least since ancient Chinese Eastern Han Dynasty (25-220 AD) (cf Zhang's seismometer: http://en.wikipedia.org/wiki/Zhang_Heng). At a smaller scale so-called micro-seismic methods provide similar analysis with greater resolution (due to the deployment of more of sensors) for analysis of hydrocarbon production and $C0_2$ sequestration monitoring. A very interesting set of results in this context has been presented by [78], in which correlated temporal and spatial variations in both micro-seismic source locations (i.e., the sites of active deformation) and shear-wave splitting of the seismic waves produced by the events (indicating the anisotropy along the propagation path from the sources to the receivers) have been observed, which seem to be linked to some underling geomechanical processes associated with the extraction of hydrocarbons nearby. This aspect will not be discussed further other than to note that to understand the observations again requires improvements in our understanding an modelling of the geomechanical phenomena and associated wave propagation.

5 Conclusions

This paper has discussed, from the authors own viewpoint, the potential of some geophysical elastic-wave methods to be used to characterise geomechanical properties and processes at laboratory and geologic-structure scales. A brief theoretical background for elastic wave propagation in geomaterials and their measurement has been presented. The key points in this discussion being that the primary data considered are travel-times and amplitudes of the propagating waves (plus, to a lesser, extent frequency and wavelength) and that the wave propagation is controlled by a material's elastic properties (and density). These elastic properties are functions of the summed contributions due to grain elasticities, compliant discontinuities (including cemented or uncemented grain contacts) and inter- or intra-granular cracks, fractures and layers, which can all be anisotropic or can produce an anisotropic aggregate material. Examples have been presented of geophysical characterisation, at the laboratory and hydrocarbon reservoir scales, of these different geomechanical attributes through anisotropy analyses, time-lapse measures of deformation and "full-field" velocity imaging (tomography). Whilst the different examples do highlight the significant potential of the different methods, they also raise a lot more questions and challenges for the future in terms of how best to understand/interpret the observations. For example, the results on the detection and mapping of seismic anisotropy through the subsurface that have been presented clearly show a sensitivity to underlying geomechanical processes, but which ones is difficult to say. In fact the more pertinent questions are how the different probable effects contribute to the observed anisotropies and how they may be unravelled to draw useful conclusions to understand the geomechanics (and hydromechanics). The same questions must be posed for the analysis of the time-lapse deformation observations. To be able to better respond to such questions requires some advances in the analysis techniques, but most importantly advances are needed in the geomechanical modelling linked to the possible observables, such as seismic travel-times, amplitudes and anisotropies. Currently little geomechanical modelling really considers changes in elastic properties due to deformation processes and so it is difficult to make links between measured data and geomechanical simulations. Damage mechanics is one of the few approaches that does directly and routinely consider such aspects, but still the links are weak. Therefore there is still a long way to go in order to bring together the worlds of geomechanical simulation (plus associated theory and laboratory analysis) and geophysical imaging, but the possibilities are large, especially in the context of the current interest in time-lapse monitoring of hydrocarbon production and $C0_2$ sequestration using geophysical techniques. As a final comment it is worth noting that in this paper just elastic-wave methods have been discussed and other geophysical techniques also have significant potential for imaging geomechanical properties and processes; e.g., electrical resistance tomography at the laboratory-scale (e.g., [8]) and gravity imaging of subsurface reservoir changes with hydrocarbon production (e.g., [3]), but similar questions exist in terms of linking the geomechanical simulations to the geophysical data.

Acknowledgements The author would like to acknowledge all those people who have been involved in the different data examples presented including (amongst others and with apologies for omissions!): Mike Kendall, Olav Bakved, Cino Viggiani, Helen Lewis, Elma Charalampidou, Adan Realy, Colin MacBeth...

References

1. Aki, K. & Richards, P.G., 1980. *Quantitative seismology: theory and methods*. W. H. Freeman.
2. Barkved, O.I., Kristiansen, T. & Fjaer, E., 2005. The 4D seismic response of a compacting reservoir - examples from the Valhall field, Norway, *In 75th Ann. Internat. Mtg., Soc. of Expl. Geophys., Expanded Abstracts*.
3. Biegert, E., Ferguson, J. & Li, X. (Eds), 2008. Special Section - 4D Gravity Monitoring, Geophysics, 73, WA1-WA173.
4. Blackman, D. K. & Kendall, J.-M., 1997. Sensitivity of teleseismic body waves to mineral texture and melt in the mantle beneath a mid-ocean ridge, *Phil. Trans. Roy. Soc. London, A*, **355**, 217–231.
5. Budiansky, B. & O'Connell, R., 1976. Elastic moduli of a cracked solid, *Int. J. Solids Structures*, **12**, 81–97.
6. Chilingarian, G.V., Donaldson, E.C., & Yen, T.F., 1995. *Subsidence due to fluid withdrawal*, Developments in Petroleum Science, **41**, Elsevier Science.

7. Cholach, P. & Schmitt, D., 2003. Seismic anisotropy of shales, *CSEG Recorder*, **28**, 39–42.
8. Comina, C., Foti, S., Musso, G., & Romero, E., 2008. EIT Oedometer: an advanced cell to monitor spatial and time variability in soil with electrical and seismic measurements, *Geotechnical Testing Journal*, **31**, 1-9.
9. Crampin, S., 1984. Effective anisotropic elastic constants for wave propagation through cracked solids, *Geophys. J. Roy. Astr. Soc.*, **76**, 135–145.
10. Daigle, M., Fratta, D., & Wang, L.B., 2005. Ultrasonic and X-ray Tomographic Imaging of Highly Contrasting Inclusions in Concrete Specimens, *GeoFrontier 2005 Conference*, Austin, TX.
11. Debski, W. & Young, R. P., 1999. Enhanced velocity tomography: Practical method of combining velocity and attenuation parameters, *Geophys. Res. Lett.*, **26**, 3253–3256.
12. Falls, D., Young, R.P., Carlson, S.R. & Chow, T., 1992. Ultrasonic Tomography and Acoustic Emission in Hydraulically Fractured Lac du Bonnet Grey Granite, *J. Geophys. Res.*, **97**, 6867–6884.
13. Fielding, E.J., Blom, R.G., & Goldstein, R.M., 1998. Rapid subsidence over oil fields measured by SAR interferometry: *Geophysical Research Letters*, **25**, 3215–3218.
14. Fortin, J., Y. Gueguen, & A. Schubnel, 2007, Effects of pore collapse and grain crushing on ultrasonic velocities and V_p/V_s. *J. Geophys. Res.*, **112**, B08207.
15. Gueguen, Y. & Schubnel, A., 2003. Elastic wave velocities and permeability of cracked rocks, *Tectonophysics*, **370**, 163–176.
16. Guilbot, J. & Smith, B., 2002. 4-D constrained depth conversion for reservoir compaction estimation: Application to Ekofisk field: *Leading Edge*, **21**, 302–308.
17. Gray, D., Roberts, G., & Head, K., , 2002 Recent advances in determination of fracture strike and crack density from p-wave seismic data. *Leading Edge*, **21**, 280.
18. Hatchell, P. & Bourne, S.J., 2005. Measuring reservoir compaction using time-lapse timeshifts: *In 75th Ann. Internat. Mtg., Soc. of Expl. Geophys., Expanded Abstracts*
19. Hall, S.A, 2006. A methodology for 7D warping and deformation monitoring using time-lapse seismic data *Geophysics*, **71**, O21–O31.
20. Hall, S.A., 2000. *Rock fracture characterisation and seismic anisotropy: application to ocean bottom sesmic data*. PhD thesis, University of Leeds, Leeds, England.
21. Hall, S.A., Kendall, J-M., Fisher, Q., & Maddock, J., 2008. Crack density tensor inversion for analysis of changes in rock frame architecture, *Geophys. J. Int.*, **173**, 577-592.
22. Hall, S.A, Lewis, H., & Macle, X., 2007. Seismic identification of faults and inter-fault damage, in *Damage and Localization*, Geological Society, London, Special Publication, Couples, G.D., & Lewis, H. (eds), pp. 187-207.
23. Hall, S.A., Viggiani, G., Charalampidou, E., Bésuelle, P., and Rousseau, C., 2007. Caractérisation de lendommagement localisé dans les géomatériaux laide des ondes ultrasonores en conditions de laboratoire, *18ème Congrès Franais de Mcanique*, Grenoble.
24. Hall, S.A, MacBeth, C., Barkved, O.I. & Wild, P., 2005. Cross-matching with interpreted warping of 3D streamer and 3D OBC data at Valhall for time-lapse assessment, *Geophysical Prospecting*, **53**, 283-297.
25. Hall, S.A., Lewis, H., Viggiani, G. and Lenoir, N., (2005). Damage assessment in geomaterials using elastic waves, *7th International Workshop on Bifurcation, Instabilities and Degradation in Geomechanics*, Chania, Greece.
26. Hall, S.A, & Kendall, J.-M., 2003. Fracture characterization at Valhall: Application of P-wave amplitude variation with offset and azimuth (AVOA) analysis to a 3D ocean-bottom data set *Geophysics*, **68**, 1150–1160.
27. Hall, S.A, Kendall, J.-M., & Barkved, O.I., 2002. Fractured reservoir characterization using P-wave AVOA analysis of 3D OBC data *The Leading Edge*, **21**, 777-781.
28. Hall, S.A, Kendall, R., Kendall, J.-M., & Sondergeld, C., 2001. Analysis of anisotropic velocities in a core sample and AVOA from a fractured vuggy carbonate reservoir, in *Anisotropy 2000: Fractures, Converted Waves, and Case Studies*, Soc. of Expl. Geophys., pp. 257–269.
29. Han, J., Schmitt, D., Collis, D., & Escartin, J., 2004. Laboratory determination of velocity anisotropy, in *Proceedings of the 2004 CSEG National Convention*.

30. Hoenig, A., 1978. Elastic moduli of a non-randomly cracked body, *Int. J. Solids Structures*, **15**, 137–154.
31. Hornby, B., 1998. Experimental laboratory determination of the dynamic elastic properties of wet, drained shales, *J. Geophys. Res.*, **103(B12)**, 29945–29964.
32. Hudson, J., 1980. Overall properties of a cracked solid, *Math. Proc. Camb. Phil. Soc.*, **88**, 371–384.
33. Hudson, J., 1981. Wave speeds and attenuation of elastic waves in material containing cracks, *Geophys. J. Roy. Astr. Soc.*, **64**, 133–150.
34. Hudson, J. & Liu, E., 1999. Effective elastic properties of heavily faulted structures, *Geophysics*, **64**, 479–485.
35. Jack, I.G., 1997. *Time-lapse seismic in reservoir management*, Society of Exploration Geophysicists, Tulsa, OK.
36. Jones, M.E., Ledra, M.J.,& Addis, M.A.,1987 *Reservoir compaction and surface subsidence due to hydrocarbon extraction*, HMSO.
37. Kachanov, M., 1980. Continuum model of medium with cracks, *J. Eng. Mech. (ASCE)*, **106**, 1039–1051.
38. Kendall, J-M., Fisher, Q. J., Covey Crump, S., Maddock, J., Carter, A., Hall, S.A., Wookey, J., Valcke, S., Casey, M., Lloyd, G., & Ben Ismail, W., 2007. Seismic anisotropy as an indicator of reservoir quality in siliciclastic rocks, in *Structurally Complex Reservoirs*, Jolley, S. J., Barr, D., Walsh, J. J. & Knipe, R. J. (eds), Geological Society, London, Special Publication **292**.
39. Kendall, R. R. & Kendall, J-M., 1996. Shear-wave amplitude anomalies in south-central Wyoming, *The Leading Edge*, **15**, 913-920.
40. King, M., Chaudhry, N., & Shakeel, A., 1995. Experimental ultrasonic velocities and permeability for sandstones with aligned cracks, *Int. J. Rock Mech. Mining Sci.*, **32**, 155–163.
41. Lahaie, F. & Grasso, J. R., 1999. Loading rate impact on fracturing pattern: Lessons from hydrocarbon recovery, Lacq gas field, France: *J. Geophys. Res.*, **104**, 17941–17954.
42. Lee, J.S., Fernandez, A.L. & Santamarina, J.C., 2005. S-Wave Velocity Tomography: Small-Scale Laboratory Application, *Geotechnical Testing Journal*, **28**, 1–9.
43. Lo, T., Coyner, K., & Toksoz, M., 1986. Experimental determination of elastic anisotropy in Berea sandstone, *Geophysics*, **51**, 164–171.
44. Lockner, D.A. & Byerlee, J.D., 1977a. Acoustic emission and fault formation in rocks. In *Proc. 1st Conf. on Acoustic Emission/Microseismic Activity in Geol. Structures and Materials* (ed. H.R. Hardy & F.W. Leighton), pp. 99-107. Trans-Tech. Publications, Clausthal-Zellerfeld.
45. Lockner, D.A. and Byerlee, J.D., 1977b. Hydrofracture in Weber sandstone at high confining pressure and differential stress. *J. Geophys. Res.* **82**, 2018-2026.
46. Lynn, H. B. & Thomsen, L., 1990. Reflection shear-wave data collected near the principal axes of azimuthal anistropy. *Geophysics*, **55**, 147–156.
47. Lynn, H.B., Simon, K.M., Bates, C., & Van Dok, R., 1996. Azimuthal anisotropy in P-wave (multiazimuth) data. *Leading Edge*, **15**, 923–928.
48. MacBeth, C. & Schuett, H., 2007. The stress dependent elastic properties of thermally induced microfractures in aeolian rotliegend sandstone, *Geophysical Prospecting*, **55**, 323-332.
49. Maddock, J., Fisher, Q., Kendall, J.-M., & Lloyd, G., 2004. Petrofabric analysis of seismic anisotropy in siliciclastic sedimentary rocks, in *66th Mtg.: Eur. Assn. Geosci. Eng., Extended Abstracts*, p. E11.
50. Mainprice, D., 1990. An efficient fortran program to calculate seismic anisotropy from the lattice preferred orientation of minerals, *Computers and Gesosciences*, **16**, 385–393.
51. Mavko, G., Mukerji, T. & Dvorkin, J., 1998. *The rock physics handbook: tools for seismic analysis in porous media*, Cambridge University Press, New York.
52. Meglis, I.L., Chow, T., Martin, C.D. & Young, R.P., 2005 Assessing in situ microcrack damage using ultrasonic velocity tomography *Int J. Rock Mechanics & Mining Sciences*, **42**, 25–34.
53. Nur, A., 1971. Effects of stress on velocity anisotropy in rocks with cracks, *J. Geophys. Res.*, **76**, 2022–2034.

54. Nur, A. & Simmons, G., 1969. Stress-induced velocity anisotropy in rock: an experimental study, *J. Geophys. Res.*, **74**, 6667–6674.
55. Olofsson, B., Probert, T., Kommedal, J.H. & Barkved,O.I., 2003 Azimuthal anisotropy from the Valhall 4C 3D survey, *The Leading Edge*, 22, 1228-1235.
56. Parker, J., Bertelli, L., & Dromgoole, P., 2003. 4D seismic technology special issue, *Petroleum Geoscience*, **9**.
57. Pattillo, P.D., Kristiansen, T.G., Sund, G.V.,& Kjelstadli, R.M., 1998. Reservoir compaction and seafloor subsidence at Valhall, *SPE*, 47274.
58. Pendrel, J., 2006. The New Reservoir Characterization, *CSEG Recorder*, **31**, 105-109
59. Potters, J.H.H.M., Groenendaal, H.J.J., Oates, S.J., Hake, J.H., & Kalden, A.B., 1999. The 3D shear experiment over the Natih field in Oman. Reservoir geology, data acquisition and anisotropy analysis, *Geophysical Prospecting*. **47**, 637-662.
60. Pratt, R.G, 1999, Seismic waveform inversion in the frequency domain, Part 1: Theory, and verification in a physical scale model, *Geophysics*, **64**, 888–901.
61. Pratt, W.E. & Johnson, D.W., 1926. Local subsidence of the Goose Creek oil field, *J. Geol.* **34**, 577–590.
62. Rasolofosaon, P., 1998. Stress-induced seismic anisotropy revisited, in *Proc. 8th Int. Worksh. Seismic Anisotropy, Rev. Inst. Franc. Petr., 53*, pp. 679–692.
63. Rasolofosaon, P., Rabbel, W., Siegesmun, S., & Vollbrecht, A., 2000. Characterization of crack distribution: fabric analysis versus ultrasonic inversion, *Geophys. J. Int.*, **141**, 413–424.
64. Sammonds, P.R., Ayling, M.R., Meredith, P.G., Murrell, S.A.F., & Jones, C., 1989. A laboratory investigation of acoustic emission and elastic wave velocity changes during rock failure under triaxial stresses, in: *Rock at great depth*, eds: Maury, V., & Fourmaintraux, D., pp. 233-240, A.A. Balkema
65. Santamarina, J.C. & Fratta, D., 2005. *Introduction to Discrete Signals and Inverse Problems in Civil Engineering*, ASCE Press, VA.
66. Santamarina, J.C, Klein, K. & Fam, M., 2001. *Soils and Waves*, J. Wiley and Sons, Chichester, UK.
67. Santamarina, J.C. & Cascante, G., 1996. Stress anisotropy and wave propagation - A micromechanical view, *Canadian Geotechnical Journal*, **33**, 770–782.
68. Sayers, C., 2002. Stress-dependent elastic anisotropy of sandstones, *Geophysical Prospecting*, **50**, 85–95.
69. Sayers, C. & Kachanov, M., 1991. A simple technique for finding effective elastic constants of cracked solids for arbitrary crack orientation statistics, *Int. J. Solids Structures*, **27**, 671–680.
70. Sayers, C. & Kachanov, M., 1995. Microcrack-induced elastic wave anisotropy of brittle rocks, *J. Geophys. Res.*, **100**, 4149–4156.
71. Schoenberg, M. & Sayers, C., 1995. Seismic anisotropy of fractured rock, *Geophysics*, **60**, 204–211.
72. Schoonbeek, J.B., 1976. Land subsidence as a result of natural gas extraction in the province of Groningen *SPE AIME*.
73. Scott T.E., & Abousleiman Y., 2004. Acoustical imaging and mechanical properties of soft rock and marine sediments, *Final Technical Report #15302, Department of Energy (DOE Award Number: DEFC26-01BC15302)*
74. Segall, P. & Fitzgerald, S.D., 1998. A note on induced stress changes in geothermal and hydrocarbon reservoirs *Tectonophysics*, **289**, 117–128.
75. Stanchits, S., Vinciguerra, S., & Dresen, G., 2006. Ultrasonic Velocities, Acoustic Emission Characteristics and Crack Damage of Basalt and Granite, *Pure and Applied Geophysics*, **5-6**, 975-994.
76. Stanchits, S., Lockner, D., & Ponomarev, A., 2003. Anisotropic changes in P-wave velocity and attenuation during deformation and fluid infiltration of granite, *Bull. Seismological Society of America*, **93**, 1803–1822.
77. Stevens, D.F., 2007. *Stress redistribution in Berea sandstone samples using acoustic emission tomography in the laboratory*, MSC thesis, Virginia Polytechnic Institute and State University.

78. Teanby, N., Kendall, J-M., Jones, R.H., & Barkved, O.I, 2004. Stress-induced temporal variations in seismic anisotropy observed in microseismic data *Geophys. J. Int.*, **156**, 459–466.
79. Teufel, L. W. & Rhett, D. W., 1992. Failure of chalk during water-flooding of the Ekofisk field *SPE* 24911.
80. Thomsen, L., 1986. Weak elastic anisotropy *Geophysics*, **51**, 1954–1966.
81. Toksoz, M. N., Cheng, C. H., & Timur, A., 1976. Velocities of seismic waves in porous rocks, *Geophysics*, **45**, 621–645.
82. Valcke, S., Casey, M., Lloyd, G., Kendall, J.-M., & Fisher, Q., 2006. Lattice preferred orientation and seismic anisotropy in sedimentary rocks, *Geophys. J. Int.*, **166**, 652–666.
83. Viggiani, G., & Atkinson, J.H., 1995. Stiffness of finegrained soil atvery small strains, *Géotechnique*, **45**, 249–265.
84. Viggiani, G. & Hall, S.A., 2008 Full-field measurements, a new tool for laboratory experimental geomechanics. In *Proceedings of the 4th International Symposium on Deformation Characteristics of Geomaterials* (Burns S.E., Mayne P.W and Santamarina J.C. (Eds.)), 3-26, IOS Press.
85. Withers, P.J., 2008. Strain measurement by Digital Image Correlation, *Strain*, **44**, 421–422.
86. Zang, A., Wagner, F.C., Stanchits, S., Dresen, G., Andresen, R. & Haidekker, M.A., 1998. Source analysis of acoustic emissions in Aue granite cores under symmetric and asymmetric compressive loads *Geophys. J. Int.*, **135**, 1113–1130.

Fracture of Ice and other Coulombic Materials

Erland M. Schulson

Abstract In this paper we review the fracture of ice on scales small (laboratory) and large (arctic sea ice cover), with emphasis on brittle compressive failure under multiaxial loading and on underlying physical processes, with reference to rock and minerals. Topics include: the structure of terrestrial ice; the brittle compressive failure envelope; failure modes; deformation features; Coulombic shear faulting; frictional sliding; physically based, micromechanical models of the initiation of Coulombic faults and of the ductile-to-brittle transition; and non-frictional/plastic shear faulting. An important point is that the physics of fracture appears to be scale independent. The paper is a synopsis of a more complete treatment given in *Creep and Fracture of Ice* (Schulson and Duval, 2009).

1 Introduction

Ice exhibits two kinds of inelastic behavior. When loaded slowly, it creeps: the material exhibits strain-rate hardening and thermal softening, and obeys the classical power-law constitutive relationship. Polar glaciers and ice sheets, for instance, deform at rates of $\sim 10^{-9}$ s^{-1} and lower (under deviatoric stresses of ~ 0.1 MPa and lower) and accumulate, through dislocation slip, volume-conserving strains in excess of unity. Similarly, meter-thick sheets of arctic sea ice buckle plastically when pushed by wind and ocean currents at speeds of around a few meters/hour against the sides of engineered structures. When loaded rapidly, on the other hand, ice fractures, at inelastic strains as low as 0.1 or lower. Strain-rate hardening is initially extinguished—indeed, strain-rate softening is evident over several decades of strain rate—but then hardening reappears under dynamic or impact conditions (Jones 1997; Shazly et al. 2006), characterized by a sensitivity lower by about a

Erland M. Schulson
Thayer School of Engineering, Dartmouth College, Hanover, NH 03755, e-mail: erland.schulson@dartmouth.edu

factor of two than that within the ductile regime; i.e., $\sigma_{brit} \propto \dot{\varepsilon}^{1/7}$ vs. $\sigma_{duc} \propto \dot{\varepsilon}^{1/3}$. Within the regime of brittle behavior, the hydrostatic component of the stress state impedes deformation, implying that frictional sliding across deformation-induced cracks and not dislocation slip plays a major role in compressive fracture. Brittle failure is both technologically and climatically important: it marks the point where ice loads on offshore structures reach a maximum (Sanderson 1988), generating ice forces in excess of 100-year wave forces (API 1995), and it dominates the deformation of the winter ice cover on the Arctic Ocean, generating openings, termed leads, through which heat and moisture enter the atmosphere (Maykut 1982; McPhee et al. 2008) and upon which new ice forms and salt is rejected to the ocean.

In this paper, we focus on the fracture of ice and on those aspects of that are most novel; namely, brittle compressive failure and the ductile-to-brittle transition. Tensile failure is less remarkable and can be well understood within established theories of crack nucleation and crack propagation, as discussed elsewhere (Schulson 2001). We limit our discussion to polycrystals of low-pressure ice or ordinary terrestrial ice, denoted ice Ih, and consider their behavior under terrestrial conditions; i.e., at temperatures above about -40°C at rates of loading within the so-called quasi-static regime. The discussion is based mainly upon measurements and observations in the laboratory, but enriched in the final section through reference to fracture of the ice cover on the Arctic Ocean and to compelling evidence of scale-independent fracture physics. Where relevant, we note the similarity between ice and rock and show how lessons from one material enlighten our understanding of the other. A more complete account of the subject of this paper is given in a new book, entitled: *Creep and Fracture of Ice* (Schulson and Duval 2009).

2 Structure of Ice

2.1 Crystal Structure

Ice Ih possesses an hexagonal crystal structure, reflected in the shape of snow-flakes. The structure is not closely packed, but open: it collapses upon melting, thereby accounting for its density (917 kg m^{-3} at -10°C) being lower than the density of water. The unit cell has dimensions c = 0.7357 nm and a = 0.4510 nm at -20°C (Rottger et al. 1994), and the ratio c/a = 1.633, independent of temperature and close to the ideal ratio for hard spheres closely packed, even though the Ih structure is an open one. The elastic behavior of ice Ih is moderately anisotropic (Gammon et al. 1983), but the plastic behavior is highly anisotripic owing to a large difference (approximately 60 times) in the critical resolved shear stress for basal versus non-basal slip (Duval et al. 1983). Dislocations glide on the basal plane in one of the three **a**-directions, creating two independent modes of deformation. Dislocation climb out of the basal plane produces two more modes. Thus, polycrystals of ice Ih have the four independent modes of deformation necessary for volume-conserving flow, albeit not necessarily uniform strain (Hutchinson 1977). Both glide and climb occur sluggishly in ice. That characteristic, plus very low resistance to crack propagation (fracture toughness $K_{Ic} \approx 0.1$ MPa $m^{0.5}$, relatively independent of temperature) accounts for the common observation of brittle behavior under moderate deformation rates right up to the melting point.

2.2 Microstructure

Polycrystalline ice is generally classified as either granular or columnar. Granular ice refers to an aggregate of equiaxed and randomly oriented grains, typically 1–10 mm in diameter. It is the kind of material from which glaciers and polar ice sheets are made, through the consolidation of snow (Paterson 2000). It usually contains porosity of which as little as ~0.5% by volume imparts a cloudy hue. A variant, termed frazil ice, consists of randomly oriented, round or needle-shaped crystals that nucleate from super-cooled water and then aggregate under turbulent conditions, forming, for instance, blockages on trash racks to hydro-electric power stations (Michel 1978) and certain layers within some regions of both the Arctic and the Antarctic sea ice covers (Gow et al. 1987, Jefferies et al. 1997). Barring the development of a deformation-induced texture, granular/frazil ice exhibits isotropic mechanical properties, even though the individual grains are plastically anisotropic.

Columnar ice, in comparison, refers to an aggregate of columnar-shaped grains, again typically 1–10 mm in diameter but many times greater in length. It is the kind of ice from which floating sheets of sea ice and river ice are most often made, through the unidirectional solidification of water. At freezing rates commonly imposed, air is generally rejected, and so columnar ice that grows via the solidification of fresh-water is highly transparent: a plate slowly grown in the laboratory to a thickness of 30 cm (over ≈ two weeks) is very clear indeed. Columnar ice possesses a crystallographic growth texture. The most common, denoted S2 (Michel and Ramseier 1971), finds the crystallographic c-axes confined more or less to the horizontal plane of the cover, but randomly oriented within that plane. Less common are two other textures, denoted S1 and S3, in which the c-axes are oriented either in the vertical plane (S1) or in a specific direction within the horizontal plane (S3). In describing the mechanical behavior of columnar ice, it is convenient to define a Cartesian material coordinate system X_i where directions X_1 and X_2 are perpendicular to the long axes of the columns and X_3 is parallel to the long axes. We consider below only those cases of loading where the loading axes and the material axes coincide. In such cases, the term σ_{ij} denotes the component of the stress tensor acting on the face normal to direction X_i in the X_j direction.

Sea ice contains additional phases. Although comprised primarily (~95% by volume) of fresh-water ice, it also contains a roughly 50/50 mixture of entrapped air and brine. These phases are present in the form of sub-mm pores distributed along grain boundaries. Within columnar sea ice, also termed congelation ice, the brine pockets are also distributed intragranularly, forming a sub-structure that consists of plate-like arrays spaced by $\sim$1 mm and parallel to the basal planes of the ice Ih crystal lattice. Unlike fresh-water ice, sea ice is optically opaque, owing to the scattering of light from the pores. Brine pockets lower both creep resistance (Cole et al. 1998; de La Chapelle et al. 1998) and tensile strength (Richter-Menge and Jones 1993), but, at the level present within sea ice, have either little or no effect on brittle compressive strength (Schulson and Gratz 1999), at least at -10°C.

2.3 Texture-Induced 2D Behavior

Owing to the presence of crystallographic texture, or to its absence as the case may be, polycrystalline ice exhibits either 2D or 3D inelastic deformation, within the regimes of both ductile and brittle behavior. S2 columnar ice deforms in a 2D manner when compressed principally across the columns: it expands in the orthogonal across-column direction, but hardly at all in the along-column direction. Granular ice, on the other hand, deforms inelastically in a 3D manner: when compressed longitudinally, the materials develop axially symmetric transverse strains. The implication is that biaxial loading has no effect at all on the load-bearing ability of texture-free, granular ice, such that the uniaxial and biaxial compressive strengths are indistinguishable (Weiss and Schulson 1995); only triaxial loading raises its strength. In comparison, both the ductile (Melton and Schulson 1997) and the brittle (more below) compressive strengths of S2 columnar ice biaxially loaded across the columns increase as the confining stress increases. In many respects, biaxially loaded, S2 ice behaves in a manner analogous to the behavior of granular ice loaded triaxially.

3 Brittle Compressive Failure

In the discussion below, we focus on S2 columnar ice. Owing to its texture, cracks that nucleate during deformation grow preferentially on planes that are oriented parallel to the long axis of the grains, both intergranularly and transgranularly. This means that when viewed along the columns, cracks do not shield each other. In granular ice, they do. Given that crack mechanics—nucleation, growth, interaction and particularly frictional sliding—are fundamental to brittle compressive failure, the opportunity presented by transparent and coarsely-grained columnar ice to allow direct and unimpeded viewing by the unaided eye has helped greatly to elucidate some of the basic physical processes at play.

3.1 Failure Envelope and Failure Modes

Figure 1 shows the compressive part of the failure envelope for fresh-water S2 ice, plotted in principal-stress space in terms of the applied normal stresses $\sigma_{11} = \sigma_1$ and $\sigma_{22} = \sigma_2$. We plot the data twice about the line $\sigma_{11} = \sigma_{22}$ to illustrate the isotropic character of the material when loaded within the $X_1 - X_2$ plane. Compressive stresses are taken to be positive. (To convert to shear stress $\tau = (\sigma_1 - \sigma_2)/2$ vs. normal stress or pressure $\sigma_n = (\sigma_1 + \sigma_2)/2$, rotate Fig. 1 by 45 (.) The envelope was obtained from experiments in the laboratory (Iliescu and Schulson 2004) in which square plates of ice 150 × 150 mm^2 × 25-100 mm thick (along-column direction) were proportionally loaded biaxially across the columns, at -10°C at a strain rate of $\dot{\varepsilon} = 5 \times 10^{-3}$ s^{-1}. The envelope is almost identical to one obtained from first-year sea ice (Section 6), and is relevant to the loading of a floating sheet when pushed against the sides of a wide, offshore structure.

Two branches are evident, one rising and the other descending. The rising branch indicates confinement-strengthening; the descending branch, confinement-weakening. The transition from strengthening to weakening is specific to columnar ice loaded across the columns, and sets in once the degree of confinement, defined by the stress ratio $R_{21} = \sigma_2/\sigma_1$, reaches a critical level R_c. For the conditions underlying Figure 1, $R_c \approx 0.2$. This value is governed by the coefficient of internal friction (Section 5).

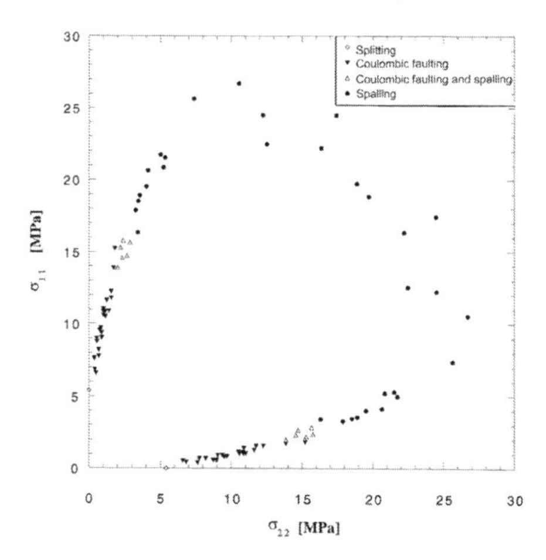

Fig. 1 Brittle compressive failure envelope for columnar-grained S2 fresh-water ice of 6±2 mm column diameter, biaxially loaded across the columns at -10°C at $\dot{\varepsilon}_{11} = 4 \times 10^{-3}$ s^{-1}. The data were obtained from plates (150 × 150 × 25 mm^3) where the smallest dimension was parallel to the long axis of the columns. The failure mode is denoted by the shape of the points. From (Iliescu and Schulson 2004).

Terminal failure along the rising branch is marked by the development of one or more Coulombic shear faults. The faults are oriented at 25°–30° to the direction of shortening and parallel to the no-load direction. Occasionally they form as conjugate sets, Figure 2. Near zero confinement ($R_{21} < 0.01$) faults transform to axial splits that are aligned more or less with the direction of loading (Wachter et al. 2009). Under higher confinement ($R > R_c$), faults transform to splits that cut across the columns, leading to across-column cleavage or spalling out of the X_1-X_2 loading plane.

Fig. 2 (a) Coulombic shear faults in columnar-grained S2 fresh-water ice biaxially loaded across the columns to terminal failure under moderate confinement, under the conditions described in the caption to Figure 1, as viewed along the columns. (b) A fully developed set of across-column cracks in the same kind of ice biaxially loaded across the columns under higher confinement, showing a set of thin plates that eventually collapse, leading to spalling out of the loading plane, as viewed across the columns. In each image, the maximum stress was applied in the vertical direction. From (Iliescu and Schulson 2004).

Within this lower confinement/Coulombic faulting regime ($0 < R < R_c$), the brittle compressive strength, taken as the value of the maximum principal stress, may be expressed by the linear equation:

$$\sigma_{1f} = \sigma_0 + q\,\sigma_2 \tag{1}$$

where the subscript f denotes terminal failure, σ_0 denotes the unconfined compressive strength and q, the slope of the envelope. The slope, like R_c, is governed by the coefficient of internal friction (Section 3.2) and decreases with increasing temperature, falling from q = 5.7±0.3 at -10°C to q = 3.48±0.21 at -3°C (Schulson et al. 2006a). Within the higher confinement/spalling regime ($R_c < R \leq 1$) the strength may be described approximately by the equation:

$$\sigma_{1f} = C - \sigma_2 \tag{2}$$

where C is a constant whose value at -10°C is 32±3 MPa. Across-column spalling appears to result from a combination of two processes, across-column cleavage-crack growth and Euler buckling: both processes are driven by the two normal components of the applied stress tensor, thus accounting for confinement weakening (Iliescu and Schulson 2004).

3.2 Confinement-Strengthening and Frictional Sliding

Underlying confinement-strengthening is frictional sliding (Jaeger and Cook 1979; Sammonds et al. 1998; Schulson et al. 2006). This may be seen as follows: Imagine an incipient macroscopic failure plane inclined by the angle θ to the direction of maximum principal stress, across which act a shear stress τ and a normal stress σ_n. Coulomb's (1773) criterion for the failure of brittle bodies dictates that terminal failure sets in when the shear stress exceeds the sum of material cohesion τ_0 and the frictional resistance to sliding $\mu_i\sigma_n$, where the parameter μ_i denotes the internal friction coefficient; i.e.

$$\tau = \tau_0 + \mu_i\sigma_n \quad . \tag{3}$$

This criterion in terms of the principal stresses may re-written:

$$|\tau| - \mu_i\sigma_n = \frac{1}{2}(\sigma_1 - \sigma_2)(\sin 2\theta + \mu\cos 2\theta) - \frac{1}{2}\mu_i(\sigma_1 + \sigma_2) = \tau_0 \quad . \tag{4}$$

Upon differentiating Equation (4) with respect to θ and setting the result to zero, the function attains its maximum value when (Jaeger and Cook 1979):

$$\tan 2\theta = 1/\mu_i \quad . \tag{5}$$

The maximum stress at terminal failure is then given by:

$$\sigma_{1f} = \frac{2\tau_0 + \sigma_2\left[(\mu_i^2+1)^{1/2} + \mu_i\right]}{\left[(\mu_i^2+1)^{1/2} - \mu_i\right]} \quad . \tag{6}$$

Thus, the slope of the low-confinement segment of the brittle compressive failure envelope is given solely in terms of the friction coefficient:

$$q = \frac{d\sigma_1}{d\sigma_2} = \frac{(\mu_i^2+1)^{1/2} + \mu_i}{(\mu_i^2+1)^{1/2} - \mu_i} = \left[(\mu_i^2+1)^{1/2} + \mu_i\right]^2 \quad . \tag{7}$$

From Equation (7) and from the slope of failure envelope shown in Figure 1, $\mu_i = 0.98 \pm 0.04$ for the relevant experimental conditions.

When account is taken of the effects on friction of temperature and sliding speed, the internal friction coefficient derived from the failure envelope has almost the same value as the coefficient of friction measured directly for sliding across fully-developed Coulombic faults (Fortt and Schulson 2007), Table 1. This suggests that the frictional processes underlying terminal failure and post-terminal failure are similar.

Table 1 Friction coefficients for S2 fresh-water ice at a sliding speed of $\approx 8\times 10^{-4}\mathrm{m\,s^{-1}}$ (from Schulson et al. 2006a; Fortt and Schulson 2007)

Temperature (°C)	Internal friction coefficient from failure envelope	Coefficient of friction for sliding across Coulombic faults
-10	0.98±0.04	0.89±0.10
-3	0.66±0.44	0.69±0.07

The above analysis is not limited to ice. Judging from the behavior of granite and other brittle materials (Byerlee 1967; Scott and Nielsen 1991a; Scott and Nielsen 1991b; Schulson et al. 2006a; Costamagna et al. 2007), it appears to apply to rock as well.

3.3 Micromechanical Processes Underlying Coulombic Faulting

Figure 3 shows the evolution of a Coulombic fault within S2 columnar fresh-water ice loaded biaxially across the columns. The long axis of the columns is perpendicular to the page. During the early stages of deformation, within the pseudo-elastic regime, grain-sized primary cracks nucleate along boundaries inclined by $\sim 45°$ to the direction of maximum principal stress, creating in the process a field of damage that is distributed globally more or less uniformly. Secondary cracks of two kinds stem from the primary cracks and curve towards the direction of shortening, Figure 4. One kind, termed an extensional or wing crack, stems from opposite ends of the primary/parent crack and serves to relax tensile stresses that develop there as a result of frictional sliding across the parent. The other kind of secondary, termed a comb crack, sprouts from one side of the sliding parent crack and relaxes tensile stresses that develop on that side through non-uniform sliding (Cooke 1997). As the applied load increases, more primary cracks nucleate, while the secondary cracks lengthen via stable mode-I crack propagation as additional sliding opens their mouths. The cracks interact and, near terminal failure, link up, creating a fault. The ones shown in Figure 3 and Figure 4 are bordered by zig-zag (i.e. wing crack) edges.

Terminal failure appears to result not simply from the linkage of only those cracks created earlier in the deformation of the body, but also from the creation and propagation of new cracks (Wachter et al. 2009) right up to the point of material collapse, in keeping with earlier hypotheses (Moore and Lockner 1995; Gupta and Bergstrom 1998). What seems to happen, judging from high-speed (10k fps) digital images (Fortt and Schulson 2009), is that right up to terminal failure cracks develop at various sites along the imminent fault, ostensibly randomly but probably related more to the orientation of grains and grain boundaries than to randomness. Destabi-

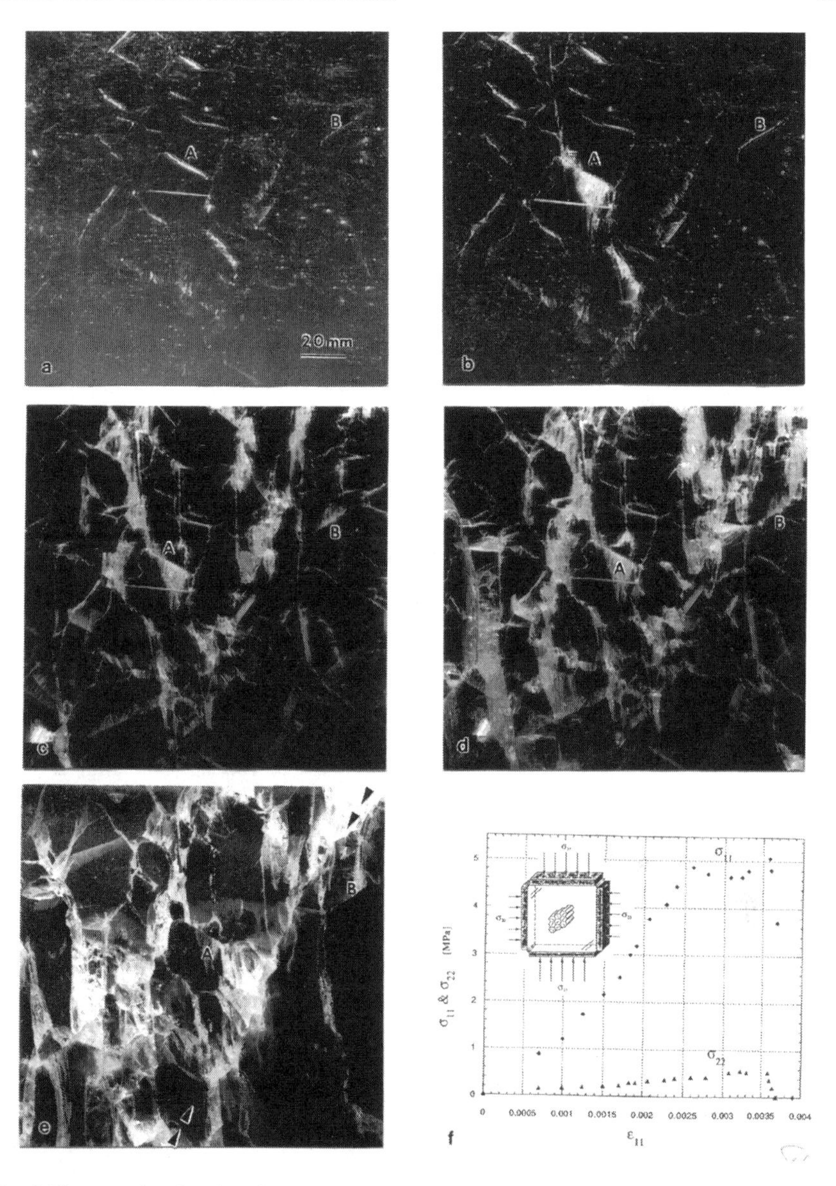

Fig. 3 Photographs showing the sequential development (a–e) of a Coulombic shear fault in columnar-grained S2 fresh-water ice proportionally pulse-loaded biaxially across the columns under moderate confinement ($\sigma_2/\sigma_1 = 0.1$) at -10°C at $\dot{\varepsilon}_{11} = 4.8 \times 10^{-3}\,\mathrm{s}^{-1}$, as viewed along the columns. The maximum principal stress was applied in the vertical direction. (f) shows the corresponding stress-strain curve. From (Schulson et al. 1999).

lization sets in, we imagine, when the remaining bridges of material can no longer support the applied load.

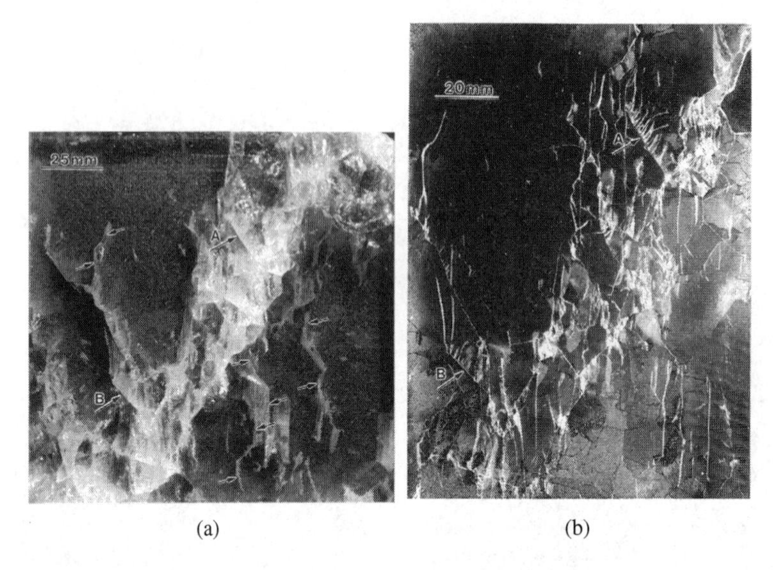

Fig. 4 Photographs of a Coulombic shear fault in the ice described in the caption to Figure 3, as viewed along the columns. (a) As seen within the specimen of full thickness (25 mm). Note the zigzag edge. The arrows point to wing cracks (b) The whitish region marked A in (a) at higher magnification as seen in a thin ($\sim$1 mm) section. Note the set of comb-like, secondary cracks that formed on one side of an inclined parent crack. From Schulson et al. (1999).

Details concerning the very final stages of terminal failure are not yet clear, but may involve some aspects of the failure of granular materials. What is clear is that a key feature in the initiation of the fault is the comb crack (Schulson et al. 1999; Schulson 2001; Renshaw and Schulson 2001). In effect, the secondary cracks that constitute this feature create sets of closely spaced micro-plates, fixed on one end and free on the other, Figure 5. We imagine that under the combination of frictional drag across their free ends and axial load, material at the root of the micro-plates experiences mixed mode-I and mode-II loading. As load increases and the stress intensity factors reach the critical level, the crack propagates and the plates "break", analogous to the failure of teeth in a comb under the action of a sliding thumb. Load is transferred to adjacent micro-plates/teeth and the process repeats itself.

Before quantifying the mechanics of fault initiation, we note that wing cracks and comb cracks also form within granular ice. They are more difficult to detect, however, owing in part to the crack shielding already noted. The other reason is the location of the parent crack itself. Unlike primary cracks within columnar ice, which can propagate all the way through the thickness along grain boundaries that thread the specimen, those within granular ice are restricted to the grains in which

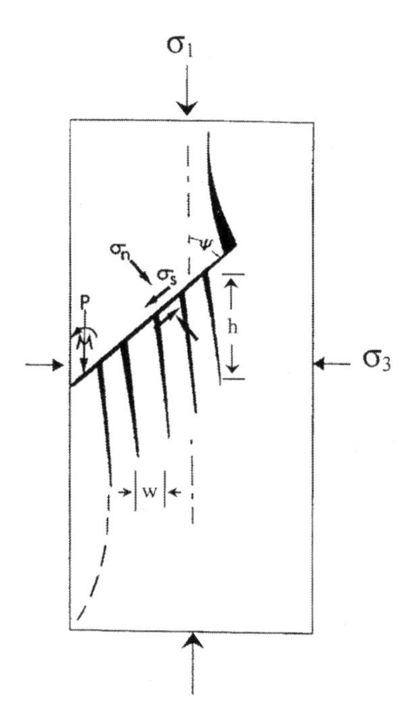

Fig. 5 Schematic sketch of a comb crack. Secondary cracks initiate from one side of a parent crack that is inclined by an acute angle ψ to the maximum (most compressive) principal stress. The secondary cracks create sets of slender micro-plates of length h and width w, fixed on one end and free on the other. Axial loading produces a load/unit depth, P, on the plates, and frictional drag across their free ends produces a moment/unit length on them, M. The moment is of greater importance than the axial load and eventually becomes large enough to "break" the micro-plates via mixed mode-I and mode-II crack propagation. From Schulson (2001).

they form. As a result, they are often buried and can only be clearly seen within thin sections that happen to contain the parent crack. This may be true of rock as well.

The triggering of a fault by micro-plate failure, incidentally, precedes work on ice. Peng and Johnson (1972), Ashby and Hallam (1986), Sammis and Ashby (1986) and Bazant and Xiang (1997), for instance, invoked the concept in their theory of brittle compressive failure. There, however, failure occurs by elastic buckling of plates fixed on both ends, formed, for instance, between adjacent wing cracks. The problem with the buckling model is that when realistic slenderness ratios are employed (measured to be $h/w \approx 3$–7 for ice) the applied stress to "break" the micro-plate is estimated to be two to three orders of magnitude greater than the measured strength of the ice (Schulson et al. 1999).

3.4 The Comb Crack Model of Coulombic Faulting

Renshaw and Schulson (2001) modeled the comb crack mechanism in terms of a number of physical parameters and the degree of confinement. Accordingly, upon incorporating frictional resistance to sliding across both the free ends and against the sides as micro-plates brush against their neighbors and upon assuming that the critical stress intensity factors for mode-I and mode-II crack growth in ice are equal (i.e., $K_{IIc} = K_{Ic}$), for which there is experimental some evidence (Shen and Lin 1986), they obtained for the micro-plate failure stress the equation:

$$\sigma_f = \frac{2K_{Ic}}{\sqrt{c}\left[\left(1+\left(1-\mu_i\frac{1+R}{1-R}\right)^{2/3}\right)^{1/2}-1\right]^{1/2}\left(1+3\mu_i^2\alpha^2(1-R)^2\right)^{1/2}} \tag{8}$$

where c denotes the half-length of primary cracks and α is the microplate slenderness ratio given by $\alpha = h/w$, where h and w, respectively, are the length and thickness of the plates. Equation (8) is subject to the limitations $0.05 \leq R < [(1+\mu_i^2)^{1/2}+\mu_i]^{-2}$: the lower limit reflects the fact that h/w is derived from two other ratios h/c and w/c that follow from an analysis of wing crack mechanics (Ashby and Hallam 1986) and comb-crack mechanics (Renshaw and Schulson 2001) within which context $\alpha \to \infty$ as $R \to 0$; and the upper limit defines the level of confinement above which the Coulomb stress is zero (i.e., $\tau - \mu_i\sigma_n = 0$) and frictional sliding is completely suppressed (Section 5).

Figure 6 compares Equation 8 with experimental measurements of the terminal failure stress of S2 ice. Over the range of confinement for which the model applies, there is relatively good agreement. However, the model under-predicts the strength somewhat, possibly because it estimates the stress to initiate or trigger the fault as opposed to the stress to lead to its full development. The upswing in the curve reflects the fact that once the confinement and hence the normal stress across the fault becomes large enough to suppress frictional sliding, a new process begins to operate. We consider this process again in Section 5 within the context of non-frictional faulting.

3.5 Application to Other Coulombic Materials

Although the above model was proposed initially for ice (Schulson et al. 1998), its development (Renshaw and Schulson 2001) is not limited to that material. Comb-cracks mechanics can also account for the brittle compressive strength of a variety of rock and minerals, Figure 7. Within those materials, comb-like features, there termed feather cracks, horse-tail cracks and splay-cracks have been reported for many years, on scales both small (Conrad and Friedman 1976; Ingraffea 1981; Wong 1982; Gottschalk et al. 1990) and large (Rispoli 1981; Segall and Pollard 1983; Granier 1985; Davies and Pollard 1986; Martel et al. 1988; Cruikshank et al.

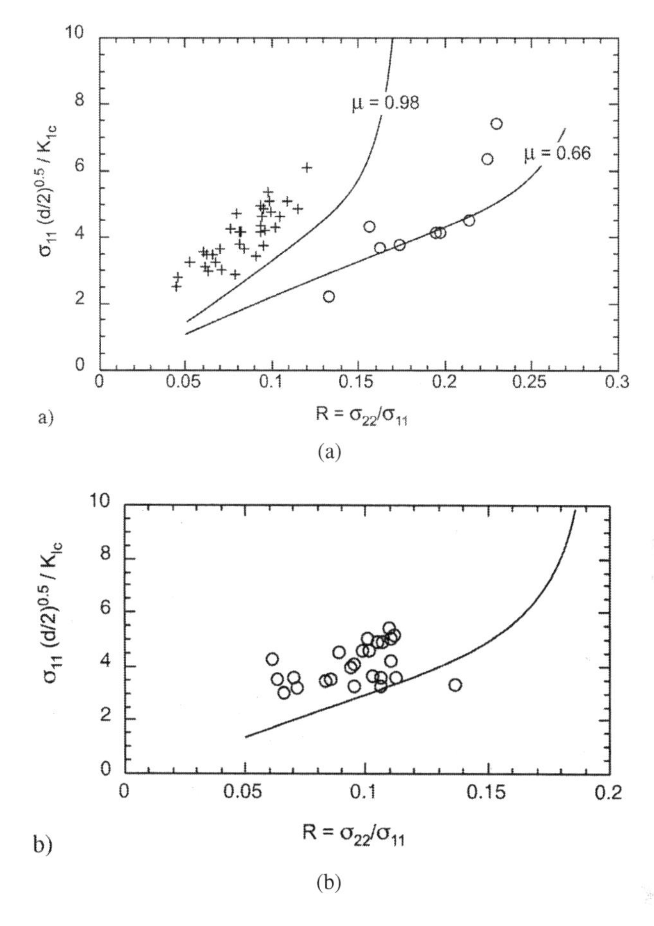

Fig. 6 Brittle compressive strength (in non-dimensionalized units) (a) of S2 fresh-water ice at -3°C (lower curve) and at -10°C and (b) of S2 first-year arctic sea ice at -10°C vs. confinement ratio $R = \sigma_{22}/\sigma_{11}$. The data were obtained from Figures 1 and 11 and from Schulson et al. (2006a,b).

1991; Mutlu and Pollard 2008; Misra et al. 2009). The novel point about ice and the subsequent development of the model is the linking of the mechanics of comb-cracking to terminal failure.

4 Ductile-to Brittle Transition

As mentioned in the Introduction, ice undergoes a ductile to brittle transition when deformed at a rate above a critical level. (In terrestrial ice mechanics, the DB transition is usually described in terms of a critical strain rate, because the rate of deformation varies by a much greater factor than does temperature.) Under compressive

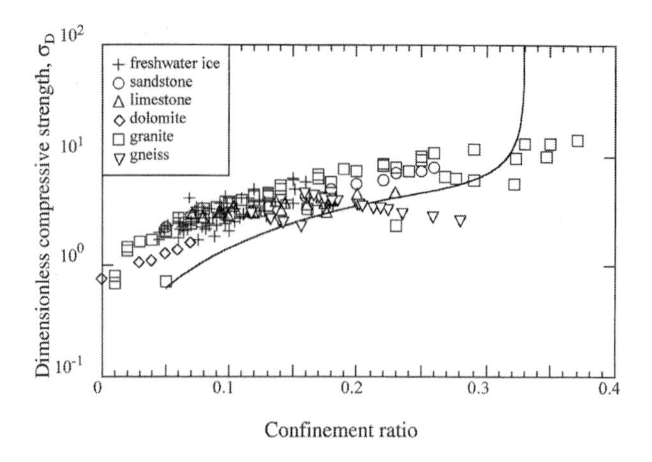

Fig. 7 Brittle compressive strength of a variety of rocks and minerals vs. confinement ratio $R = \sigma_3/\sigma_1$. The curve was computed from Equation 8 using parametric values listed by Renshaw and Schulson (2001). The non-dimensional stress is defined as $\sigma_f\sqrt{c}/K_{Ic}$.

loading the transition strain rate is $\dot{\varepsilon}_{tc} \approx 10^{-4}\,s^{-1}$, for relatively small (~1 m), unconfined bodies of fresh-water ice of ~5 mm grain size loaded at temperatures around -10°C. The transition strain rate increases by about an order of magnitude upon reducing the grain size to ~1 mm; is about an order of magnitude greater for sea ice owing to its lower creep resistance; is somewhat higher under moderate confinement; decreases by about an order of magnitude upon lowering temperature to -40°C; and is lower still within larger bodies (Schulson 2001). The DB transition is both technologically and scientifically important: as already noted, it marks the point where ice loads on offshore structures reach a maximum; and, when localization is considered, it accounts for the fractured condition of the arctic sea ice cover (Marsan et al. 2004), slowly though the ice appears to deform when viewed on large spatial and temporal scales (Kwok 2001).

The transition corresponds not to the loading conditions under which cracks first form. We eliminate that possibility, because ice can be riddled with short, deformation-induced cracks (that render the body milky white in appearance) and still exhibit macroscopically ductile behavior. Nor does the transition correspond to an increasing preference for intergranular cracks as opposed to transgranular cracks (Gold 1997). That criterion predicts an effect of grain size opposite to the one observed. Instead, the DB transition marks the conditions under which cracks begin to propagate (Schulson 1990, 2001), albeit in a stable manner, and can be explained in terms of a competition between crack-tip creep and crack propagation.

4.1 Competition Between Creep and Fracture

Located at the tips of secondary cracks is a creep zone of radius r_c. We imagine (Schulson 1990; Renshaw and Schulson 2001) that when the size of this zone exceeds a certain fraction f of the length of the primary crack, secondary cracks are blunted and so do not propagate. As a result, the material exhibits macroscopically ductile behavior. When the creep zone is smaller than that fraction of the parent crack size, on the other hand, secondary cracks do propagate and brittle behavior ensues. The DB criterion may then be expressed as:

$$r_c = f_c \,. \tag{9}$$

The DB transition, in other words, may be viewed as a localized competition between creep and fracture: creep dominates at lower deformation rates (or, alternatively, at higher temperatures) where time/temperature is sufficient to relax crack-tip stresses below those required for crack propagation; at higher rates (lower temperatures) fracture dominates.

4.2 Micro-Mechanical Model of the Ductile-Brittle Transition

Schulson (1990) and Renshaw and Schulson (2001) quantified this picture. Upon invoking the Riedel-Rice (1980) analysis of crack-tip creep for non-interacting cracks loaded externally under tension, equating crack size to grain size ($2c = d$), upon invoking secondary creep only and upon deriving the value $f \approx 1/6\pi$ from considerations of crack-tip stresses, they obtained for the transition strain rate, in dimensionless form, the equation:

$$\dot{\varepsilon}_D = \frac{\dot{\varepsilon}_{tc} d^{3/2}}{BK_{Ic}} = \frac{25}{(1-R) - \mu(1+R)} \tag{10}$$

subject to the limitation $R < (1-\mu)/(1+\mu)$, where B is a temperature-dependent constant in the power-law creep equation ($\dot{\varepsilon} = B\sigma^n$ where n=3) and the parameters K_{Ic} and R have the same meaning as above; μ we take not to be the coefficient of internal friction, but the coefficient of friction for an earlier stage of deformation, of the kind described by Kennedy et al. (2000) for sliding across relatively smooth surfaces. The model dictates several points that agree with observation: that the transition strain rate scales as (grain size)$^{-3/2}$, as observed (Batto and Schulson 1993); that the transition strain rate is an order of magnitude greater for sea ice than for fresh-water ice, given that $B_{\text{seaice}} \approx 10B_{\text{freshwaterice}}$ (de La Chapelle and Duval 1999; Cole et al. 1998) and that both the fracture toughness and the coefficient of friction of the two materials have similar values (Schulson and Duval 2009), again in agreement with observation (Nickolayev and Schulson 1995); and that the transition strain rate increases with increasing confinement, in quantitative agreement with experimental measurements on both fresh-water ice (Schulson and Buck 1995) and salt-water ice (Schulson and Nickolayev 1995). The other point to note is that the micro-mechanical model incorporates only physical parameters (i.e., no adjustable parameters) that can be/ have been measured independently.

Sodhi et al. (1998) developed a model that also incorporates creep and fracture. As discussed by Schulson and Duval (2009), that model is more phenomenological in nature and specific to deformation under indentation. It includes several parameters unique to that loading configuration, but excludes frictional sliding. In excluding friction, the model neglects a fundamental step in brittle compressive failure.

4.3 Application to Other Coulombic Materials

Once again, a model developed for ice works for a variety of rock and minerals. Upon inserting into Equation (10) appropriate values (obtained from the literature) of the materials parameters, derived experimentally and tabulated in the original paper, Renshaw and Schulson (2001) obtained the result shown in Figure 8. Closed symbols denote ductile behavior and open symbols, brittle behavior. For all materials examined, including ice, the model, shown by the curve through the data, quite accurately distinguishes the two kinds of behavior.

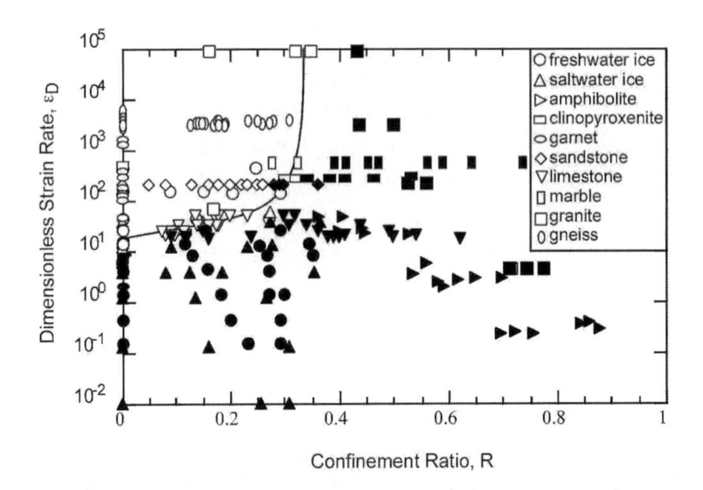

Fig. 8 Dimensionless ductile-to-brittle transition strain rate vs. compressive stress ratio $R = \sigma_3/\sigma_1$, for a variety of polycrystalline rocks and minerals. Open symbols indicate brittle behavior and closed symbols indicate ductile behavior. From Renshaw and Schulson (2001).

5 Coulombic vs. Plastic Faulting

So far in this discussion, we have focused on inelastic behavior of ice rapidly deformed (albeit within the regime of quasi-static behavior) and loaded under moderate confinement. There, frictional sliding operates, because the Coulomb stress is non-zero; correspondingly, the material exhibits pressure-hardening. Under higher confinement, where frictional sliding is suppressed, ice can still deform inelastically, because the deviatoric component of the stress tensor is non-zero. However, the character of the deformation is different. Dislocation-based plasticity now governs, characterized by pressure-independence, (Jones 1982; Rist and Murrell 1994). Unlike the low-rate, dislocation-based deformation that operates within the regime of ductile behavior, the higher-rate plasticity is macroscopically localized. The localization is manifested in the development of a second kind of shear fault, termed a non-frictional or plastic fault, inclined by $\sim 45^\circ$ to the directions of maximum and minimum principal stress; i.e., along planes close to those on which the applied shear stress reaches a maximum, Figure 9. Plastic faulting occurs as suddenly as Coulombic faulting and, for that reason, could be viewed as brittle failure. Plastic faults are comprised not of microcracks, but of recrystallized grains, indicative of a combination of plastic flow and deformation-induced heat. Adiabatic heating, where thermal softening dominates both strain and strain-rate hardening, may be at play here (Schulson 2002). Plastic faulting mode of failure operates within both fresh-water ice (Durham et al. 1983; Rist and Murrell 1994; Gagnon and Gammon 1995; Meglis et al. 1999) and salt-water ice (Sammonds et al. 1989, 1998; Gratz and Schulson 1997), at temperatures as low as -196 to -114°C (Durham et al. 1983).

Plastic faulting operates only under triaxial loading. We can show from Equation (6) that this failure mode is expected to set in just before the degree of confinement reaches the level defined by the equation (Schulson 2002):

$$R = \frac{\sigma_3}{\sigma_1} = \frac{(1+\mu_i^2)^{1/2} - \mu_i}{(1+\mu_i^2)^{1/2} + \mu_i} = \left[(1+\mu_i^2)^{1/2} + \mu_i\right]^{-2} = \frac{1}{q} \tag{11}$$

where σ_3 denotes the smallest (least compressive) principal stress. At temperatures between -10 and -3°C, for instance, $\mu_i \approx 0.8$ and this specifies a confinement of $R \approx 0.2$ to activate the process, in reasonable agreement with observation (Golding et al. 2009).

Practical implications? From the perspective of the interaction between an offshore engineered structure and a floating ice sheet, plastic faulting may limit the pressure that develops towards the more highly confined, central region of the contact zone. There, zones of "blue ice" have been observed (Frederking 1990), comprised of recrystallized grains.

Again ice mimics rock and minerals. In a number of materials (granite, quartzite, anorthosite and diabase) there is evidence, summarized by Renshaw and Schulson (2004), of sudden, non-frictional faulting under higher confinement. As in ice, thermal softening could again dominate strain and strain-rate hardening (Hobbs and Ord 1988).

Fig. 9 Photograph of a thin section from S2 columnar fresh-water ice, as observed through polarizing filters. The parent specimen was proportionally loaded to just beyond terminal failure (to $\varepsilon_{11} = 0.024$) along the path $\sigma_{11} : \sigma_{22} : \sigma_{33} = 1 : 0.5 : 0.2$ at -10°C at $\dot{\varepsilon}_{11} = 6 \times 10^{-3}\ \mathrm{s}^{-1}$. Note the plastic fault and the recrystallized grains within. The short, white lines are reflections from non-propagating, across-column cracks. Scale: smallest division=1 mm. From Golding et al. (2009).

6 Scale-Independent Fracture Physics: Fracture of the Arctic Sea Ice Cover

There is growing evidence that the mechanisms underlying the brittle compressive failure of ice are independent of spatial scale, and that fracture of the Arctic sea ice cover, noted in the Introduction, can be understood within the context of frictional crack sliding and crack growth (Schulson and Hibler 1991; Hibler and Schulson 2000; Weiss 2001a, b; Weiss 2003; Marsan et al. 2004; Weiss and Marsan 2004; Schulson 2004; Weiss et al. 2007; Rampal et al 2007; Weiss and Schulson 2009). The cover is a floating body, $\approx$ 1–3 m thick, that encompasses an area of $\approx 10 \times 10^6$ km^2. It is loaded, predominantly under biaxial compressive stresses (Richter-Menge and Elder 1998; Richter-Menge et al. 2002), that develop via a combination of wind and ocean current (Thorndike and Colony 1982). The cover fails through a number of different failure processes, including rafting, ridging and strike-slip-like faulting (Marko and Thomson 1977). Faults have been more broadly termed linear kinematic features (Kwok 1996; 2001), many of which lace through the entire Arctic Basin and some of which form intersecting or conjugate sets (Marko and Thomson 1977; Erlingsson 1988, Walter and Overland 1995, Schulson 2004) of intersection angle $2\theta \approx 20$–$50°$.

The evidence for scale independence, stated briefly, is as follows:

(i) *Scale-independent fracture patterns:* Comparison of the fracture pattern within some regions of the winter ice cover on the Beaufort Sea with the pattern of fracture observed within specimens of S2 sea ice harvested from the winter cover and then loaded in the laboratory to terminal failure under biaxial, across-column compression at a rate sufficiently great to impart brittle behavior reveals remarkable similarity, Figure 10. On both the geophysical and the laboratory scales, patterns exhibit features such as wing-like cracks, comb-like cracks and Coulombic-like shear faults punctuated with rhomboidal-shaped openings (Schulson and Hibler 1991; Schulson 2004). The openings indicate either left-lateral or right-lateral sliding, analogous to strike-slip faulting in Earth's crust.

(ii) *Failure stress:* At the site of the SHEBA field experiment (Perovich 1999), for instance, the maximum principal stress (i.e., most compressive stress) within the cover, measured in-situ mid-November 1997, was $\sigma_c \approx 30$ kPa (Richter-Menge et al. 2002). This is in fair agreement with the stress calculated from Ashby-Hallam (1986) wing-crack mechanics and from the geometry of cracks that formed at the SHEBA site (see Fig. 9d of Schulson 2004, and the attendant discussion); i.e.

$$\sigma_c = \frac{K_c\sqrt{L}}{\sqrt{\pi c}0.23(1-\mu_i)} \approx 50 \text{ kPa} \tag{12}$$

where $K_c \approx 250 \text{ kPa m}^{1/2}$ (Dempsey 1996) is an apparent fracture toughness that includes energy dissipated through creep, $L = l/c \approx 3$ km/3.5 km ≈ 0.9, where l denotes the length of the wing crack, and c and $\mu_i \approx 0.8$ have the same meaning as above.

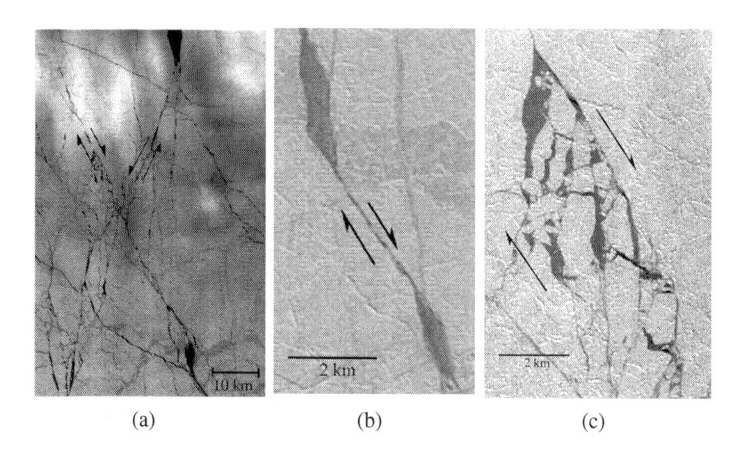

Fig. 10 (a) Landsat-7 images of Arctic sea ice cover showing (a) conjugate set of Coulombic shear faults, (b) wing-like cracks and (c) comb-like cracks. From (Schulson 2004).

(iii) *Scale-independent shape of failure envelopes:* In keeping with the "look-alikeness" of fracture features on different scales, the failure envelope for the winter ice cover has the same shape as the envelope for specimens harvested from the cover and then biaxially compressed to terminal failure in the laboratory (Weiss et al. 2007; Schulson et al. 2006), Figure 11. However, the field stresses are smaller, by around two to three orders of magnitude. This difference is caused by stress concentrators in the form of structural defects within the ice sheet (e.g., parent sliding cracks that formed, perhaps, through differential thermal contraction) whose linear dimensions are around five orders of magnitude larger than grain-sized cracks within laboratory specimens; i.e., kilometers vs centimeters. Barring healing, the failure strength, to a first approximation, scales inversely with (defect length)$^{1/2}$.

(iv) *Scale-independent friction coefficient:* The fact that the slope of the failure envelope is scale-independent implies that the coefficient of internal friction is scale independent. The value derived from the envelopes, $\mu_i \approx 0.8$, dictates from the phenomenological theory of brittle compressive failure (Equation 5) that the angle of intersection between conjugate faults is $\approx 50°$. The nature of this apparent scale-independence of the friction coefficient is beyond the scope of this paper. Suffice it to say that it appears to reside not in the interlocking of asperities, but in self-similar geometrical characteristics of fault gouge (Weiss and Schulson 2009).

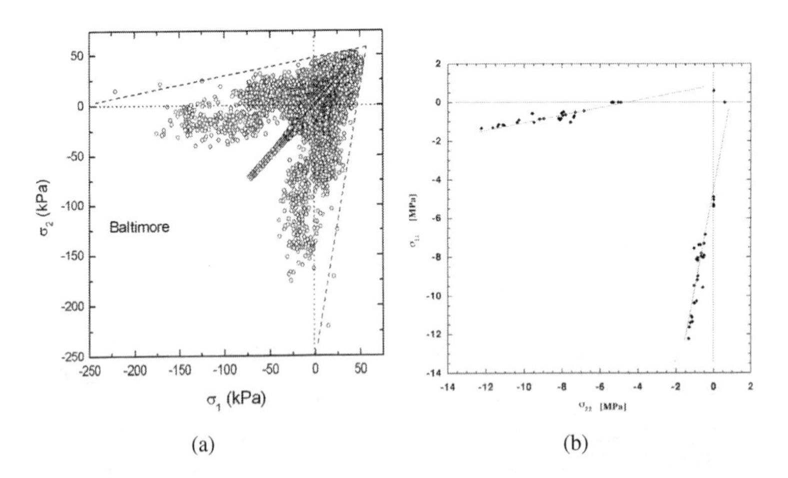

Fig. 11 (a) Stress states within the winter sea ice cover on the Arctic Ocean during the SHEBA experiment, measured in-situ at the "Baltimore" site by Richter-Menge et al. (2002) and plotted in principal stress space. From Weiss et al. (2007), (b) Coulombic branch of the brittle compressive failure envelope for columnar-grained S2 first-year sea ice. The material was harvested from the winter cover on the Arctic Ocean and then was loaded to terminal failure in the laboratory by compressing biaxially across the columns at -10°C at $\dot{\varepsilon}_1 = 1.5 \times 10^{-2}\,\mathrm{s}^{-1}$. Note that the slope of the laboratory envelope is the same as the slope of the field envelope (dotted line) that bounds the in-situ stresses shown in Figure 9(a). Data from (Schulson et al. 2006b).

(v) *Ductile-brittle transition:* Based upon the fact that deformation within the sea ice cover is highly localized, Marsan et al. (2004) calculated from a multi-fractal analysis the cumulative probability of observing a given deformation rate at different spatial scales. They found that at the spatial scale of 1 meter on a time scale of 3 days about 15% of the deformation is accommodated within the regime of brittle behavior (i.e., $\dot{\varepsilon} > 10^{-4}\,\mathrm{s}^{-1}$), corresponding to about 0.2% of the surface area of the ice cover. A similar result was obtained by Rampal et al. (2007): on a spatial scale of 100-1000 meters and a time scale of 3 hours about 20% of the deformation occurred at a strain rate $> 10^{-4}\,\mathrm{s}^{-1}$. In other words, the brittle behavior on the larger scale appears to originate from highly localized, high-strain-rate deformation on the smaller scale.

Much of this evidence, limited though it is, has been obtained during the past five years; i.e., since McNutt and Overland (2003) questioned the idea of scale-independence and argued instead that it is the degree of disconnectedness between scales that accounts for the organization of the system. That view, based upon hierarchy theory of complexity (for review, see O'Neill et al. 1986), holds that there are emergent properties that represent discontinuities in scale, just as there are in biological systems where a cell does not govern the mechanics of an organ nor an organ the mechanics of the body as a whole. Sea ice transitions, for example, are specified as somewhere around the "floe scale" of 1 km, around the "multi-floe scale" of 2-10 km, and around the "aggregate scale" of 10-75 km. To support their view, McNutt and Overland (2003) cited the observation that the indentation pressure of arctic ice sheets decreases as the size of the sheet/indenter increases, from $\sim$10 MPa at $10^{-4}\,\mathrm{m}^2$ to $\sim$0.01 MPa at $10^5\,\mathrm{m}^2$ (Sanderson 1988). A problem with the hierarchy argument is that the different scales are defined rather arbitrarily. Also, hierarchy theory defies the results of fractal analyses (Weiss 2001, 2003) which reveal neither a characteristic length in the fragmentation pattern of ice in nine orders of magnitude, from 10^{-4} m to 10^5 m, which includes the proposed transition scales, nor a characteristic length in the deformation of the winter sea ice cover (Marsan et al. 2004). As to the difference in strength, that observation could be explained, as already mentioned, in terms of the difference in size of structural defects that weaken the ice.

Our sense at this juncture, therefore, is that the evidence for scale-independent fracture physics is more compelling than the argument against.

7 Summary

To summarize, we have reviewed the fracture of ice on scales small (laboratory) and large (arctic sea ice cover), with emphasis on brittle compressive failure under multiaxial loading and with reference to rock and minerals.

When loaded under low to moderate confinement, terminal failure occurs through the development of one or more Coulombic shear faults. The process involves the nucleation, growth and interaction of both primary and secondary cracks as well as frictional sliding across crack faces. The coefficient of internal friction governs fault orientation, and controls as well the slope of the confinement-strengthening branch of the brittle compressive failure envelope. A micromechanical model, based upon frictional drag across the ends of fixed-free micro-plates created through the development of comb-like secondary cracks, accounts reasonably well for the terminal failure stress. When modified to include creep-based stress relaxation at the tips of secondary cracks, the model accounts as well for the ductile-to-brittle transition, not only of ice but also of rock and minerals. Observations of the failure of the arctic sea ice cover, few though they are, reveal fracture features very much like those seen on the smaller scale and a failure envelope similar in shape, but smaller in size than generated in the laboratory, suggesting that the physics of brittle compressive failure is scale independent.

The story changes under higher confinement. There, under triaxial loading, frictional sliding is suppressed. Failure now occurs via non-frictional or plastic faulting, a macroscopically brittle-like process but one that is based upon dislocation mechanics as opposed to crack mechanics, augmented by dynamic recrystallization. Less is known of this failure mode, although preliminary analysis suggests that adiabatic heating may be at play here.

References

1. American Petroleum Institute (1995) Recommended practice 2N — planning, designing and constructing structures and pipelines for arctic conditions
2. Ashby MF and Hallam SD (1986) The failure of brittle solids containing small cracks under compressive stress states. Acta Metall 34:497–510
3. Batto RA and Schulson EM (1993) On the ductile-to-brittle transition in ice under compression. Acta metall Mater 41:2219–2225
4. Bazant ZP and Xiang Y (1997) Size effect in compression fracture: Splitting crack band propagation. J Eng Mech-ASCE February:162–172
5. Byerlee JD (1967) Frictional characteristics of granite under high confining pressure. J Geophys Res 72:3639–3648
6. Cole DM, Johnson RA and Durell GD (1998) Cyclic loading and creep response of aligned first-year sea ice. J Geophys Res 103:21751–21758
7. Conrad RE, II and Friedman M (1976) Microscopic feather fractures in the faulting process. Tectonophysics 33:187–198
8. Cooke ML (1997) Fracture localization along faults with spatially varying friction. J Geophys Res 102:24425–24434
9. Costamanga R, Renner J and Bruhns OT (2007) Relationship between fracture and friction for brittle rocks. Mech Mater 39:291–301
10. Coulomb CA (1773) Sur une application des règles de maximis et minimis à quelques problèmes de statique relatifs à l'architecture. Acad Roy Sci Mémoires de Mathématiques et de Physique par divers savants 7:343–382
11. Cruikshank KM, Zhao G and al e (1991) Analysis of minor fractures associated with joints and faulted joints. J Struct Geol 13:865–886
12. Davies RK and Pollard DD (1986) Relations between left-laterial strike-slip faults and right-lateral kink bands in granodiorite, Mt. Abbot quadrangle, Sierra Nevada, California. Pure Appl Geophys 124:177–201
13. de LaChapelle SD, Duval P and Baudelet B (1995) Compressive creep of polycrystalline ice containing a liquid phase. Scripta Metall Mater 33:447–450
14. de la Chapelle SD, Milsch H, Castelnau O et al (1999) Compressive creep of ice containing a liquid intergranular phase: Rate-controlling processes in the dislocation creep regime. Geophys Res Lett 26:251–254
15. Dempsey JP (1996) Scale effects on the fracture of ice. In Proceedings of The Johannes Weertman Symposium. Arsenault R J et al (eds) Anaheim, California
16. Durham WB, Heard HC and Kirby SH (1983) Experimental deformation of polycrystalline H_2O ice at high pressure and low temperature: Preliminary results. J Geophys Res 88:B377-B392

17. Duval P, Ashby MF and Anderman I (1983) Rate-controlling processes in the creep of polycrystalline ice. J Phys Chem 87:4066–4074
18. Erlingsson B (1988) Two-dimensional deformation patterns in sea ice. J Glaciol 301–308
19. Fortt AL and Schulson EM (2007) The resistance to sliding along coulombic shear faults in ice. Acta Mater 55:2253–2264
20. Fortt AL and Schulson EM (2009, unpublished results)
21. Frederking R, Jordaan IJ and McCallum JS (1990) Field tests of ice indentation at medium scale, Hobson's choice ice island. In Proceedings of IAHR Symposium on Ice. Espoo, Finland
22. Gagnon RE and Gammon PH (1995) Triaxial experiments on iceberg and glacier ice. J Glaciol 41:528–540
23. Gammon PH, Kiefte H, Clouter MJ et al (1983) Elastic constants of artificial ice and natural ice samples by brillouin spectroscopy. J Glaciol 29:433–460
24. Gold LW (1997) Statistical characteristics for the type and length of deformation-induced cracks in columnar-grain ice. J Glaciol 43:311–320
25. Golding N, Schulson EM and Renshaw CE (2009, unpublished results)
26. Gottschalk RR, Kronenberg AK, Russel JE et al (1990) Mechanical anisotropy of gniess: failure criterion and textural sources of directional behavior. J Geophys Res 95:613–621
27. Gow AJ, Ackley SF, Buck KR et al (1987) Physical and structural characteristics of Weddell Sea pack ice. CRREL Report 87–14
28. Granier T (1985) Origin, damping and pattern development of faults in granite. Tectonics 4:721–737
29. Gratz ET and Schulson EM (1997) Brittle compressive failure of columnar saline ice under triaxial compression. J Geophys Res 102:5091–5107
30. Gupta V and Bergström JS (1998) Compressive failure of rocks by shear faulting. J Geophys Res 103:23875–23895
31. Hibler WDI and Schulson EM (2000) On modeling the anisotropic failure and flow of flawed sea ice. J Geophys Res 105:17105–17120
32. Hobbs BE and Ord A (1988) Plastic instabilities: Implications for the origin of intermediate and deep focus earthquakes. J Geophys Res 93:10521–10540
33. Hutchinson JW (1977) Creep and plasticity of hexagonal polycrystals as related to a single-crystal slip. Metall Trans A8:1465–1469
34. Iliescu D and Schulson EM (2004) The brittle compressive failure of fresh-water columnar ice loaded biaxially. Acta Mater 52:5723–5735
35. Ingraffea AR (1981) Mixed-mode fracture initiation in Indiana limestone and westerly granite. Proc US Symp Rock Mechanics 22:196–191
36. Jaeger JC and Cook NGW (1979) Fundamentals of rock mechanics. Chapman and Hall, London
37. Jeffries MO, Worby AP, Morris K et al (1997) Seasonal variations in the properties and structural composition of sea ice and snow cover in the Bellinghausen and Amundsen seas, Antarctica. J Glaciol 41:138–151
38. Jones SJ (1997) High strain-rate compression tests on ice. J Phys Chem B 101:6099–6101
39. Jones SJ (1982) The confined compressive strength of polycrystalline ice. J Glaciol 28:171–177
40. Kennedy FE, Schulson EM and Jones D (2000) Friction of ice on ice at low sliding velocities. Phil Mag A 80:1093–1110
41. Kwok RE (1998) Analysis of SAR data of the polar oceans. IN: The radarsat geophysical processor system Tsatsoulis C and R Kwok (eds), Springer-Verlag, Berlin
42. Kwok R (2001) Scaling laws in ice mechanics. IN: Dempsey J P and H H Shen (eds) Deformation of the arctic ocean sea ice cover between November 1996 and April 1997: A qualitative survey, Kluwer Academic Publishing.
43. Marko JR and Thomson RE (1977) Rectilinear leads and internal motions in the ice pack of the western arctic ocean. J Geophys Res 82:979–987
44. Marsan D, Stern H, Lindsay R et al (2004) Scale dependence and localization of the deformation of arctic sea ice. Phys Rev Lett 93:178501

45. Martel SJ, Pollard DD and Segall P (1988) Development of simple strike-slip fault zones in granitic rock, mount abbott quadrangle, Sierra Nevada, California. Geol Soc Am Bull 100:1451–1465
46. Maykut GA (1982) Large-scale heat exchange and ice production in the central arctic. J Geophys Res 87:7971–7984
47. McNutt SL and Overland JE (2003) Spatial hierarchy in arctic sea ice dynamics. Tellus Series A-Dynamic Meterology and Oceanography 55:181–191
48. McPhee MG, Morison J and Nilsen F (2008) Revisiting heat and salt exchange at the ice-ocean interface: Ocean flux and modeling considerations. J Geophys Res 113:doi 10/1029/2007JC004383
49. Meglis IL, Melanson PM and Jordaan IJ (1999) Microstructural change in ice: Ii. Creep behavior under triaxial stress conditions. J Glaciol 45:438–448
50. Melton JS and Schulson EM (Year) The ductile deformation of columnar (S2) saline ice under triaxial compression. In Proceedings of 7th ISOPE 1997. ISOPE Honolulu, Hawaii
51. Michel B (1978) Ice mechanics. Laval University Press
52. Michel B and Ramseier RO (1971) Classification of river and lake ice. Can Geotech J 8:36–45
53. Misra S, Mandal N, Dhar R et al (2009) Mechanisms of deformation localization at the tips of shear fractures: Findings from analogue experiments and field evidence. J Geophys Res 114:doi: 10.1029/2008JB005637
54. Moore DE and Lockner DA (1995) The role of microcracking in shear-fracture propagation in granite. J Struct Geol 17:95–114
55. Mutlu O and Pollard DD (2008) On the patterns of wing cracks along an outcrop scale flaw: A numerical modeling approach using complementarity. J Geophys Res 113:doi: 10.1029/2007JB005284
56. Nickolayev OY and Schulson EM (1995) Grain-boundary sliding and across-column cracking in columnar ice. Phil Mag Lett
57. O'Neill RV, DeAngelis DL, Waide JB et al (1986) A hierarchical concept of ecosystems. Princeton University Press, Princeton 72:93–97
58. Paterson WS (2000) Physics of glaciers. Butterworth Heineman
59. Peng S and Johnson AM (1972) Crack growth and faulting in cylindrical specimens of chelmsford granite. Int J Rock Mech Min Sci 9:37–86
60. Perovich DK, Andreas EL, Curry JA et al (1999) Year on the ice gives climate insights. EOS Trans 481, 485–486
61. Rampal P, Weiss J, Marsan D et al (2007) Scaling properties of sea ice deformation from buoy dispersion analysis. J Geophys Res 113:C03002
62. Renshaw CE and Schulson EM (2001) Universal behavior in compressive failure of brittle materials. Nature 412:897–900
63. Renshaw CE and Schulson EM (2004) Plastic faulting: Brittle-like failure under high confinement. J Geophys Res 109:1–10
64. Richter-Menge JA and Elder BC (1998) Characteristics of pack ice stress in the alaskas beaufort sea. J Geophys Res-Oceans 103:21817–21829
65. Richter-Menge JA and Jones KF (1993) The tensile strength of first-year sea ice. J Glaciol 39:609–618
66. Richter-Menge JA, McNutt SL, Overland JE et al (2002) Relating arctic pack ice stress and deformation under winter conditions. J Geophys Res-Oceans 107:art. 8040
67. Riedel H and Rice JR (1980) Tensile cracks in creeping solids. ASTM-STP 7700:112–130
68. Rispoli R (1981) Stress fields about strike-slip faults from stylolites and tension gashes. Tectonophysics 75:T29-T36
69. Rist MA and Murrell SAF (1994) Ice triaxial deformation and fracture. J Glaciol 40:305–318
70. Rottger K, Endriss A, Ihringer J et al (1994) Lattice constants and thermal expansion of H_2O and D_2O ice Ih between 10 and 265 k. Acta Crystallograph, Sect B 50:644–648
71. Sammis CG and Ashby MF (1986) The failure of brittle porous solids under compressive stress states. Acta metall 34:511–526
72. Sammonds PR, Murrell SAF and Rist MA (1989) Fracture of multi-year sea ice under triaxial stresses: apparatus description and preliminary results. Trans ASME 111:258–263

73. Sammonds PR, Murrell SAF and Rist MA (1998) Fracture of multi-year sea ice. J Geophys Res 103:21795–21815
74. Sanderson TJO (1988) Ice mechanics: Risks to offshore structures. Graham & Trotman, London
75. Schulson EM (1990) The brittle compressive fracture of ice. Acta metall mater 38:1963–1976
76. Schulson EM (2001) Brittle failure of ice. Engineering Fracture Mechanics 68:1839–1887
77. Schulson EM (2004) Compressive shear faults within the arctic sea ice cover on scales large and small. J Geophys Res 109:1–23
78. Schulson EM and Buck SE (1995) The ductile-to-brittle transition and ductile failure envelopes of orthotropic ice under biaxial compression. Acta metall Mater 43:3661–3668
79. Schulson EM and Duval P (2009) Creep and fracture of ice. Cambridge University Press, Cambridge
80. Schulson EM and Gratz ET (1999) The brittle compressive failure of orthotropic ice under triaxial loading. Acta Mater 47:745–755
81. Schulson EM and Hibler WD (1991) The fracture of ice on scales large and small: Arctic leads and wing cracks. J Glaciology 37:319–323
82. Schulson EM and Nickolayev OY (1995) Failure of columnar saline ice under biaxial compression: failure envelopes and the brittle-to-ductile transition. J Geophys Res 100:22383–22400
83. Schulson EM, Iliescu D and Renshaw CE (1999) On the initiation of shear faults during brittle compressive failure: A new mechanism. J Geophys Res 104:695–705
84. Schulson EM, Fortt A, Iliescu D et al (2006a) On the role of frictional sliding in the compressive fracture of ice and granite: Terminal vs. Post-terminal failure. Acta Mater 54:3923–3932
85. Schulson EM, Fortt A, Iliescu D et al (2006b) Failure envelope of first-year arctic sea ice: The role of friction in compressive fracture. J Geophys Res 111:doi: 10.1029/2005JC003234186
86. Schulson EM (2002) Compressive shear faulting in ice: Plastic vs. Coulombic faults. Acta Mater 3415–3424
87. Scott TE and Nielsen KC (1991) The effects of porosity on the brittle-ductile transition in sandstones. J Geophys Res 96:405–414
88. Scott TE and Nielsen KC (1991) The effects of porosity on fault reactivation in sandstones. J Geophy Res-Solid Earth and Planets 96:2352–2362
89. Segall P and Pollard DD (1983) Nucleation and growth of strike-slip faults in granite. J Geophys Res 88:555–568
90. Shazly M, Prakash V, Lerch BA (2006) High-strain rate compression testing of ice. NASA STI Program Office, Report No. 200–213866
91. Shen W and Lin SZ (Year) Fracture toughness of bohai bay sea ice. In Proceedings of 5th International Offshore Mech and Arctic Eng Symp. V. J. Lunardini Y S W, A. O. Ayorinde, D. V. Sodhi (eds) OMAE-AIME
92. Sodhi DS, Takeuchi T, Nakazawa N et al (1998) Medium-scale indentation tests on sea ice at various speeds. Cold Reg Sci Technol 28:161–182
93. Thorndike AS and Colony R (1982) Sea ice motion in response to geostrophic winds. J Geophys Res-Oceans Atmos 87:5845–5852
94. Wachter LM, Renshaw CE and Schulson EM (2009) Transition in brittle failure mode in ice under low confinement. Acta Mater 57:345–355
95. Walter BA, Overland JE and Turet P (1995) A comparison of satellite-derived and aircraft-measured regional surface sinsible heat fluxes over the beaufort sea. J Geophys Res 100:4584–4591
96. Weiss J (2001) Fracture and fragmentation of ice: A fractral analysis of scale invariance. Eng Fracture Mech 68:1975–2012
97. Weiss J (2003) Scaling of fracture and faulting of ice on earth. Surv Geophys 24:185–227
98. Weiss J and Marsan D (2004) Scale properties of sea ice deformation and fracturing. C R Phys 735–751
99. Weiss J and Schulson EM (1995) The failure of fresh-water granular ice under multi-axial compressive loading. Acta metall mater 43:2303–2315

100. Weiss J and Schulson EM (2009) Coulombic faulting from the grain scale to the geophysical scale: Lessons from ice. J Physics D: Appl Physics (in press)
101. Weiss J, Schulson EM and Stern HL (2007) Sea ice rheology in-situ, satellite and laboratory observations: Fracture and friction. Earth Planet Sci Lett 255:1–8
102. Wong T-F (1982) Micromechanics of faulting in westerly granite. Int J Rock Mechanics and Mining Sciences 19:49–64

Experimental studies of the viscoplasticty of ice and snow

Jacques Meyssonnier, Armelle Philip, Laura Capolo, and Philippe Mansuy

1 Introduction

Polar ice sheets are fundamental elements of the climate system. Their extent and elevation have a direct influence on the global atmospheric circulation (through the albedo and wind intensity and direction) while the fresh water input from ice shelves and icebergs influence the ocean circulation. Furthermore the paleoclimatic records found in the deep ice cores drilled in polar ice sheets are an important source of information on the Earth climate mechanisms. As regards valley glaciers, as found in alpine regions, because they are very sensitive to temperature fluctuations they are nowadays considered as key indicators of present climate change. A good estimate of ice sheets mass balance response to climate change is needed to understand past sea level changes, while the mass balance of alpine glaciers is directly connected to present sea level rise. In this context it is very important to model the slow flow of glaciers and ice sheets, and to do so as well as possible, to improve our knowledge on the mechanical properties of ice. On the other hand, a wide range of engineering problems involve ice and snow mechanics, many of which in the high velocity or/and strain rate regimes (as for instance ice-structure interaction or snow avalanches). Since these later topics are exposed in other contributions to this book

---

J. Meyssonnier
Laboratoire de Glaciologie et Géophysique de l'Environnement (LGGE), Université Joseph Fourier (UJF) – CNRS , BP 96, F-38402 Saint-Martin d'Hères Cedex, France. e-mail: meyssonnier@lgge.obs.ujf-grenoble.fr
A. Philip
LGGE, UJF – CNRS , BP 96, F-38402 Saint-Martin d'Hères Cedex, France. e-mail: armelle.philip@lgge.obs.ujf-grenoble.fr
L. Capolo
LGGE, UJF – CNRS , now at Grenoble Sciences, 715, rue de la Houille Blanche BP 53 - 38041 Grenoble Cedex 9. e-mail: laura.capolo@ujf-grenoble.fr
Ph. Mansuy
LGGE, UJF – CNRS , now at Centre de Technologies Michelin Ladoux, 63040 Clermont-Ferrand, France. e-mail: philippe.mansuy@fr.michelin.com

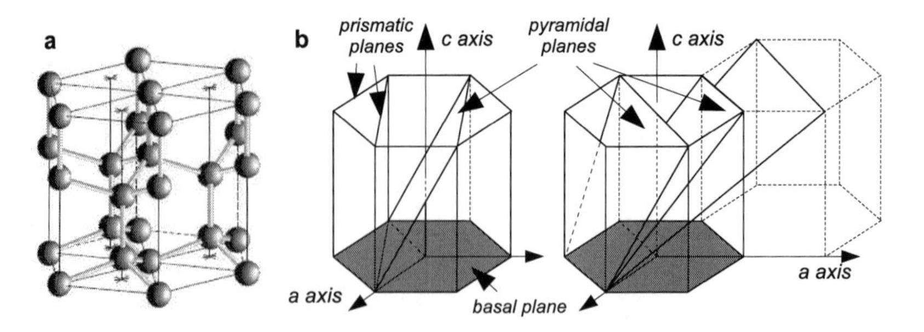

Fig. 1 Structure of hexagonal ice. **a** schematic of the oxygen atoms lattice. **b** definitions.

(see e.g. E. Schulson and M. Schneebeli) we will restrict ourselves to the very slow deformation of ice and snow in what glaciologist call the "viscoplastic regime".

2 Viscoplasticity of ice

To explain what motivates the experimental studies described in the following sections it is necessary to give a short overview of the basics of ice viscoplasticity.

2.1 Ice is a crystalline material

Under the conditions of pressure and temperature that prevail on Earth water crystallizes into "hexagonal" ice. Since the structure of the water molecule is preserved, each oxygen atom is linked to 2 hydrogen atoms by strong covalent bonds (those of the water molecule) and 2 other hydrogens (which belong to other water molecules) by hydrogen bonds energetically weaker. Thus each oxygen atom is the centroid of a tetrahedron whose vertexes are the oxygen atoms of the neighbouring molecules. However, for a given tetrahedron there are 6 possible arrangements of the hydrogens, so that the hexagonal symmetry concerns only the arrangement of the oxygen atoms. The hexagonal symmetry axis is called the c axis (Fig. 1a). It is perpendicular to the basal plane of the ice crystal. The 3 symmetry axes in the basal plane are the a-axes. The crystallographic planes perpendicular to the basal plane are the prismatic planes. Any other crystallographic plane is denoted as pyramidal (Fig. 1b).

Ordinary ice is usually a polycrystalline material. Each grain of the polycrystal is a "single crystal". In general the grains can be distinguished from each other by their crystallographic orientations; they are separated from each other by grain boundaries.

2.2 Dislocations in ice

The slow viscoplastic deformation of the ice (single) crystal is made possible owing to the presence of dislocations, like for other crystalline materials. Dislocations are linear defects that produce a local discontinuity of the crystal lattice. The displacement of dislocations (dislocation glide) allows the shearing of crystalline planes which results in a permanent deformation (see Fig. 2). A dislocation is characterized by the direction of its line **l** and its Burgers vector **b** which gives the amplitude and direction of the local distortion (see Fig. 2): it allows to re-close an initially closed circuit joining atoms in the perfect crystal that would be opened when surrounding the dislocation line. The two special cases are the "edge" dislocation with **b** perpendicular to **l**, and the "screw" dislocation with **b** parallel to **l**. The other dislocations are "mixed" dislocations, i.e. a combination of edge and screw dislocations. The local distortion of the lattice induced by a dislocation is associated to a strain energy that can be considered as elastic since the (oxygen) atoms come back to their equilibrium position after the passage of the dislocation. Under the hypotheses that the theory of elasticity is applicable (which is true far from the dislocation line) and of isotropic elasticity, it is possible to achieve expressions for the dislocations energies per unit length of the dislocation line [43].

A dislocation can move by dislocation glide with its line remaining in the same crystalline plane. Its screw component may shift on the crystalline plane of an other family that contains its Burgers vector : this is the cross slip mechnism. At high temperature the dislocation can move out of its glide plane while keeping the same glide system (same family plane) by dislocation climb. Since this process implies the diffusion of vacancies (missing atoms), which are very few in ice, its probability is very low for ice [20].

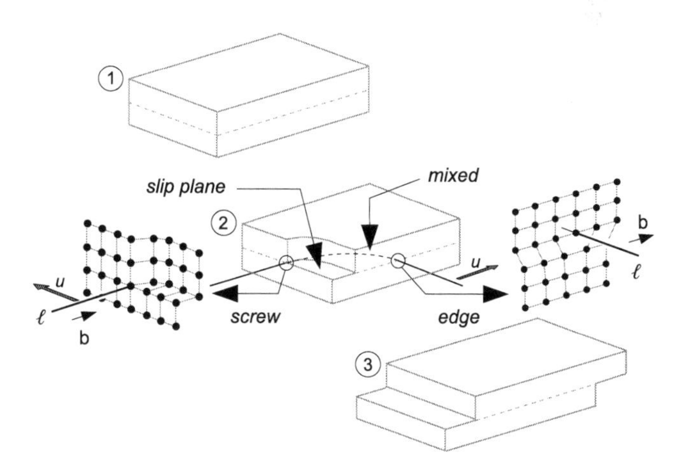

Fig. 2 Screw, edge and mixed dislocations. The permanent deformation from 1 to 3 results from the motion of the dislocation lines **l** along **u** in the slip plane; **b** is the Burgers vector of the dislocations.

It has been established for long [15] that the viscoplastic deformation of the ice single crystal occurs mainly owing to the gliding of basal dislocations. These basal dislocations can move under very low (<0.01 MPa) basal shear stresses [20]. Their velocity is proportional to the shear stress and their mobility M is temperature dependent following an Arrhenius law of the form

$$M(T) = M_0 \exp -\frac{Q}{R}(\frac{1}{T} - \frac{1}{T_0}), \quad (1)$$

where T_0 is a reference temperature and the activation energy Q varies from 60 to 90 kJ mol^{-1} according to different authors [12, 42, 36, 37, 21]. (The universal gas constant is R = 8,314 $\times$ 10^{-3} kJ mol^{-1}K^{-1}.) This mobility is a few μms^{-1}MPa^{-1} at -10°C.

2.3 Single crystal vs. Polycrystal

The two degrees of freedom provided by basal dislocation glide are not enough to accommodate any given strain state in the single crystal. Dislocation glide on other crystalline planes (prismatic or pyramidal) does not contribute significantly to the deformation, and neither dislocation climb or cross slip have been observed [38]. As a consequence of the predominant basal dislocation glide, the ice single crystal exhibits a very strong anisotropy of its viscoplastic behaviour. The strain rate differs by many orders of magnitude according to whether basal glide is easy or difficult [8] (at -10°C shearing parallel to the basal plane is 10^4 to 10^5 faster than in any other configuration of the crystal in the range 0.1 to 10 MPa). The typical creep curve for a single crystal sheared parallel to its basal plane is shown on Fig. 3a. The primary transient stage appears as accelerating creep (softening) owing to the multiplication of dislocations and is followed by steady creep. During this stationary stage the shear strain rate parallel to the basal plane $\dot{\gamma}_b$ can be expressed as

$$\dot{\gamma}_b = A(T)\tau_b^n \text{ , with } \quad A = A_0 \exp -\frac{Q}{RT} \text{ ,} \quad (2)$$

where τ_b is the basal resolved shear stress and the stress exponent n is close to 1.5 [17, 18] (although most authors adopt the value of 2). The "fluidity" parameter A is strongly dependent on the temperature T. It follows an Arrhenius law (1) with an activation energy Q of about 60 kJ mol^{-1}.

As concerns the ice polycrystal, the typical creep curve (Fig. 3b) consists of a primary transient stage of decelerating creep (hardening), followed by a very short stage of secondary creep, if any, and by a tertiary creep stage (accelerating creep). It differs from that of other materials, as metals or ceramics for instance, in that the secondary creep stage is reduced to an inflexion point on the creep curve (the "minimum creep rate"). At the very beginning of loading the dislocations can move almost freely inside each grain. However the grain boundaries are an obstacle to the movement of most of dislocations, so that during the primary stage an internal stress field develops inside each grain, which results in strain hardening. Since laboratory tests on ice are made at a very high temperature relative to the melting point T_m (in general $T/T_m > 0.9$; at lower temperatures the tests would be much too long) the high stresses generated at the grain boundaries initiate recrystallization: new grains free from dislocations are nucleated to replace the highly distorted areas of the grains. The onset of recrystallization corresponds to the inflexion point on the creep curve. In the following of the creep test the ice polycrystal experiences constant generalized recrystallisation, which can lead eventually to a constant creep rate (the initial microstructure of the polycrystal is completely lost). For a polycrystal which is macroscopically isotropic (made of a large number of grains with random crystallographic orientations), the minimum strain rate – stress relation is expressed as Norton-Hoff's law, called "Glen's law" by glaciologists (since [14]). This constitutive law links the strain rate **d** to the stress deviator **s** as

$$d_{ij} = \frac{B(T)}{2}\tau^{n-1}s_{ij}, \tag{3}$$

where $\tau^2 = s_{ij}s_{ij}/2$ is the second invariant of **s**, and the fluidity parameter B depends on the temperature T according to Arrhenius law (1) with an activation energy Q= 78 kJ mol^{-1} below -10°C [29]. For polycrystalline isotropic ice tested in the laboratory under relatively high deviatoric stresses ($\tau > 0.1$ MPa) the stress exponent n is about 3 and the reference parameter at T_m is B_m=70 MPa^{-3}a^{-1} ($\approx 2.2\times 10^{-6}$MPa^{-3}s^{-1}) according to [29]. Glen's law (3) is generally adopted for modelling tertiary creep: the exponent n keeps the value of 3, and to account for the softening of ice B_m is multiplied by a factor between 3 and 4 [29]. The isotropic ice polycrystal deforms slower than the single crystal (at -10°C a single crystal sheared parallel to its basal plane deforms 10^2 to 10^3 faster than the polycrystal under 1 MPa).

Figures 3a and b show that the single crystal of ice and the polycrystal have a different behaviour. The viscoplastic behaviour of the polycrystal follows a power

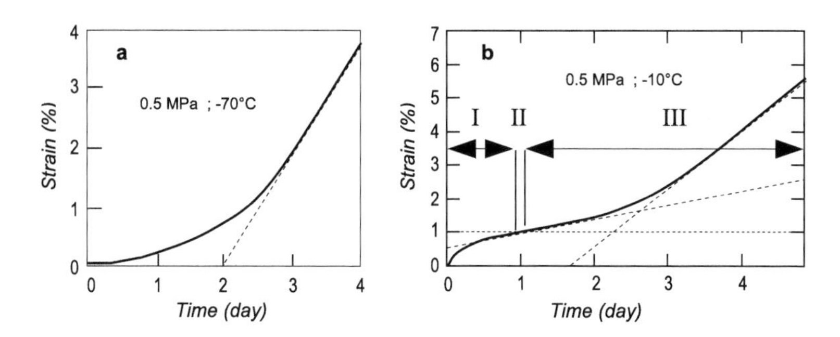

Fig. 3 Typical creep curves for ice. **a** single crystal. **b** polycrystal. Note that the curves are for different temperatures: the time to reach the polycrystal minimum strain rate would be approximately 30 years at -70°C.

law with stress exponent $n = 3$ in (3), whereas for the single crystal the value of n in (2) is 1.5. This difference indicates that the types of physical mechanisms involved in the deformation are different (this has strong consequences as regards for instance the construction of self-consistent models). It results essentially from the very strong viscoplastic anisotropy of the single crystal and from the fact that grain boundaries act as barriers to the movement of dislocations. In a polycrystal there are first strong strain incompatibilities between the grains (inter-granular heterogeneity) which give rise to high stress concentrations near the grain boundaries. As the deformation proceeds a heterogeneous strain field develops inside each grain (intra-granular heterogeneity).

3 Experimental study of the heterogeneous deformation of ice

To achieve a better knowledge of the processes involved in the deformation of polycrystalline ice, and in particular the mechanisms responsible for intragranular heterogeneity, two series of creep tests have been carried out on specially designed specimens of laboratory grown ice. The first experiments were performed in the laboratory by observing ice under polarized light (because of the anisotropy of hexagonal ice a thin section placed between cross polarisers shows the grains with different c axis orientations in different colours). The second series were performed under X ray synchrotron radiation to observe the deformation of ice near a grain boundary or a triple junction of grains. In both situations we studied either single crystals or "multi-crystals", i.e. polycrystals with a very limited number of grains.

3.1 Observations under polarized light

Because of its strong anisotropy it is very difficult to perform a mechanical test on a single crystal of ice. In the very early stage of the deformation two kinds of strain heterogeneity develop. First, as basal dislocation glide begins there is only a discrete distribution of basal planes which are activated and, as creep proceeds, the number of activated planes increases (see Fig. 4). This kind of heterogeneity, studied by [33] long ago has been recently revisited by [32]. We are not interested here in this mechanism since we will see that at the grain scale the resulting deformation may be considered as homogeneous. The second type of heterogeneity is strain localization induced by the loading conditions at the boundaries of the specimen. A single crystal compressed at angle with respect to its basal plane first exhibits a bending of the basal planes (visible on Fig. 4), followed ultimately by the formation of sub-boundaries (i.e. a kind of recrystallization: the initially single crystal becomes a polycrystal). To lessen this effect the specimens (either single or multi-crystals) were embedded in a matrix of fine-grained ice (see Fig. 5). The matrix grain size was much smaller than that of the specimens and could be considered as

Fig. 4 Compression of a single crystal. The crystal is a plate 10 mm thick. Its basal planes are at 45° from the compression axis (black arrow). Left: activated slip lines appear from the beginning of the test; right: after 33% strain (86400 s under 0.5 MPa) bending of the basal planes becomes visible.

macroscopically isotropic (owing to the very large number of its grains oriented at random). Doing so, the stress concentrations at the ice-press platen contacts were removed far away from the studied multicrystals.

Different shapes and configurations of multicrystals were tested to study the influence of the size, of the crystallographic orientation, and of the geometry of its grains. We will report here on only one type which consists of a large disk-shaped single crystal 30 to 50 mm in diameter, referred to as the central grain in the following, surrounded by ice grains of different sizes and orientations (see Fig. 5c). We will compare the observations made on this type of multicrystal to that made on a circular shaped single crystal inclusion (i.e. central grain alone).

The specimens had the overall shape of a thick plate, 210 mm high, 140 mm wide and 8 mm thick. Each single or multi-crystal was grown in the laboratory from monocrystalline seeds such that the c-axes of the grains lay in the plane of the plate. The grain size of the matrix (< 1 mm) was much smaller than the specimen thickness and the central grain size.

The creep experiments were performed in a cold room at -10°C. The ice plate was maintained in a vertical plane by two glass plates disposed on either side of its largest faces. The compression stress was exerted on the two horizontal edges, while the two vertical edges were free surfaces. The observations were done by using two cross polarisers placed each side of the ice plate. During a test photographs were taken at regular time intervals to follow the evolution of the shape and of the lattice orientation of the grains, as well as strain localization.

At the beginning of the creep tests, basal slip bands appeared across the central grain (or across the single crystal inclusion) as an alternation of well defined coloured stripes parallel to the direction of the basal planes, about 1 mm wide. After a plate deformation of about 1% the slip bands transformed into well-defined dark slip lines. During the starting stage, other bands, more diffuse (much less marked), less numerous and larger (2 to 5 mm wide) than the slip lines appeared perpendicular to the direction of the basal plane.

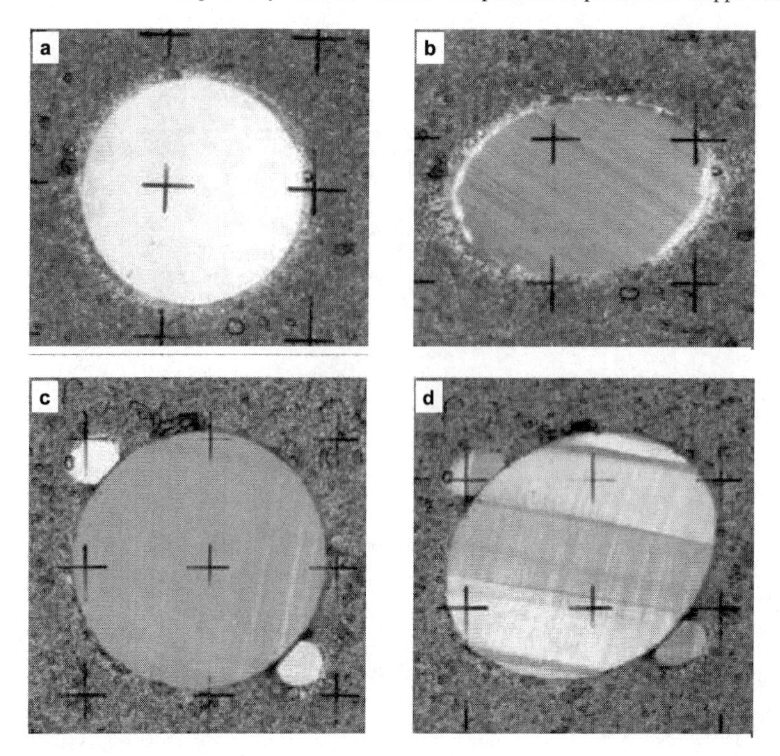

Fig. 5 Mono and multi- crystalline inclusions. The c axes of the central grains are in the plane of the specimen. **a** monocrystalline inclusion 30 mm in diameter with basal planes at 45° from the compression axis (vertical). **b** monocrystalline inclusion (a) deformed up to 6% under 0.5 MPa; **c** monocrystalline inclusion 50 mm in diameter with two adjacent grain (7 mm in diameter) and with its basal planes at 10° from the compression axis; **d** inclusion (c) deformed up to 19% under 0.75 MPa. (Note that the fine grained matrix dimensions are much larger than shown on the photographs.)

The circular monocrystalline inclusions exhibited a homogeneous distribution of the activated basal planes, with a slip lines spacing between 0.1 and 1 mm (see Fig. 5b). They did not show any glide system apart from basal, nor any hint of strain localization apart from the slip lines. This was observed even for a crystal with its basal plane at 45° from the compression axis deformed up to 50%. During the tests the circular shape crystal evolved towards an ellipse and the basal planes rotated towards the direction perpendicular to the compression axis (see Fig. 5b) as predicted by Eshelby inclusion theory [9]. The long axis of the ellipse rotated in the same direction as the basal planes but not at the same rate. The diffuse bands formed perpendicular to the basal planes disappeared with increasing strain, except for one or two which remained stable near the ends of the ellipse long axis without transforming into lines of localization.

By comparison, the deformation of multicrystals was not homogeneous and different kinds of strain localization phenomena occurred inside the grains (see Fig. 5c and d). Some of the diffuse bands evolved into well developed localization features, whereas some other disappeared. The two main localization phenomena observed in the multicrystals were kink bands (Fig. 6a) and bending bands (Fig. 6b), accompanied by recrystallization.

The kink bands formed initially perpendicular to the direction of the basal plane and were initiated at triple junctions on the boundary of the central grain. Each band delimited two areas of the crystal with the same basal plane orientations. Inside the kink band, about 0.5 to 1 mm wide, the basal planes remained parallel to each other, but at high angle from the adjacent areas, up to 60° for a total strain of 10% (see Fig. 6a). These kink bands acted as shear bands, providing an additional degree of freedom for the deformation of the inclusion. During the tests, the kink bands did not remain perpendicular to the basal planes and with the basal planes deviation from the compression axis increasing, other kink bands formed perpendicular to the basal planes, becoming more and more marked until the test was stopped. The localization process associated to kinking did not seem to be dependent on the crystallographic orientations of the surrounding grains which were randomly oriented: a kink band was initiated at each triple junction.

Bending bands were characterized by a progressive change of the orientation of the basal slip lines from one side of the band to the other, corresponding to progressive bending of the basal planes over a relatively wide area. The curvature of the basal planes suggested that the central grain was submitted locally to a flexion about an axis normal to the plane of the ice plate. These bands delimited two regions of the inclusion with different crystallographic orientations. Some specimens exhibited wide bending areas, delimited either by two narrower and almost parallel kink bands crossing through the inclusion, or by well marked sub-boundaries (see Fig. 5d). For some specimens the variation in the orientation of the basal planes from side to side of the bending area was about 15 °. Other evidences of bending of the basal planes were observed at the grain boundaries between the central grain and the surrounding grains. This type of accommodation mechanism was likely to occur at all grain boundaries, but was clearly visible only in the largest crystals. Indication of the evolution of the progressive change of orientation towards the formation of sub-boundaries can be seen in Fig. 6b.

Recrystallization was observed in the fine-grained matrix, close to the multicrystal boundary. This form of strain accommodation was a consequence of the discontinuous conditions at the interface with the fine-grained matrix, leading to a local increase in the stored energy near the boundary. Recrystallization was also observed in the bulk of the matrix (after a large amount of deformation only a minority of the initial grains was still recognizable) and in the grains surrounding the central grain after a deformation of about 10%. Only a few small crystals appeared sometimes in the central grain at its boundary. Therefore, the size of the grains seemed to influence the recrystallization rate, this rate being slower the larger the grains.

The observed localization features are strongly related with the dynamics of dislocations inside the grains. According to [19], kink band formation can be explained by the rearrangement into dislocation walls of a rapid supply of dislocations in the basal planes when dislocation climb is difficult, which is the case for ice

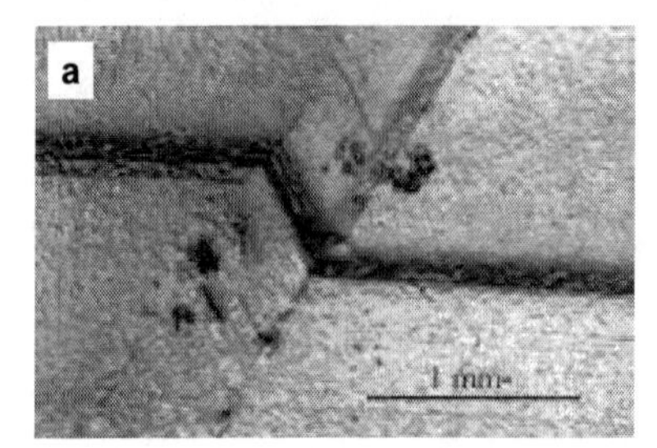

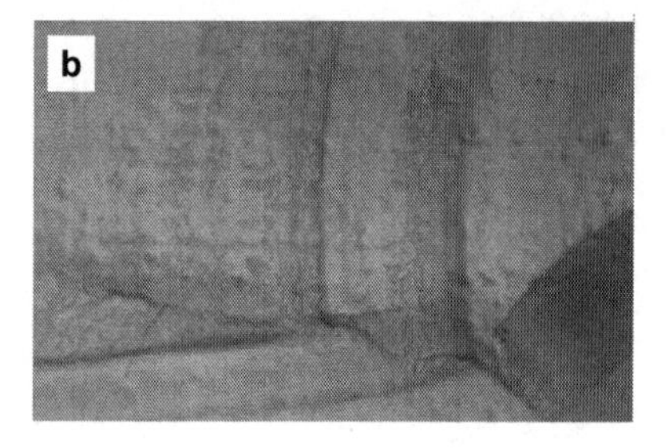

Fig. 6 Two examples of localization features observed in the specimens. **a** kink band. **b** bending of the basal planes.

[1, 41, 11, 22]. Bending bands allow the accommodation of strain gradients by geometrically necessary dislocations. The excess of dislocations of one sign rearranges into walls, leading ultimately to the formation of sub-boundaries. It was already observed by [3] and [45]. Our observations suggest that the occurrence of the diffuse bands is closely related to stress concentrations at the matrix-multicrystal interface. Since the size of the grains was found to influence the occurrence of recrystallization, the disappearance of these bands in the monocrystalline inclusions could be explained by the recrystallization of the small adjacent grains of the matrix.

3.2 Observations under X-ray radiation

The high stress concentrations which take place in the vicinity of grain boundaries in polycrystalline ice lead to heterogeneous intragranular strain fields, therefore to a heterogeneous lattice distortion field in each grain. In order to understand the evolution of the dislocation density during the deformation of ice, in-situ compression tests have been performed under X-rays at the ID-19 beam line of the European Synchrotron Radiation Facility (ESRF) in Grenoble using diffraction topography techniques.

Mainly two types of ice specimens have been studied: single crystals and tri-crystals in order to enlighten the role of grain-boundaries and triple junctions. The specimens were thin sections ($21 \times 17 \times 1\mathrm{mm}^3$) of ice grown in the laboratory. They were cut from batches of columnar ice (lake ice type with column shaped grains) grown by freezing deionized water in a vertical temperature gradient. By cutting thin sections parallel to each other this allows to obtain specimens with very close crystallographic orientations. The thin sections were compressed in their plane using a specially designed small press fit for being installed into a transparent refrigerated cell. The applied compressive stress was in the range 0.02 MPa to 0.1 MPa. To avoid buckling and sublimation during the tests, the ice thin sections were inserted between two plastic plates transparent to X-rays (see Fig. 7a).

a

b
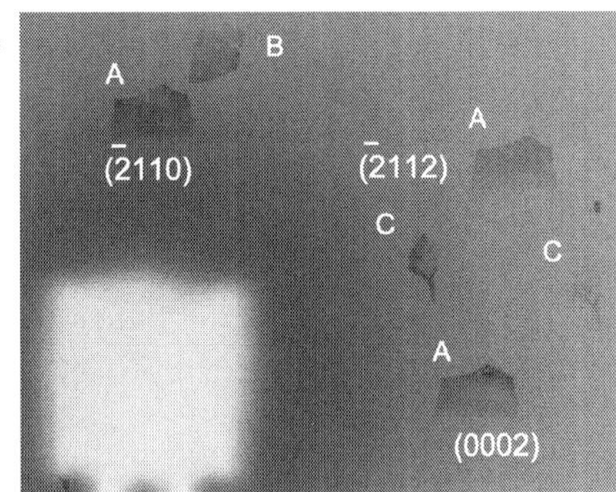

Fig. 7 Photograph of a tri-crystal studied by x-ray topography **a** and typical topogram **b** showing the simultaneous diffractions spots of the grains.

The basics of X-ray diffraction topography is Bragg's law which links the lattice spacing of the family of diffracting planes d to the X-ray wavelength λ as

$$n\lambda = 2d\sin\theta \ , \tag{4}$$

where θ is the incidence angle of the X-ray beam with respect to the diffracting plane and n is an integer (the "order" of the diffraction). X-ray polychromatic light ("white beam") was used for this study because it provides reflections on different families of crystallographic planes for the same crystal (considering two different d there is always two λ in the beam spectrum that allow fulfilling Bragg's condition (4)). This allows to obtain more information on the observed dislocation features. Also, when studying tri-crystals, for particular relative crystallographic orientations of the grains the white beam may diffract simultaneously for more than one grain on the detector (film), which is particularly interesting for grain boundary and triple junction studies. X-ray topograms were recorded on films. A typical example is shown on Fig. 7b.

Our first experiments on ice single crystals showed very large variations in the dislocation density of the undeformed laboratory grown crystals. Since a too high dislocation density prevents the observation of individual dislocations, a long time was dedicated to assessing several growth conditions in order to identify a protocol ensuring a low initial dislocation density (each crystal growth lasts about 1 month). The crystals "quality" was assessed on one hand by counting the number of dislocations visible on topograms obtained using a polychromatic beam and on the other hand by using the "re-focused hard X-rays" technique [5]. At the end, the growth method which was selected provides us with crystals exhibiting very low initial dislocation densities, as low as $10^6 \mathrm{m}^{-2}$ (the dislocation density is given in terms of length of dislocation line per unit volume of the material).

The observation of individual dislocations is only possible in the early stage of the deformation. During compression, it was possible to follow the displacement of individual dislocations. Their velocity was found to be between 0.5 and 1 $\mu\mathrm{ms}^{-1}$ at $-10°$C under a compression stress of 0.3MPa, which is in agreement with the order

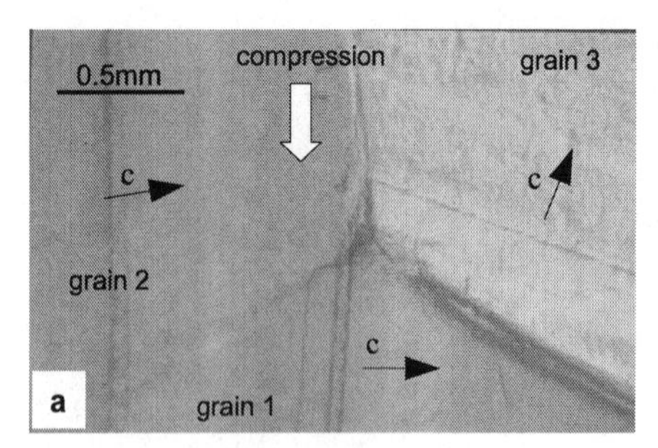

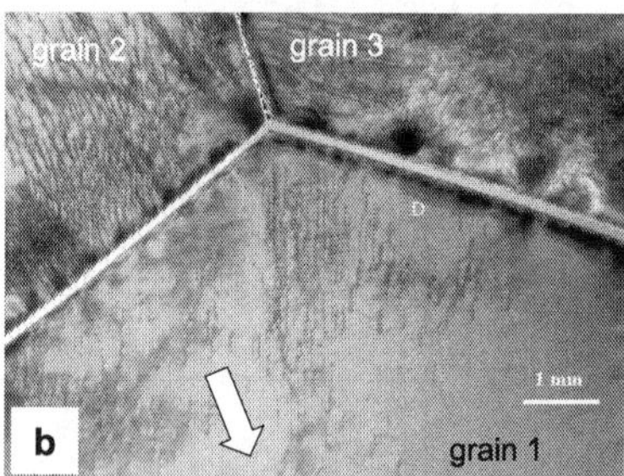

Fig. 8 Photograph under polarized light and x-ray topograms of two tri-crystal with the same initial structure. **a** the tricrystal as been loaded under 0.9 MPa for 10 h, the misorientation between the grains c-axes are c_1–c_2: 45°, c_2–c_3: 55°, c_3–c_1: 10°, and between the a-axes : a_1–a_2: 0°, a_2–a_3: 50°, a_3–a_1: 20° ; **b** Topograms of the $(1,0,\bar{1},2)$ planes of the 3 grains of the tri-crystal (compressive stress 0.3 MPa during 15 minutes).

of magnitude of the mobility of dislocations in ice found in the literature (e.g. [37]). During loading the dislocation density increases and it was possible to observe its evolution in the grains qualitatively. Grain boundaries are the place of stress concentrations that appear as black areas limited by white loops in the topograms. Figure 8 shows the topograms obtained simultaneously of the grains of a tri-crystal. At the beginning of loading, the black regions seem to be related to the internal elastic strain since they disappear when the external load is removed. However, after the specimen has been loaded for some time then unloaded, the topographs still exhibit black areas at the grain boundaries: these must be related to the energy stored in the dislocations that accumulate due to the strain incompatibility between grains. The observations made during compression tests performed under polarized light on a tri-crystal with the same structure (cut from the same batch) are in agreement with that obtained under X-rays. Figure 8 shows that slip lines pass directly through the grain boundary from grain 1 to grain 2. The topography study, which allows to measure the crystallographic orientation of the grains, shows that these two grains have a-axes in the same direction and are then "strain-compatible" (the Burgers vectors of the basal dislocations in both grains are parallel to each other). The topograms show practically no distortion at the grain 1 – grain 2 boundary. The topography measurements show that the dislocation Burgers vectors of grains 1 and 3 are not compatible: the energy stored at the boundary needs to be released by recrystallization.

When the deformation is too high and the dislocations too numerous to be observed individually it is still possible to estimate the density of "geometrically necessary dislocations" (GND) [35, 2, 43] from the measure of the distortion of the lattice planes. These dislocations allow to accommodate the viscoplastic strain gradients, either locally or at the grain scale. They arise from the build-up of single-sign dislocations at obstacles to dislocation motion, such as grain boundaries and can be described by an effective Burgers vector allowing accommodating the deformation in an energetically favourable way.

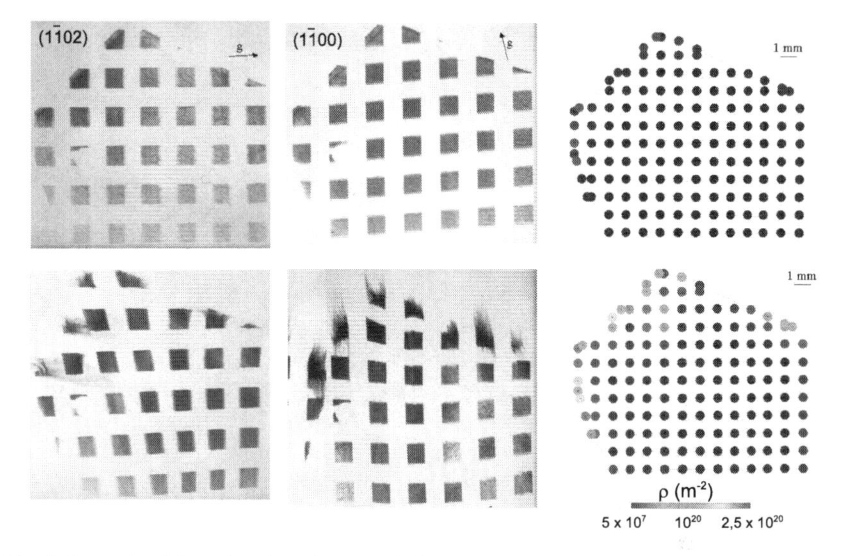

Fig. 9 Example of the estimation of geometrically necessary dislocation densities (GND) using the reticulography method. Left: topograms obtained by using a Tungsten grid of a pyramidal plane and of a prismatic plane before compression and after 1 hour under 0.6MPa; right: GND density estimated from the analysis of the x-ray topograms.

Following [35] a schematic two-dimensional crystal initially rectangular needs a GND density $\rho = 1/(Rb)$ to accommodate a homogeneous bending which leads to a radius of curvature R (b is the length of the dislocations Burgers vector). This simple example shows that strain gradients, GNDs and lattice curvature are intimately associated.

A generalization to a three dimensional geometry is made possible by using the reticulography method [27, 28]. This method consists in placing a grid of absorbent material (Tungsten) in front of the specimen in order to split the incident beam in a number of sub-beams. Each diffraction spot is thus split into sub-regions. By pointing the grid nodes on the topograms of different lattice planes (at least 2) it is possible to obtain the components of the diffraction vectors (normal to the diffracting planes) at each node and then to reconstruct the crystallographic orientation of the crystal (i.e. the crystal reference frame) at each node. From this field of local orientations it is possible to derive the crystal curvature tensor, then to access the density of GNDs. This can be done at different stages of the deformation. An example (extracted from L. Capolo's PhD thesis [6, 7]) of topograms obtained with 1mm $\times$ 1mm split beams is shown on Fig. 9a for a grain of a tricrystal before compression and after one hour under 0.6 MPa. Figure 9b shows the corresponding distribution of the GNDs density calculated from the mapping of the crystallographic orientations. These results show the heterogeneous evolution of the distortions. Due to the strain incompatibilities between the grains the GNDs density is higher in the vicinity of the grain boundaries and at the triple junction. Such results are very promising, however the grid size must be much reduced in order to detect the strong strain gradients that occur in ice.

4 Experimental study of the micromechanics of dry snow

Snow consists of ice grains linked by bonds. Its porosity is filled by air containing a certain amount of water vapour. Below 0°C snow is called dry snow.

Three mechanisms have been identified as possibly active in the deformation of dry snow [40]: breaking of bonds, grain boundary sliding (GBS) and intragranular deformation. GBS and grain deformation are very sensitive to temperature (Arrheniuslaw with activation energies Q below -15°C of about 50 kJ mol^{-1} for GBS [16] and 60 kJ mol^{-1} for intracrystalline creep of monocrystalline ice [17]. Since snow is a porous medium, local stress concentrations arise at the grain contacts where the local cross sectional area is minimum. These stress concentrations are thought to play a significant role [26] and this explains why snow is often described as a granular material e.g. [34], which is questionable. On one hand, because ice exhibits a very strong viscoplastic anisotropy in the temperature range relevant for engineering problems (ratio T/T_m between 0.92 and 1) the intragranular deformation of the snow grains cannot be ruled out (Note that the very few snow models that account for the deformation of the grains assume that ice behaves as an isotropic ice polycrystal, e.g. [25]). On the other hand the fact that GBS is often presented as the dominant mechanism in the deformation of dry snow needs to be assessed (GBS is certainly an active mechanism in the case of wet snow whose grains are linked to each other by capillarity forces). GBS has been presented as an active deformation mechanism in ice by [16], however this result concerned very small grain sizes (less than 100 μm). Besides there is little direct evidence of GBS occurrence in ice or snow in the literature (even the experiments of [23] on bicrystals sheared parallel to their grain boundary have shown that GBS is not a dominant mechanism).

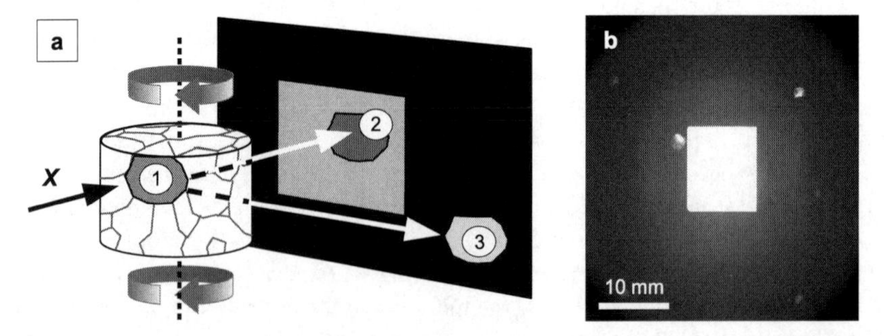

Fig. 10 Principle of the Diffraction Contrast Tomography technique. **a** schematic of the experimental setting; **b** typical radiograph showing the transmitted beam (usable as an absorption contrast tomography radiograph) and the diffraction spots from the grains which are under Bragg's condition.

Fig. 11 Photograph of the large grained snow studied to assess the application of DCT to snow **a** and **b** three-dimensional DCT image of the specimen with the crystalline orientations indicated by different colours.

To investigate the relative influences of intragranular deformation and GBS in the deformation of snow in the ductile regime (thus putting the breaking of bonds out of scope) we started an experimental study 2 years ago. More generally this study aims at improving our knowledge of the influence of the snow microstructure on its mechanical behaviour by observing how snow deforms at the grain scale in the viscoplastic regime.

The experimental work consists in performing compression tests on snow under X-ray radiation. The tests are monitored by microtomography in absorption mode, i.e. the "classical" micro computed tomography technique (μCT), and using a new technique still under development called Diffraction Contrast Tomography (DCT) [30, 24]. In addition to the geometrical information supplied by μCT, the DCT method provides the crystallographic orientations of the snow grains, which is essential in view of understanding how snow deforms and of achieving realistic micro-mechanical simulations of the experiments because of the anisotropy of ice.

The principle of DCT is as follows. The experimental set-up is similar to the one used for μCT : the snow specimen is placed in a refrigerated cell on a rotation stage and irradiated by a monochromatic synchrotron X-ray beam (see Fig. 10a). Since the snow grains are made of ice crystals each grain happens to be in Bragg's diffraction condition from time to time during the specimen rotation (i.e. one of the grain crystalline planes, characterized by the lattice spacing d, is illuminated with incidence angle θ such that Bragg's law (4) is fulfilled). Each of these events produces a diffracted beam which corresponds to an extinction spot in the transmitted beam (loss of intensity). However only the planes that diffract at small angles θ can be captured by the detector whose size is physically limited. Recording the diffracted spots implies that the cross section of the transmitted beam (i.e. the dimensions of the sample) is less than that of the detector (see Fig. 10b). The diffracted beams are recorded as two-dimensional diffraction spots. The analysis of these diffraction spots allows to determine the crystallographic orientation and three dimensional shape of the grains in the specimen. In addition a three dimensional image can be

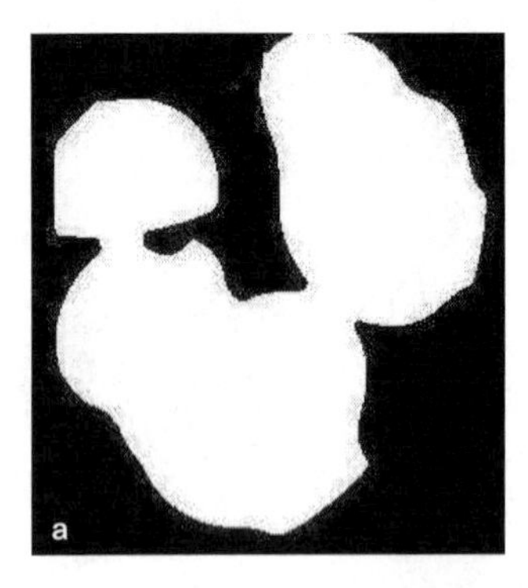

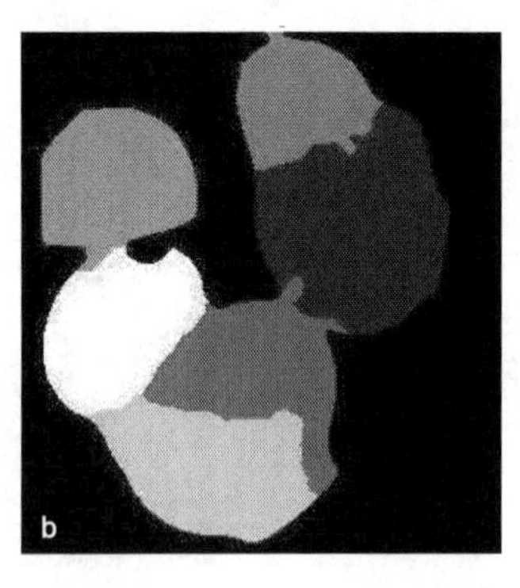

Fig. 12 Snow grains identification by DCT. **a** horizontal cross section of a specimen as obtained by conventional microtomography (snow is white). **b** the same horizontal cross section obtained by DCT show that some snow grains are made of several ice crystals.

obtained by analysing the intensity of the transmitted beam, as done when using μCT.

The quality of the DCT reconstruction depends on the number of diffraction spots that can be recorded (the more, the better). To collect a large number of diffraction spots, the active area of the detector must be relatively large compared with the specimen dimensions to allow as large as possible angles θ. Since the camera definition is fixed (2048×2048 pixels), the actual dimensions of the detector are function of the camera optics: with an effective pixel size of 20 μm the detector area is about 40×40 mm^2. Because of the refrigerated cell the detector cannot be approached less than 65 mm from the specimen. This fixes the limit incidence angle θ for possibly detected spots. The other parameter that can be tuned is the X-ray wavelength. By setting the beam energy to 20 keV we are able to record the spots which correspond to the basal plane $\{0,0,0,2\}$, the two principal prismatic planes $\{1,\bar{1},0,0\}$ and $\{1,1,\bar{2},0\}$ and pyramidal planes $\{1,1,\bar{2},2\}$ and $\{2,0,\bar{2},1\}$.

The snow specimens are cylinders whose dimensions depend on the mean grain size (because the resolution of the camera optics must be adapted to the grain size; the detector area depends on the camera optics; and the direct image of the specimen must cover only a small fraction of the detector). The first experiments were done with grains larger than 1mm in diameter (Fig. 11a) in order to limit their number and to obtain large diffraction spots convenient for testing the applicability of the DCT. With the 20 μm optics, the specimens were 10 mm high and about 10 mm in diameter.

The results obtained for an undeformed snow sample imaged using the DCT technique is shown on Fig. 11b. On this image, the colours of the grains are representative of the grain c-axis vector orientation. This shows that the DCT input will be very valuable for further micro-mechanical simulations of snow. The same kind of three-dimensional image can be obtained with deformed specimens. However we must keep in mind that if a significant part of the deformation is intragranular, then the deformation inside the grains will become more and more heterogeneous, which will set the limits of the DCT technique because to be usable the diffraction spots must remain separable from the background of the radiographs.

Another information given by DCT concerns the identification of the grains. The classic μCT method allow to define a grain from its geometric shape. Looking to Fig. 12 one can see that what a snow grain may be composed of a few ice crystals (i.e. some snow grains are multi-crystals).

At last, by analysing the recorded diffraction spots it is possible to obtain information about the evolution of the crystalline quality of the grains. A perfect crystal should give an entire diffraction spot in a very narrow interval of rotation of the specimen. In reality the distortion of the crystal lattice has the effect of spreading the diffraction spot over an interval of rotation angle. By recording the size of these intervals for each grain at different stages of deformation of the specimens we have found that the distortion of the grain lattices increased during the compression test, which is a good indication that intragranular deformation does occur during the deformation of snow.

5 Conclusion

The isotropic power-law (3) is quite convenient for simulating the flow of temperate valley glaciers whose ice, close to the melting point, experiences continuous recrystallization. However the observations of deep ice cores drilled in polar ice sheets have shown preferential c axes orientations of the grains. As a consequence, owing to the very strong anisotropy of the ice crystal, polar ice exhibits a strain-induced macroscopic anisotropy. Modelling its behaviour is a complex matter because laboratory studies under the conditions existing in polar ice sheets, i.e. very low deviatoric stresses and temperature, are not possible (the time required to reach a significant amount of deformation is much too long). The only possibility which is left is to assess a rheological model by using field data, which requires to simulate the large scale flow of ice. This has been on going in the recent years by proposing a simplified model for the ice grain behaviour [31] which has been integrated in a homogenization model for anisotropic polycrystalline ice [13]. Nevertheless the problem of modelling properly the ice grain behaviour still remains. The three experiments described in the present paper are a small step towards a better understanding of the mechanisms involved. They are a very small part of the experimental work which has been started long ago (e.g. [33]) and is actually ongoing as concerns ice (e.g. [44]) or snow (e.g. [39, 4]). This experimental work needs to be continued and improved. Since its output consists only in observations which need to be interpreted, it must be accompanied by an important effort for modelling the dynamics of dislocations (e.g. [10]) and the localization of the deformation.

Acknowledgements The authors are particularly grateful to Prof. J. Baruchel, Dr. P. Bastie and Dr. W. Lüdwig for their help with the X-rays experiments at the European Synchrotron Radiation Facility (ESRF) and Laue-Langevin Institute (ILL) in Grenoble. We greatly acknowledge the financial support of the French Agence Nationale de la Recherche through the contract ANR-06-BLAN-0396 for the "Snow-White" project. We also thank our colleagues from the LGGE, CEN, 3Sr, and Cemagref laboratories in Grenoble who are part of this ANR project, namely C. Brutel-Vuilmet, B. Chareyre, F. Darve, F. Donzé, A. Dufour, F. Flin, L. Gillibert, J. Kozicki, B. Lesaffre, F. Nicot, J-M. Panel, E. Pougatch, Ph. Puglièse, and S. Rolland du Roscoat.

References

1. Ahmad, S., Withworth, R. (1988) Dislocation motion in ice : a study by synchrotron x-ray topography. Phil. Mag. A 57(5): 749–766.
2. Ashby, M.F. (1970) The deformation of plastically non homogeneous material. Phil. Mag. 13: 399–424.
3. Azuma, N., Higashi, A. (1985) Formation processes of ice fabric patterns in ice sheets. Ann. Glaciol. 6: 130–134.
4. Baker, I., Obbard, R., Iliescu, D., Meese, D. (2007) Microstructural characterization of firn. Hydrol. Process. 21: 1624-1629.
5. Bastie, P. , Hamelin, B. (1996) La méthode de Laue refocalisée à haute énergie: une technique d'étude en volume des monocristaux. Journal de Physique IV, colloque C4 6: 13–21.
6. Capolo, L., Philip, A., Meyssonnier, J. (2005) Etude de la déformation de la glace sous rayonnement X, Proc. 17ème Congrès Français de Mécanique, Troyes, 29 Août - 2 Sept. 2005.
7. Capolo, L. (2007) Contribution à l'étude des hétérognités de déformation viscoplastique de la glace Ih mono et multi cristalline: essais de compression in-situ sous rayonnement X Thèse de doctorat de l'Université Joseph Fourier (Grenoble 1), 13 Juillet 2007.
8. Duval, P., Ashby, M. F., Andermann, I. (1983) Rate-controlling processes in the creep of polycrystalline ice. J. Phys. Chem. 87(21): 4066–4074.
9. Eshelby, J.D. (1957) The determination of the elastic field of an ellipsoidal inclusion, and related problems. Proc. Roy. Soc. A 241: 376–396.
10. Roberston C.F., Fivel, M.C. (1999) A study of the submicron indent-induced plastic deformation. J. Mater. Res. 14: 2251–2258.
11. Fukuda, A., Higashi, A. (1973) Dynamical behaviour of dislocations in ice crystals. Crystal Lattice Defects 4(4): 203–210.
12. Fukuda, A., Hondoh, T., Higashi, A. (1987) Dislocation mechanisms of plastic deformation of ice. J. de Physique, Paris colloque C1, supplément au n° 3 48(3): 163–173.
13. Gillet-Chaulet, F., Gagliardini, O., Meyssonnier, J., Zwinger, Th., Ruokolainen, J. (2006) Flow-induced anisotropy in polar ice and related ice-sheet flow modelling. J. Non-Newtonian Fluid Mech. 134: 33-43.
14. Glen, J. W. (1955) The creep of polycrystalline ice. Proc. Royal. Soc. London A (228) : 519–538.
15. Glen, J.W., Perutz, M.F. (1954) The growth and deformation of ice crystals. J. Glaciol. 2(10): 397–403.
16. Goldsby, D. L. and Kohlstedt, D. L. (2001) Superplastic deformation of ice: experimental observations J. Geophys. Res. 106 (B6): 11017-11030.
17. Higashi, A., Koinuma, S., Mae, S. (1964) Plastic yielding in ice single crystals. Jap. J. Appl. Phys. 3(10) :610–616.
18. Higashi, A., Koinuma, S., Mae, S. (1965) Bending creep of ice single crystals. Jap. J. Appl. Phys. 4(8) :575–582.
19. Honeycombe, R. W. K. (1984) The plastic deformation of metals. (E. Arnold Eds, 2nd ed.)
20. Hondoh, T. (1992) Glide and climb processes of dislocations in ice. in Phys. and Chem. of Ice, Hokkaido University Press, Sapporo, Japan
21. Hondoh, T. (2000) Nature and behavior of dislocations in ice. in Physics of Ice Core Records, Hokkaido University Press, Sapporo: 3–23.

22. Hondoh, T., Iwamatsu, H., Mae, S. (1990) Dislocation mobility for non-basal glide in ice measured by in situ X-ray topography. Phil. Mag. A 62(1): 89–102.
23. Ignat, M., Frost, H. J. (1987) Grain boundary sliding in ice J. Phys. C1, suppl.3 48: pp.189-195.
24. Johnson, G., King, A., Hoennicke, M., Marrow, T., Lüdwig, W. (2008) X-ray diffraction contrast tomography: a novel technique for for three-dimensional grain mapping of polycrystals II. The combined case J. Appl. Cryst. 41(2): 310-318.
25. Johnson, J. B. and Hopkins, M. A., (2005) Identifying microstructural deformation mechanisms in snow using discrete-element modeling J. Glaciol. 51 (174): 432-442.
26. Kry, P.R. (1975) The relationship between the visco-elastic and structural properties of fine-grained snow J. Glaciol. 14(72) : 467-477.
27. Lang, A.R., Makepeace, A.P.W. (1996) Reticulography : a simple and sensitive technique for mapping misorientations in single crystals. J. Synchrotron Rad. 3: 313–315.
28. Lang, A.R., Makepeace, A.P.W. (1999) Synchrotron X-ray reticulography : principles and applications. J. Phys. D : Appl.Phys. 32: A97–A103.
29. Lliboutry, L., Duval, P. (1985) Various isotropic and anisotropic ices found in glacier and polar ice caps and their corresponding rheologies. Annales Geophysicæ 3(2): 207–224.
30. Lüdwig, W., Schmidt, S., Lauridsen, E.M., Poulsen, H.F. (2008) X-ray diffraction contrast tomography: a novel technique for three-dimensional grain mapping of polycrystals. I. Direct beam case J. Appl. Cryst. 41(2): 302-309.
31. Mansuy, Ph., Meyssonnier, J., Philip, A. (2002) Localization of deformation in polycrystalline ice: experiments and numerical simulations with a simple grain model. Computational Materials Science 25(1-2): 142–150.
32. Montagnat, M., Weiss, J., Chevy, J., Duval, P., Brunjail, H., Bastie, P., Gil Sevillano, J. (2006) The heterogeneous nature of slip in ice single crystals deformed under torsion. Phil. Mag. 86(27) : 4259–4270.
33. Nakaya, U. (1958) Mechanical properties of single crystals of ice. U.S. Snow, Ice and Permafrost Research Establishment. Research Report 28 : 1–46.
34. Nicot, F. (2004) Constitutive modelling of snow as a cohesive granular material. Gran. Matter. 6: 47-60.
35. Nye, J.F. (1953) Some geometrical relations in dislocated crystals. Acta Metallurgica 1: 153–162.
36. Okada, Y., Hondoh, T., Mae,S. (1996) Glide motion of dislocations in ice close to the melting temperature. Abstract presented at Int. Symp. on the Physics and Chemistry of Ice, Hanover (New Hampshire, USA, August 27-31, 1996 : 93–95.
37. Okada, Y., Hondoh, T., Mae,S. (1999) Basal glide of dislocations in ice observed by synchrotron radiation topography. Phil. mag. A 79(11): 2853–2868.
38. Petrenko, V.F., Whitworth, R.W. (1999) Physics of Ice. Oxford University Press, 1st edition.
39. Scapozza, C., Bartelt, P. (2003) Triaxial tests on snow at low strain rate. Part II. Constitutive behaviour. J. Glaciol. 49 (164): 91–101.
40. Shapiro, L. H., Johnson, J. B., Sturm, M., Blaisdell, G. L. (1997) Snow Mechanics: Review of the state of knowledge and applications CRREL Report 97-3.
41. Shearwood, C., Whitworth, R.W. (1989) X-ray topographic observations of edge dislocation glide on non-basal planes in ice. J. Glaciol. 35(120): 281–283.
42. Shearwood, C., Whitworth, R. W. (1991) The velocity of dislocations in ice. Phil. Mag. A 64(2): 289–302.
43. Weertman, J., Weertman, J. R. (1970) Théorie élémentaire des dislocations. Ed. Masson & Cie, Paris, 1970.
44. Weiss, J., Louchet, F. (2006) Seismology of plastic deformation. Scripta Materialia 54: 747-751.
45. Wilson, C. J. L., Zhang, Y. (1994) Comparison between experiment and computer modelling of plane-strain simple-shear ice deformation. J. Glaciol. 40(134): 46–55.

Discontinuities in granular materials: Particle-level mechanisms

J. Carlos Santamarina and Hosung Shin

Abstract Discontinuous planes often develop in soils and affect the mechanical behavior (stiffness and strength) and transport properties of sediments (fluid migration and diffusion). The fundamental understanding of the development of discontinuities in soils must recognize their inherent granular nature and effective-stress dependent behavior. We use complementary experimental, analytical and numerical methods to study particle-scale mechanisms involved in contraction-driven shear failure due to mineral dissolution, desiccation cracks, and hydraulic fractures. We show that: (1) under zero-lateral strain conditions, particle-scale volume contraction causes a stress decrease from k_0-to-k_a so that shear strain localization can develop in sediments with post-peak strain softening response; (2) the development of desiccation cracks in fine grained sediments is determined by the invasion of the air-water interface membrane and ensuing changes in particle forces and displacements; (3) hydraulic fracture results from positive feedback between changes in pore size and the associated changes in particle-level capillary forces (immiscible fluids), seepage drag forces (miscible fluids) and skeletal forces. These particle-level mechanisms are compatible with the effective stress dependent frictional behavior of soils.

1 Introduction

Soils are granular materials; the granular skeleton and pore structure determine the mechanical and hydraulic properties of the soil mass. Strain localization planes in the granular mass can develop along sedimentation planes, or by changes in boundary and loading conditions that cause positive feedback and a runaway effect. These localizations have profound practical relevance because stability and deformation characteristics of geo-structures are often controlled by the presence of discontinuities that act as either weak zones or preferential paths for gas or fluid migration.

J. Carlos Santamarina and Hosung Shin
Georgia Institute of Technology, Atlanta, GA 30328, e-mail: carlos.santamarina@ce.gatech.edu

The development of discontinuities in granular materials has attracted much research attention, and various modes of failure have been identified, including open, closed, and shear discontinuities. Specific examples include shear bands, desiccation cracks, polygonal fault systems, ice lenses, hydraulic fractures, and grouting-induced fractures.

In this article we summarize a fundamental particle and pore scale understanding of localization mechanisms associated to mineral dissolution, desiccation cracks and hydraulic fracture. The complete study is documented in Shin (2009) and in references by the authors listed herein.

2 Mineral dissolution and shear strain localization

Shear failure in sediments is generally linked with active boundary conditions, such as those imposed by tectonic stresses. Under conditions of no lateral strain, and in the absence of tectonic stress, soil mechanic theories predict a simple one-dimensional compaction in which sediment particles displace vertically without shear failure during fluid pressure diffusion. Conflicting with this theory, shear failure planes are often found in sediments that formed under near horizontal burial conditions in zones with no tectonic activity.

Various thermo-chemo-hydro-mechanical coupling mechanisms were hypothesized to explain the development of these discontinuities. At present, mineral dissolution appears as the most plausible triggering mechanisms. We used experimental, analytical and numerical methods to explore the evolution of internal stresses, conditions for localization and the ensuing deformation field.

2.1 Evolution of internal stresses

Experimental results were obtained using a soft oedometer filled with a granular mixture of salt and glass beads saturated with brine. After loading, the pore fluid concentration was decreased gradually to dissolve the salt grains. The ratio between horizontal-to-vertical effective stress decreased from the initial value $k_0 \approx 1 - \sin\phi$ (as in Jaky 1944) to a value similar to $k_a = \tan^2(45 - \phi/2)$ at the Coulomb failure condition (experimental details and results in Shin and Santamarina 2009a). If dissolution continued, the horizontal stress increased towards k_0 with episodic changes in k.

Similar results were obtained using discrete element simulations, by decreasing the particle diameter for a randomly selected set of particles (Fig. 1). While horizontal stress recovery often follows upon further dissolution, marked differences in fabric are observed between the pre and post-dissolution fabrics.

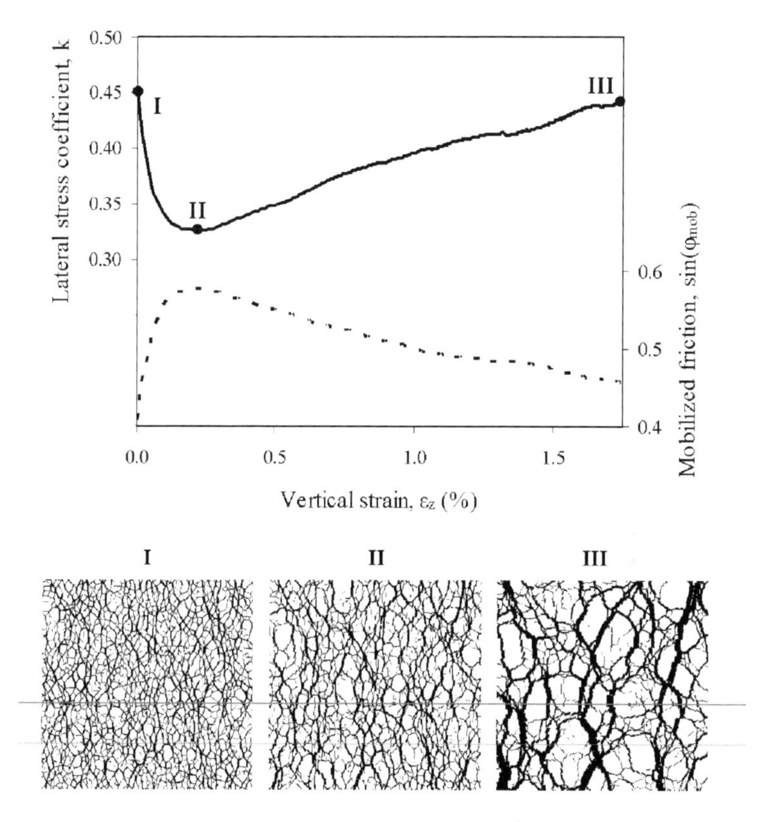

Fig. 1 Sediment evolution during particle dissolution. Discrete element simulation of a 2D packing of 9999 disks. The diameter of 20% of the particles - selected at random- is gradually reduced while keeping zero lateral strain and constant vertical stress boundary conditions. (a) Lateral stress coefficient k and mobilized friction, (b) Interparticle force networks at different stages of dissolution.

Both experimental and numerical results showed the increase in void ratio and the development of loose packing after dissociation; this observation may explain the response of some natural soils (e.g., those reported in Burland 1990).

We can conclude that grain mass loss due to mineral dissolution produces a pronounced horizontal stress drop under zero lateral strain conditions and the state of stress may reach the active shear failure k_a-condition.

2.2 Shear localization

Shear strain localization and the formation of shear bands reflect the development of energetically new beneficial deformation mechanism whereby less energy is required to deform along a few planes rather than in the homogeneous deformation of the whole sediment. Mathematically, shear bands form when the strain hardening $h = \partial\sigma_d/\partial\gamma_{pl}$ exceeds a critical value h_{cr} (where σ_d is the deviatoric stress and γ_{pl} is the plastic shear strain). The values of h_{cr} depend on material strength parameters and loading path (Rudnicki and Rice 1975). In the case of Drucker-Prager model with shear hardening, h_{cr} becomes (Perrin and Leblond 1993, Lade 2002)

$$\frac{h_{cr}}{G} = \frac{1+\nu}{9(1-\nu)}(\beta-\mu)^2 - \frac{1}{2}(1+\nu)\left[-N+\frac{1}{3}(\beta+\mu)\right]^2 \tag{1}$$

where $\mu = \tan\phi$ and $\beta = \tan\psi$ capture the peak friction angle ϕ and the angle of dilation ψ. Stress anisotropy is represented by the parameter $N = \sqrt{2}\sigma_2'/\sqrt{\sigma_1'^2+\sigma_2'^2+\sigma_3'^2}$.

In sediments that have experienced particle-level volume contraction, the stress in the vertical direction is the maximum principal stress and the two horizontal stresses are of the same magnitude. For this stress condition, shear faulting occurs in the strain softening regime in homogenous media; therefore, the formation of shear faults associated with volume contraction will take place in materials that exhibit post-peak strength softening.

A complex pattern of strain localization is anticipated because there is no preferred direction for shear planes (Desrues and Viggiani 2004). Polygonal faults emerge as a kinematically admissible fault morphology, as observed in the field (Cartwright and Dewhurst 1998).

To gain further insight into the potential development of shear strain localization during particle-level contraction, we conducted finite element simulations of a medium subjected to constant vertical stress under zero lateral strain boundary conditions using a Drucker-Prager frictional model with non-associated flow rule (details in Shin et al. 2008). The nucleation of localization was facilitated by creating a correlated random field for volume contraction. Results show diffused strain localization in perfectly plastic media $\phi_{res} = \phi_p$ (Fig. 2-a) and marked shear strain localization when the medium was modeled with post-peak strength softening $\phi_{res} < \phi_p$ (Fig. 2-b).

2.3 Deformation Field

The regions surrounding individual polygonal faults show evidence of local strain accumulations that closely match theoretically predicted near-field strains for blind normal faults (Barnett et al. 1987). Tectonic normal faults are widely considered to follow a simple scaling relationship between maximum displacement and maximum dimension (Cowie and Scholz 1992, Schultz et al. 2006). We collected data for maximum throw δ versus fault height H for 629 polygonal faults: all measurements are bound within $\delta/H = 0.045 \pm 0.016$ (Shin et al. 2009). These are surprisingly large displacements for systems that have not experienced extension in their basal

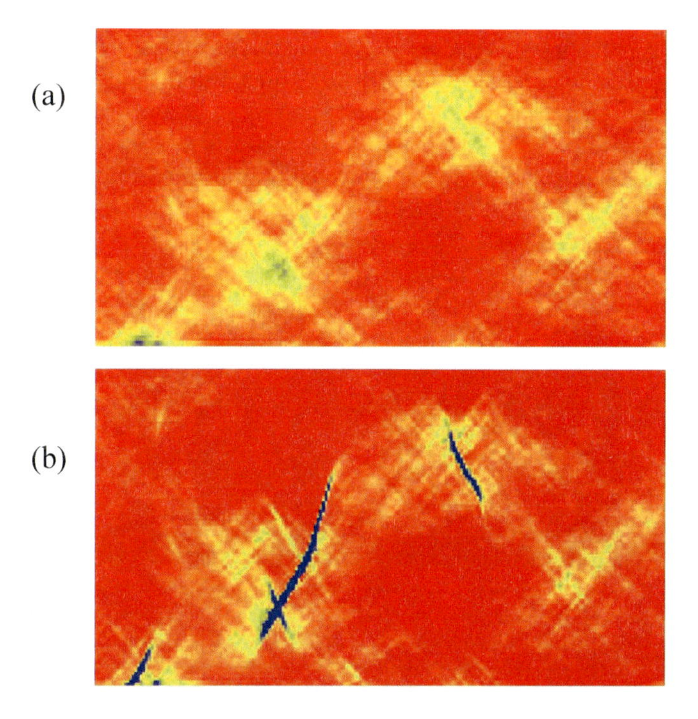

Fig. 2 Finite element simulation - Correlated random field of volume contraction within a frictional material. (a) Diffused deviatoric strain distribution in perfectly plastic medium – without softening $\phi_{res} = \phi_p = 30°$. (b) Shear strain localization is facilitated if there is post peak strain softening ($\phi_p = 30°$, $\phi_{res} = 10°$).

planes, and the δ/H values are much larger than for most tectonic normal faults (as compared to trends in Cowie and Scholz 1992 at an effective stress $\sigma' <40$ MPa).

Let's assume a shear plane at $\beta = 45° + \phi/2$ (Vermeer 1990). The stress redistribution after shear localization leads to a new horizontal stress ratio $k_r = \sigma'_{hr}/\sigma'_z$

$$k_r = \frac{\cos\beta\sin(\beta - \phi_r)}{\sin\phi_r + \cos\beta sin(\beta - \phi_r)} \quad . \tag{2}$$

The horizontal stress change from k_0 to k_r produces a contractive horizontal strain $\varepsilon_h = \Delta\sigma'_h/E$. We assume that the region where strains accumulate around the fault has a parabolic shape, and integrate horizontal strains to obtain the horizontal displacement u_h; finally, we compute the associated vertical displacement to satisfy compatibility, $u_z = u_h \tan\beta$. The maximum normalized throw δ/H predicted by this analysis is

$$\frac{\delta}{H} = \frac{(k_r - k_0)\tan\beta}{2}\,\frac{\sigma'_z}{E} \tag{3}$$

which is a function of the horizontal stress change $(k_r - k_0)$ and inversely proportional to sediment stiffness-to-stress ratio E/σ'_z.

Numerical simulations are conducted using a Drucker-Prager model where the Coulomb-type strength criterion is linearly proportional to the effective confining stress, i.e., no cohesion. Predicted volumetric strains $\varepsilon_v \, q/E$ and normalized throw δ/H are shown in Fig. 3 (Note: the volumetric strain is normalized by the ratio between the vertical load q at the burial depth and the sediment stiffness E).

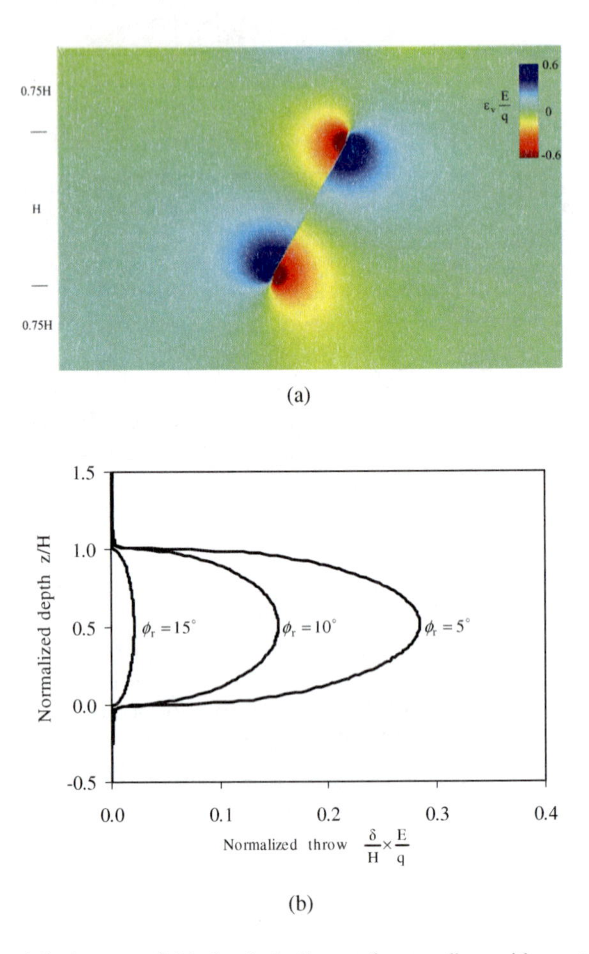

Fig. 3 Strain and displacement field after fault slippage for a medium with constant stiffness with depth. (a) Normalized volumetric strain $\varepsilon_v E/q$ (Contractive is positive - blue). (b) Normalized throw along the fault length for different residual friction angles. Model parameters: sediment friction angles $\phi = 30^\circ$, $\phi_{res} = 5^\circ$, initial state of stress $k_0 = 1 - \sin\phi = 0.5$, fracture orientation $\beta = 45^\circ + \phi/2 = 60^\circ$.

Numerical results (Fig. 3) and analytical predictions (Equation 3) match measured trends for very small values of stiffness $1.5 < E/\sigma_z' < 10$. (As a reference, standard values for sediments that have not experienced dissolution range from $E/\sigma' \approx 40$ at large strains to $E/\sigma' \approx 100$ for small strains – Santamarina, et al.,

2001). Indeed, it can be shown that the normalized equivalent skeletal compressibility E_{eq}/σ' for sediments that experienced dissolution and exhibit high porosity before shear is

$$\frac{k_r - k_a}{\alpha} > \frac{E_{eq}}{\sigma_z'} > \frac{k_r - k_a}{\alpha + 0.3(1-\alpha)C_c/(1+e_0)} \tag{4}$$

resulting in an estimate of $E_{eq}/\sigma' \approx$ 1.5-to-6 in sediments that went through α = 10% dissolution. This low E_{eq}/σ' estimate is compatible with field observations and confirms the role of dissolution as a causal mechanism.

3 Desiccation cracks in saturated fine-grained soils

It is often assumed that soils crack when the tensile stress exceeds the soil tensile strength (Lachenbruch, 1962). However, this mechanistic interpretation fails to recognize the inherent, cohesionless-yet-frictional effective stress dependent behavior of soils. Furthermore, assumptions such as zero effective stress at the crack tip may not necessarily reflect the underlying particle level mechanisms.

3.1 Fundamental Mechanism – Fine Grained Soils

A new hypothetical mechanism for desiccation crack initiation and growth is proposed herein, centered on the air-water interface membrane (Shin and Santamarina 2009b).

The sequence of events summarized in Fig. 4 has been identified from multiple experimental observations. (a) The water surface is above the sediment surface; water evaporates freely and the pore fluid pressure is positive everywhere inside the sediment. (b) Eventually, the water level reaches the sediment surface; at this instant, the pore fluid pressure on the surface is zero. (c) Further evaporation brings the air-water interface membrane against grain surfaces; the membrane resists invading the soil and capillary suction develops in the pore fluid; the effective stress increases and the soil undergoes vertical one dimensional settlement that is equal to the amount of water that evaporates. (d) The increased stiffness of the soil skeleton hinders further consolidation, and the membrane invades the largest pores. (e) Particles displace normal to and away from the air-water interface; pore size increases at the tip of the invading front so that further membrane invasion is favored at the tip signaling crack initiation.

It is important to highlight that the *effective stress remains in compression everywhere in the soil mass including at the tip of the desiccation crack.*

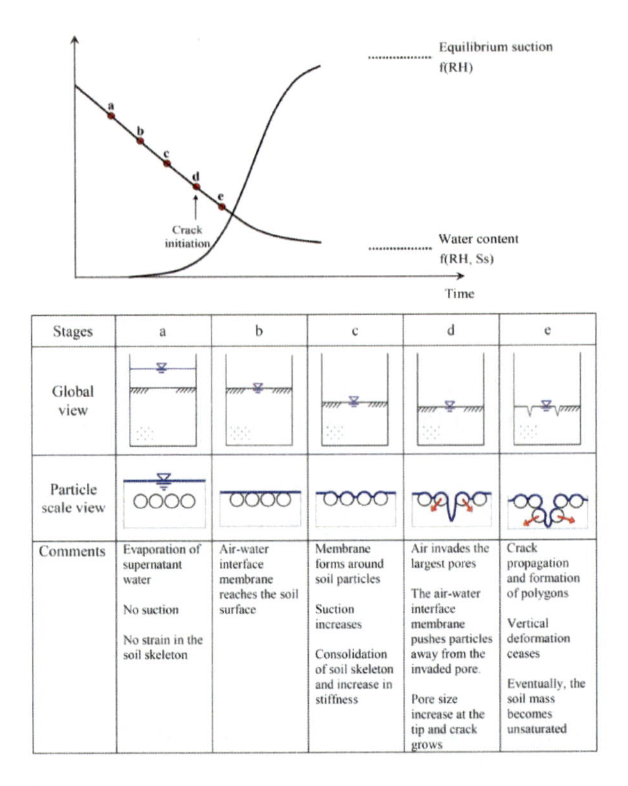

Stages	a	b	c	d	e
Global view					
Particle scale view					
Comments	Evaporation of supernatant water No suction No strain in the soil skeleton	Air-water interface membrane reaches the soil surface	Membrane forms around soil particles Suction increases Consolidation of soil skeleton and increase in stiffness	Air invades the largest pores The air-water interface membrane pushes particles away from the invaded pore. Pore size increase at the tip and crack grows	Crack propagation and formation of polygons Vertical deformation ceases Eventually, the soil mass becomes unsaturated

Fig. 4 Water evaporation, soil desiccation and crack formation - Summary of pore and particle scale processes.

3.2 Initiation – Surface Features

Topographic features and indentations are nucleation sites for desiccation cracks (see also Towner 1988, Zabat et al. 1997, Weinberger 1999). Furthermore, we notice in all our experiments that cracks always start at the bottom of the defect, not at the equator. Let's consider a Gaussian-shaped surface defect in a modified cam clay medium subjected to internal suction. The initial equilibrium condition is at $p' = 0.01$ kPa with homogeneous void ratio $e = 3.92$; suction is increased gradually to reach $p' = 100$ kPa (Fig. 5). The far field sediment experiences 1D consolidation and reaches a void ratio $e = 2.105$. However, the $e - \sigma'$ path at the tip gradually deviates from the $e - \sigma'$ path in the far field, and the void ratio converges to a value higher than anywhere else, i.e., allowing for preferential membrane invasion and crack initiation.

The air entry suction can be estimated from Laplace's equation in terms of pore size d_p or more conveniently in terms of specific surface S_s and void ratio e,

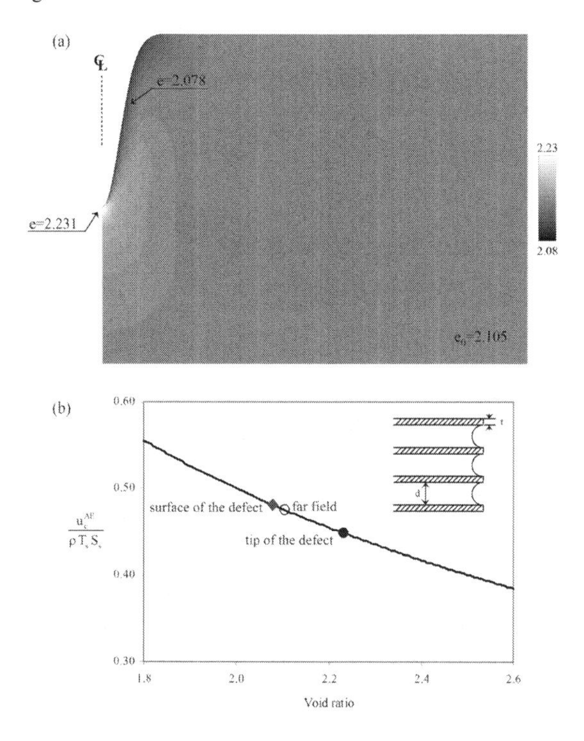

Fig. 5 Surface defects and crack initiation. Local void ratio evolution during desiccation - Finite element analysis. The domain size is 6 mm×6 mm, and the defect depth is 1 mm. (a) Void ratio when suction u = 100 kPa. (b) Required suction for air entry at the tip and on the surface as a function of the local void ratios.

$$u_c^{AE} = \frac{2T_s}{d_p} = \frac{\rho T_s S_s}{e} \tag{5}$$

where ρ [g/cm^3] is the mass density of the mineral that makes the particles and T_s = 0.072 N/m is the surface tension of water.

3.3 Implications

Fracture initiation and propagation based on the invasion of the water-air interfacial membrane can properly explain all known observations related to desiccation crack formation in soils, including (details in Shin and Santamarina 2009b): pore fluid and fabric effects on desiccation crack pattern (due to pore size distribution and air entry); slower crack propagation velocity as the crack approaches the free boundary of a pre-existing crack (due to delayed water migration and a smaller reduction in air-entry value next to the free boundary); right angle junction in crack pattern formation (caused by the rotation of the void ratio expansion contours towards the pre-existing free boundary); and frictional resistance responsible for the apparent tensile strength, yet with failure surface normal to the tensile direction (due to the increase in void ratio and membrane invasion normal to the applied extension).

4 Hydraulic fracture in granular materials

Hydraulic fractures affect a wide range of geosystems, and environmental applications (Massarsch 1978, Andersen et al. 1994, Garagash 2006, Soga et al. 2006). However, a proper understanding is still lacking, in part due to puzzling observations: on one hand, cohesionless-frictional sediments have no tensile strength so the failure mechanism must be different from the tension-dominated hydraulic fracture in brittle solids (Bohloli and de Pater 2006); on the other hand, hydraulic fractures in soils are in opening mode and perpendicular to the minor effective principal stress (e.g. X-ray images in Toshikazu et al. 2002), rather than Coulomb shear planes in the 45° orientation.

In this section we develop a new fundamental understanding of hydraulic fractures in granular materials that is compatible with the effective stress dependent frictional behavior of granular materials.

4.1 Invading and Host Fluids – Particle-Level Forces

Particle-level mechanisms in fluid-driven fractures depend on the miscibility between the invading fluid and the host fluid that saturates the granular medium. Relevant particle level forces in a medium made of grains size d are: skeletal force $F_{sk} = \sigma' d^2$ (associated to effective stress σ'), capillary force $F_c = \pi d T_s$ (when fluids are immiscible), and seepage force $F_s = 3\pi\mu v d$ (when the invading fluid of viscosity μ traverses the pore space with flow velocity v). An opening will form in the soil mass when capillary and/or seepage forces exceed the skeletal force.

4.2 Fracture Initiation and Propagation

The developing of an opening driven by the forced invasion of an immiscible fluid in a soil mass resembles the formation of desiccation cracks: the interfacial membrane invades the sediment at large voids, wedge action causes the increase in pore size and allows for further membrane invasion (Fig. 6a). Numerical simulations confirm these observations (effective stress formulation, modified cam clay model, frictional media without cohesion). Similar to results obtained for the case of desiccation cracks, void ratio changes follow the 1D normal consolidation line in the far field; however, the void ratio at the tip of defects evolves to a value higher than anywhere else creating the proper situation for positive feedback that leads to crack initiation and propagation.

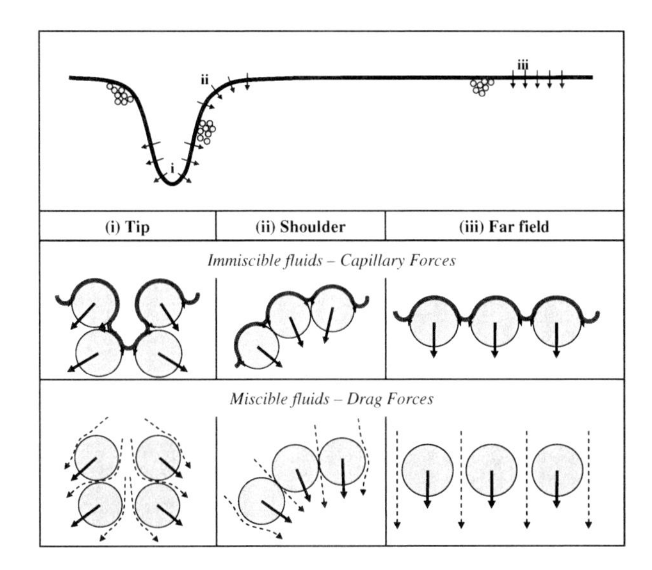

Fig. 6 Void ratio evolution around a surface defect during the forced invasion of (a) an immiscible fluid, and (b) a miscible fluid.

When miscible fluids are involved, the seepage force associated to the pore flow velocity (i.e., the imposed hydraulic gradient) generates the grain forces (i.e., effective stress field) that drive opening mode fracture propagation (Fig. 6b). In this case, the positive-feedback condition at the tip of the fracture involves: pore size opening, decreased energy loss (e.g. Hagen–Poiseuille), increase in fluid pressure, increase in seepage force and further pore opening leading to fracture propagation orthogonal to the acting seepage force.

Hydraulic wedging in both cases, i.e., with or without an interfacial membrane, is facilitated by the initial pore size distribution near the injection point.

4.3 Experimental Evidence

The initiation of hydraulic fractures on planar surfaces is experimentally studied using a high pressure oedometric cell. A Ca-montmorillonite slurry (LL = 97%; PL = 47%; initial water content w = 150%) is placed inside a cylindrical, stainless steel oedometer chamber and allowed to consolidate to 2 kPa. Either a miscible (water) or an immiscible fluid (oil) is placed on top of the sediment and the see-through top cap is assembled. Finally, the upper fluid is pressurized while allowing drainage through the bottom port. The soil surface is photographed using a digital camera operating in time-lapse mode. Images in Fig. 7 show the evolution of hydraulic fractures in both cases. Once again, surface defects play a critical role in fracture initiation. While drainage starts immediately after opening the bottom drainage port, there is a time delay for fracture initiation: fluid-driven separation starts when particle-level forces around surface pores or defects prompt grain separation and "hydraulic wedging". Hence, hydraulic fracturing is intimately linked to pressure diffusion.

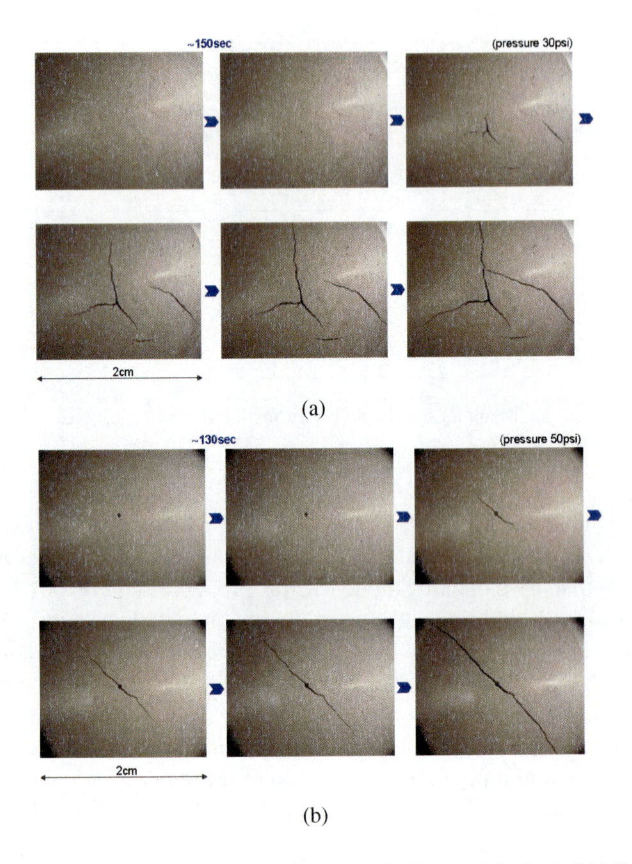

Fig. 7 Hydraulic fracture formation on the soil surface. (a) Immiscible fluid. (b) Miscible fluid.

4.4 Discussion on Pore Size Distribution

Pore size distribution plays a prevalent role on fracture initiation pressure in fluid driven cracks as large pores tend to be invaded/enlarged first. Sediments typically exhibit a log-normal pore size distribution rather than uniform distribution (Garcia-Bengochea, et al., 1979; Juang and Holtz, 1986; Tanaka, et al., 2003), and the standard deviation σ^* in logarithmic scale $\log(d_p/1\mu\text{m})$ is between $0.3 \leq \sigma^* \leq 0.7$. Fracture initiation starts at large pores that are α-standard deviations above the mean pore size $\overline{d}$; this is the characteristic pore size $d_p^* = \overline{d}\ 10^{\alpha\sigma^*}$.

This result can be used to obtain a lower estimate of the capillary pressure at fracture initiation and to obtain the spacing at fracture saturation (see details in Shin 2009). It can be concluded that the pressure at fracture initiation will be lower in soils with wider pore size distributions, at the same global void ratio.

5 Conclusions

Discontinuous planes often develop in soils. These discontinuities affect the mechanical behavior (stiffness and strength) and transport properties of sediments (fluid migration and diffusion). The fundamental understanding of the development of discontinuities in soils must be compatible with their inherent granular nature and effective-stress dependent behavior.

Dissolution is an ubiquitous phenomenon in soil diagenesis. Particle-level volume reduction in uncemented granular media under zero-lateral strain boundary conditions causes a decrease in horizontal stress, leading to internal shear failure conditions and the potential development of shear strain localization. Both the genesis of polygonal fault systems and the magnitude of shear displacements they experienced can be properly explained assuming mineral dissolution during sediment diagenesis.

While dry or drained granular materials fail in shear only (in agreement with the Coulomb criterion), saturated granular materials can couple granular behavior with fluid flow to give rise to opening type failure modes where the fracture walls are kept open by the interaction between the invading fluid and the granular medium.

Skeletal forces (effective stress dependent), capillary forces (when immiscible fluids are involved) and seepage drag-forces (associated with flow velocity) interact at surface defects or large surface pores to give rise to positive feedback mechanisms that promote fracture initiation and propagation: particle displacement around the largest pores enlarge voids at the tip to facilitate further localized invasion. Desiccation cracks are an example of localized invasion of the interfacial membrane between water and air.

Effective stresses remain in compression everywhere throughout the granular medium including at the tip of opening fractures.

Acknowledgements Support for this research was provided by the Goizueta Foundation and the National Science Foundation. These studies evolved from interactions with J. Cartwrigth (polygonal faults), M. Dusseault and L. Germanovich (hydraulic fracture), and Salvador Lazcano (desiccation cracks).

References

1. Andersen KH, Rawlings CG, Lunne TA et al (1994) Estimation of hydraulic fracture pressure in clay. Canadian geotechnical journal 31:817-828.
2. Barnett JAM, Mortimer J, Rippon JH et al (1987) Displacement geometry in the volume containing a single normal fault. American Association of Petroleum Geologists Bulletin 71:925-937.
3. Bohloli B, de Pater CJ (2006) Experimental study on hydraulic fracturing of soft rocks: Influence of fluid rheology and confining stress. Journal of Petroleum Science and Engineering 53:1-12.
4. Burland JB (1990) On the compressibility and shear-strength of natural clays. Géotechnique 40:329-378.
5. Cartwright JA, Dewhurst DN (1998) Layer-bound compaction faults in fine-grained sediments. Bulletin of the Geological Society of America 110:1242-1257.
6. Cowie PA, Scholz CH (1992) Physical explanation for the displacement length relationship of faults using a post-yield fracture-mechanics model. Journal of Structural Geology 14:1133-1148.
7. Desrues J, Viggiani G (2004) Strain localization in sand: an overview of the experimental results obtained in Grenoble using stereophotogrammetry. International Journal for Numerical and Analytical Methods in Geomechanics 28:279-321.
8. Garagash DI (2006) Propagation of a plane-strain hydraulic fracture with a fluid lag: Early-time solution. International Journal of Solids and Structures 43:5811-5835.
9. Garcia-Bengochea I, Lovell CW, Altschaeffl AG (1979) Pore distribution and permeability of silty clays. Journal of the Geotechnical Engineering Division 105:839-856.
10. Jaky J (1944) The coefficient of earth pressure at rest. Journal of the Union of Hungarian Engineers and Architects 355-358..
11. Juang CH, Holtz RD (1986) Fabric pore size distribution, and permeability of sandy soils. Journal of geotechnical engineering ASCE 112:855-868.
12. Lachenbruch AH (1962) Mechanics of thermal contraction cracks and ice-wedge polygons in permafrost. Geological Society of America, New York.
13. Lade PV (2002) Instability, shear banding, and failure in granular materials. International Journal of Solids and Structures 39:3337-57.
14. Massarsch KR (1978) New aspects of soil fracturing in clay. Journal of the Geotechnical Engineering Division ASCE 104:1109-1123.
15. Perrin G, Leblond JB (1993) Rudnicki and Rice analysis of strain localization revisited. Journal of Applied Mechanics-Transactions of the ASME 60:842-846.
16. Rudnicki JW, Rice JR (1975) Conditions for the localization of deformation in pressure-sensitive dilatant materials. Journal of the Mechanics and Physics of Solids 23:371-394.
17. Santamarina JC, Klein KA, Fam MA (2001) Soils and waves : particulate materials behavior, characterization and process monitoring, J. Wiley & Sons, New York.
18. Schultz RA, Okubo CH, Wilkins SJ (2006) Displacement-length scaling relations for faults on the terrestrial planets. Journal of Structural Geology 28:2182-2193.
19. Shin H (2009) Development of discontinuities in granular media. Georgia Institute of Technology, Atlanta.
20. Shin H, Santamarina JC (2009a) Mineral dissolution on ko effects Journal of Geotechnical and Geoenvironmental Engineering ASCE in press.
21. Shin H, Santamarina JC (2009b) Desiccation cracks in saturated fine-grained soils: Particle level phenomena and effective stress analysis under review.
22. Shin H, Santamarina JC, Cartwright JA (2008) Contraction-driven shear failure in compacting uncemented sediments. Geology 36:931-934.
23. Shin H, Santamarina JC, Cartwright JA (2009) Displacement field in contraction driven faults under review.
24. Soga K, Gafar KO, Ng MYA et al (2006) Macro and micro behaviour of soil fracturing. International Symposium on Geomechanics and Geotechnics of Particulate Media - Geomechanics and Geotechnics of Particulate Media, Yamaguchi Japan 421-427.

25. Tanaka H, Shiwakoti DR, Omukai N et al (2003) Pore size distribution of clayey soils measured by mercury intrusion porosimetry and its relation to hydraulic conductivity. Soils and Foundations 43:63-73.
26. Toshikazu H, Yoshiyuki M, Kenichi M et al (2002) Features of crack propagation by hydraulic fracturing in cohesive soil. Experimental study on seepage failure of small earth dams. Transactions of the Japanese Society of Irrigation, Drainage and Reclamation Engineering 219:383-392.
27. Towner GD (1988) The influence of sand- and silt-size particles on the cracking during drying of small clay-dominated aggregates. Journal of Soil Science 39:347-356.
28. Vermeer PA (1990) The orientation of shear bands in biaxial tests. Géotechnique 40:223-236.
29. Weinberger R (1999) Initiation and growth of cracks during desiccation of stratified muddy sediments. Journal of Structural Geology 21:379-386.
30. Zabat M, Vayer-Besançon M, Harba R et al (1997) Surface topography and mechanical properties of smectite films, Springer, Berlin.

Grain Crushing, Pore Collapse and Strain Localization in Porous Sandstone

Teng-fong Wong and Patrick Baud

Abstract In many reservoir engineering and tectonic problems, the ability to predict the occurrence and extent of inelastic deformation and strain localization hinges upon a fundamental understanding of the phenomenology and micromechanics of compaction in reservoir rock. This paper reviews recent research advances on inelastic compaction and strain localization in porous sandstone, with focus on the synthesis of laboratory data, quantitative characterization of damage, fracture mechanics modeling and discrete element simulation. Specific topics reviewed here include: phenomenology of inelastic compaction and failure, micromechanics of grain crushing and pore collapse, bifurcation analysis and discrete element modeling of compaction bands, and stress conditions for the propagation of compaction bands.

1 Introduction

A fundamental understanding of inelastic compaction and strain localization is critical to many problems in reservoir and geotechnical engineering, for which deformation and fluid transport are intimately related. In a reservoir or aquifer, extraction of fluid reduces the pore pressure and thus increases the effective stress, which can impact the stress field and hydromechanical properties, possibly leading to inelastic deformation and failure manifested by phenomena such as surface subsidence, well failure and induced seismicity (Boutéca et al. 1996; Segall 1989; Fredrich et al.

Teng-fong Wong
Department of Geosciences and Department of Mechanical Engineering, State University of New York, Stony Brook, NY 11794-2100, USA, e-mail: Teng-fong.Wong@stonybrook.edu

Patrick Baud
Istitut de Physique du Globe de Strasbourg (UMR 7516 CNRS, Universite de Strasbourg/EOST), Strasbourg, France, e-mail: Patrick.Baud@eost.u-strasbg.fr

2000; Wong et al. 2004). Significant reduction of permeability may also accompany the compaction (Zhu and Wong 1997; Vajdova et al. 2004a).

The ability to predict both the occurrence and extent of inelastic deformation and strain localization hinges upon an understanding of the phenomenology and micromechanics of compaction in porous rock. The focus of this review is on porous sandstone, with emphasis on advances in the past decade since our last review (Wong and Baud 1999). Laboratory observations on the mechanical behavior and failure mode of sandstone will be summarized first. We next review advances in microstructural characterization of damage evolution and numerical simulation of the micromehanics. The last topic is strain localization in the form of compaction band formation, which represents an important breakthrough in the conceptual understanding of the mechanics of compaction in a porous medium.

2 Phenomenology of Inelastic Compaction and Failure

In this review we will adopt the convention that compressive stresses and compactive strains (i.e., shortening and porosity decrease) are positive. The maximum and minimum (compressive) principal stresses are denoted by σ_1 and σ_3, respectively. The pore pressure will be denoted by P_p, and the difference between the confining pressure ($P_c = \sigma_2 = \sigma_3$) and pore pressure will be referred to as the "effective pressure" P_{eff}. Experimental data were all acquired under fully "drained" conditions.

When subjected to an overall compressive loading at room temperature, a porous sandstone may fail by shear localization or by distributed cataclastic flow. Under a relatively low effective pressure, the application of a deviatoric stress field induces a porous sandstone to dilate and ultimately fail by brittle faulting. In contrast, the cataclastic deformation is delocalized when the same sandstone is triaxially compressed under a relatively high effective pressure. This failure mode is usually associated with appreciable porosity decrease while the rock strain hardens (Paterson and Wong 2005).

2.1 Strain hardening and shear-enhanced compaction

This transition between two failure modes are often referred to as the "brittle-ductile transition". Its phenomenology can be illustrated by the complete set of mechanical data for Adamswiller sandstone (Wong et al. 1997) shown in Fig. 1. The top panel shows the differential stress $\sigma_1 - \sigma_3$ versus the axial strain for six conventional triaxial compression experiments at a fixed pore pressure of 10 MPa and with confining pressures maintained at 15, 30, 50, 70, 110 and 160 MPa, respectively. The bottom panel shows porosity decrease versus axial strain for the same samples.

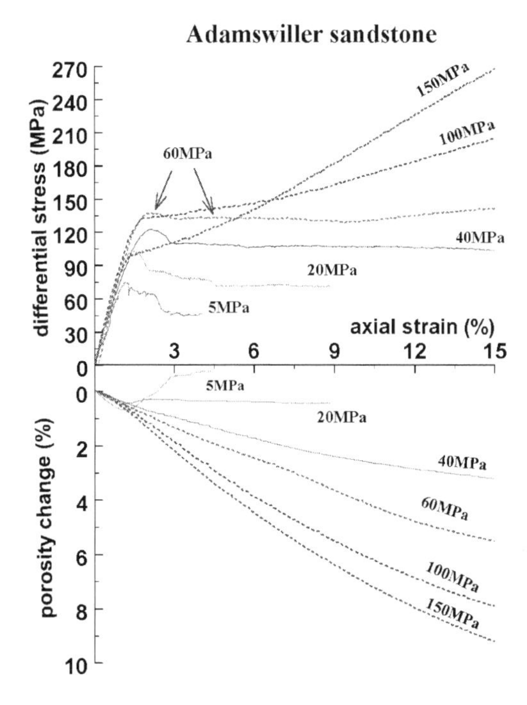

Fig. 1 Mechanical data for saturated Adamswiller sandstone (with initial porosity of 22.6%). Differential stress and porosity change were plotted versus axial strain. Effective pressures were as indicated. The solid curves are for samples which failed by shear localization, and the dashed curves are for samples which failed by cataclastic flow, with delocalized compaction and strain hardening.

The samples deformed at an effective pressures of 5 MPa and 20 MPa are representative of the brittle faulting regime. The differential stress attained a peak, beyond which strain softening was observed and the stress progressively dropped to a residual level. The peak stress shows a positive correlation with effective pressure, which is typical of Mohr-Coulomb type of brittle failure. The porosity initially decreased, but near the peak stress it reversed to an increase indicating dilation of the pore space. The dilation decreased with increasing effective pressure. Visual inspection of post-peak samples confirmed that they failed by shear localization, with a thorough-going shear band cutting across each sample.

The dashed curves (for samples deformed at effective pressures of 60, 100, and 150 MPa) are representative of the ductile regime. The slopes of the differential stress-axial strain curve were nonnegative, and the porosity decreased monotonically with deformation. Shear localization was not evident in these samples undergoing compactive cataclastic flow to an axial strain of up to 30%. The sample deformed at 40 MPa showed a peak stress and strain softening, but the porosity persistently decreased with deformation. This "transitional" mode of failure is manifested by the development of several conjugate shear bands in the sample.

Additional insights are gained by plotting the effective mean stress $(\sigma_1 + 2\sigma_3)/3 - P_p$ versus the porosity change (Fig. 2). For reference, the hydrostat is also shown (as the dashed curves). In a triaxial compression experiment, the nonhydrostatic and hydrostatic loadings are coupled together. If the porosity change is solely controlled by the hydrostatic stresses, then the triaxial data (solid curves) should coincide with the hydrostat (dashed curves) in Fig. 2. Deviations from the hydrostat would imply that additional porosity change was induced by the deviatoric stresses. In the cataclastic flow regime, the triaxial curve for a given effective pressure coincided with the hydrostat up to a critical stress state (indicated by C^* in Fig. 2), beyond which there was an accelerated decrease in porosity in comparison to the hydrostat. At stress levels beyond C^* the deviatoric stress field provided significant contribution to the compactive strain, and this phenomenon is referred to as "shear-enhanced compaction" (Curran and Carroll 1979; Wong et al. 1997). In contrast, the porosity change behavior at effective pressures of 5 and 20 MPa was such that the compaction decelerated in comparison to the hydrostat beyond critical stress states marked as C' (Fig. 2). This implies that at stress levels beyond C' the deviatoric stress field induced the pore space to undergo dilatancy.

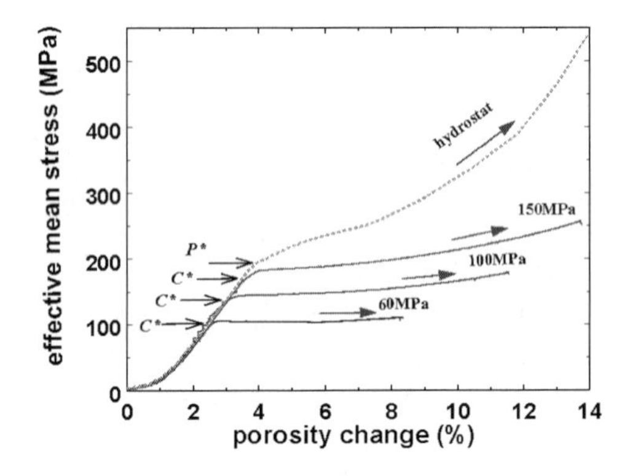

Fig. 2 Shear-enhanced compaction in saturated Adamswiller sandstone. The solid curves show the effective mean stress as a function of porosity change for triaxial compression tests at fixed effective pressures as indicated. For reference the hydrostat is shown as dashed curves. The critical stress states C' and C^* are indicated by the arrows.

2.2 *Onset of compactive yield and evolution of the yield cap*

The critical stress levels for the onset of shear-enhanced compaction varies significantly among sandstones. Data for six sandstone with porosities ranging from 15%

to 35% (from Wong et al. 1997) are shown in Fig. 3 in the stress space, with coordinates given by the effective mean stress P $(=(\sigma_1 + 2\sigma_3)/3 - P_p)$ and the differential stress Q $(= \sigma_1 - \sigma_3)$. For reference, the Mohr-Coulomb peak stresses (for brittle fracture under relatively low effective pressures) are also included as solid symbols. The C^* data (open symbols) map out the initial yield envelopes for the onset of shear-enhanced compaction, that is approximately elliptical and corresponds to a negative correlation between P and Q at the critical stress states. There is an overall trend for such an elliptical yield cap to expand with decreasing porosity or grain size (Wong et al. 1997). However, other microstructural parameters (such as cementation and clay content) may also exert significant control, as illustrated in Fig. 4 by the very different yield caps for three sandstones studied by Tembe et al. (2008) with comparable porosities (in the range of 23-25%) and grain sizes (in the range of 0.16-0.22 mm).

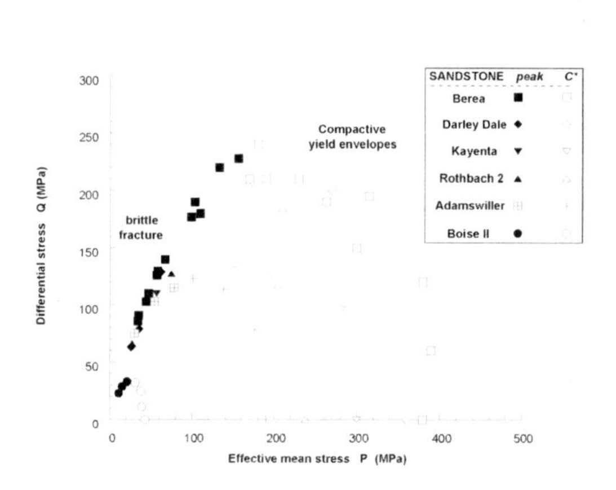

Fig. 3 Stress state C^* at the onset of shear-enhanced compaction (open symbols) and peak stress for brittle fracture (solid symbols) are shown in the P (effective mean stress) and Q (differential stress) space. Note that the compactive yield envelopes have approximately elliptical shapes with negative slopes. Data for six saturated sandstones (with porosities ranging from 15% to 35%) are shown.

These laboratory data on initial yield stresses of porous sandstone are in qualitative agreement with two classes of constitutive models: the critical state model (Schofield and Wroth 1968) which has had a profound impact in soil mechanics, and the cap model which was formulated by DiMaggio and Sandler (1971) with both soil and porous rock in mind. models. To further test the applicability of these plasticity models, Baud et al. (2006) compared sandstone data on strain hardening and evolution of subsequent yield surfaces with model predictions (taking the plastic volumetric strain as the hardening parameter). While many versions of the critical

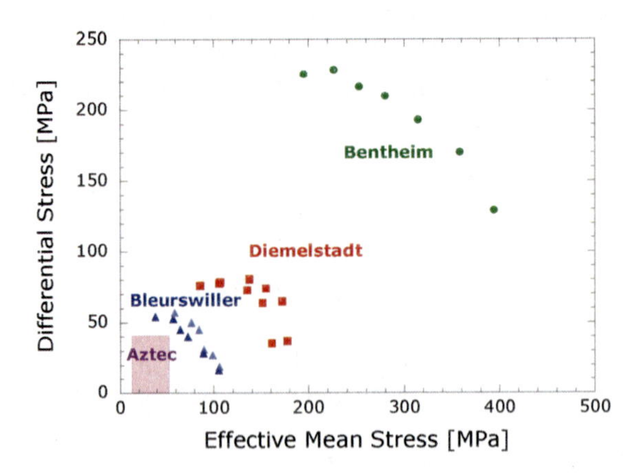

Fig. 4 Yield stresses at the onset of shear-enhanced compaction for saturated samples of three sandstones (with porosities in the range of 21-23%) that develop discrete compaction bands (after Tembe et al. 2008). The shaded rectangle is the inferred paleostress for the Aztec sandstone in Valley of Fire (Sternlof et al., 2005).

state model have been proposed, they focused on the one proposed by Carroll (1991) for porous reservoir rocks.

In their analysis of four sandstones (with porosities ranging from 13% to 23%), Baud et al. (2006) concluded that Bentheim sandstone showed the best agreement with both classes of plasticity model to relatively large strain. Figure 5 highlights such a comparison with DiMaggio and Sandler's (1971) cap model. Up to a plastic volumetric strain of 6%, the initial and subsequent caps maintained an elliptical shape with an almost constant aspect ratio of ~ 1.2. Notwithstanding the agreement in evolution of the yield cap, it should be noted that the normality condition is implicitly assumed in the formulation of the critical state and cap models. Since the shear-enhanced compaction observed in Bentheim sandstone was appreciably more than that predicted for an associative flow rule, Baud et al. (2006) concluded that it would be necessary to incorporate a non-associative model if the objective is to realistically capture the inelastic deformation of these sandstones over a broad range of effective pressures, an issue that will be discussed further in section 4 with regard to compaction localization.

3 Micromechanics of Grain Crushing and Pore Collapse

Acoustic emission (AE) and microstructural observations provide important insights into the spatial and temporal evolution of grain-scale damage during inelastic compaction. Typically the AE activity shows an accelerated increase near the compactive yield stresses and peaked after the sample has undergone a significant amount of in-

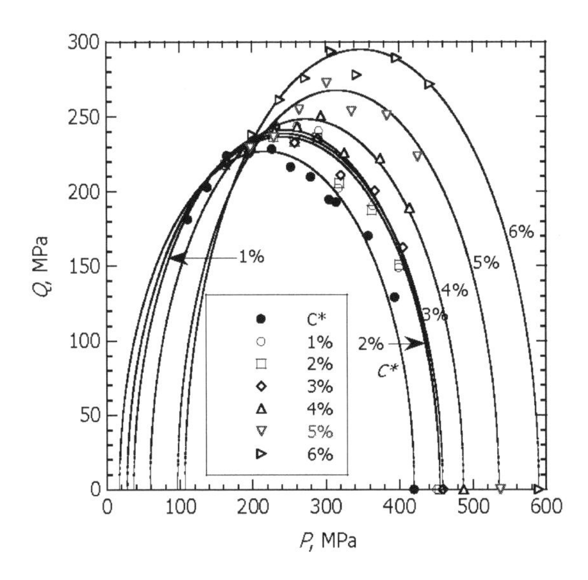

Fig. 5 Comparison between DiMaggio and Sandlers's (1991) cap model and experimental data of Bentheim limestone: Best fit of the elliptical yield surface to the initial yield data C^* and subsequent yield data for increasing plastic volumetric strain up to 6%.

elastic compaction (Zhang et al. 1990b; Wong et al. 1997; Baud et al. 2004). AE activity in rock is a proxy for damage processes that include microcracking, frictional slip, and pore collapse. The evolution of damage associated with these processes in porous sandstones have been characterized by Zhang et al. (1990a), Menéndez et al. (1996), Wu et al. (2000) and Mair et al. (2002). Intragranular cracking dominates the hydrostatic and triaxial compaction processes. The surge in AE activity at the onset of inelastic compaction marks the onset of "Hertzian fractures" that emanate from impinging grain contacts due to local tensile stress concentrations (Figs. 6a and 7a). These intragranular cracks extend across the grains, leading to grain crushing and ultimately pore collapse (Figs. 6b and 7b).

The initiation of grain crushing can be modeled by a Hertzian fracture model (Zhang et al. 1990a), in which the porous rock is idealized as a randomly packed assemblage of spherical particles of several distinct sizes. The macroscopic loading induces tensile stress concentration in the vicinity of the circle of contact between two impinging grains, with the maximum tensile stress attained at the perimeter of the contact area. Application of linear elastic fracture mechanics would predict that the critical pressure P^* decreases with increasing porosity and grain size according to a power law with an exponent of -1.5, in basic agreement with laboratory data for porous siliciclastic rocks and unconsolidated materials (including sand and glass spheres) spanning a broad range of porosities (Wong et al. 1997).

A number of damage mechanics models (e.g. Ricard and Bercovici 2003; Hamiel et al. 2004) have been developed to capture the key mechanical attributes and mi-

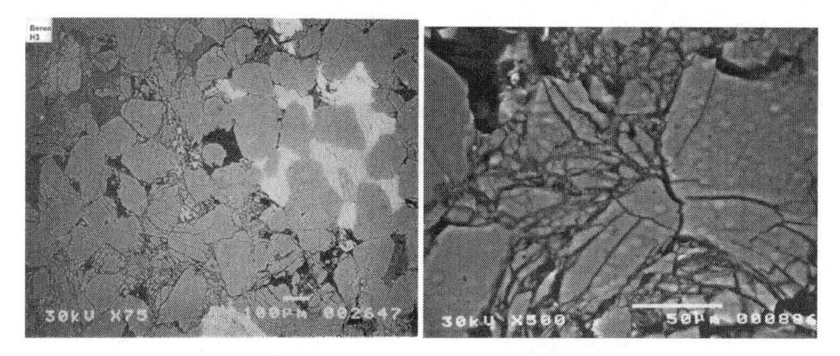

Fig. 6 Scanning electron micrographs of hydrostatically compacted samples of Berea sandstone (Menéndez et al. 1996). **a** Onset of grain crushing. **b** extensile microcracks emanating from an impinging grain contact.

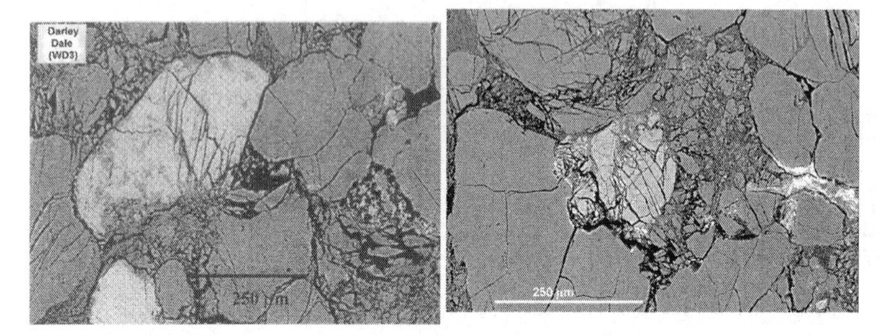

Fig. 7 Scanning electron micrographs of shear compacted samples of Darley Dale sandstone (Wu et al. 2000). **a** Onset of grain crushing. **b** Further development of grain crushing and pore collapse, as observed in the sample WD3. The σ_1 direction is vertical in these micrographs.

crostructural observations. The concept of "breakage mechanics" was proposed recently by Einav (2007) to explicitly account for the evolution of grain crushing. An alternative approach is to incorporate grain crushing into discrete element modeling. To simulate grain crushing and pore collapse, Wang et al. (2008) introduced an intragranular damage mechanism that allows for the shrinkage of an element if one of its normal contact stresses attains a critical value. Their model captures key mechanical attributes (Fig. 1) associated with brittle-ductile transition in porous sandstone, as well as damage evolution associated with compaction bands which will be discussed in the next section.

4 Onset of Compaction Localization and Propagation of Compaction Band

Strain localization is a pervasive phenomenon in the Earth's crust. It occurs over a broad spectrum of length scales, ranging from cm-sized laboratory samples up to crustal fault zones extending over hundreds of kilometers. Conventionally shear and extensile discontinuities in the forms of faults and joints are the only two modes of strain localization considered to be viable in geomaterials. Hence the discovery of a third mode of localization in the form of compaction bands (Mollema and Antonellini 1996; Olsson 1999) is a conceptual breakthrough.

Compaction bands are planar structures that have undergone inelastic compaction predominantly due to shortening in a direction subperpendicular to their planar surface with negligible shear offset. In the field such compaction bands have been observed in the Aztec sandstone from Valley of Fire, Nevada (Hill 1989; Sternlof et al. 2005) and in the Navajo sandstone of the Kaibab Monocline, Utah (Mollema and Antonellini 1996). These tabular zones of compaction typically have thickness on the order of 1-10 mm and trace lengths on the order of 1-10 m.

Laboratory studies have elucidated the mechanics of compaction localization in sandstones with porosities ranging from 13% to 28% (Olsson and Holcomb 2000; Klein et al. 2001; Tembe et al. 2008). Compactant failure at stress states in the transitional regime from brittle faulting to cataclastic flow often develops in a localized manner (Wong et al. 2001), manifested by failure modes associated with a broad spectrum of geometric complexity (Baud et al. 2004). In some porous sandstones, discrete compaction bands may develop over a broad range of stress conditions (Fig. 4). Once localization has initiated, the primary mode of accommodating further compaction is by the initiation of additional discrete bands that remain relatively narrow (Fig. 8). The geometric attributes of discrete compaction bands have been systematically characterized in laboratory deformed samples. Fairly extensive data on the thickness, tortuosity, and spacing are available from quantitative microstructural observations (Baud et al. 2004; Vajdova et al. 2004a; Tembe et al. 2008), X-ray CT-imaging (Louis et al. 2006) and AE locations (Fortin et al. 2006; Townend et al. 2008). A common observation is that the initiation and propagation of compaction bands seem to be controlled by grain-scale heterogeneities. The discrete bands made up of intensely comminuted grains typically span laterally over $\sim$ 2-3 grains, with thicknesses $<$ 1 mm. Significant reductions of permeability have been observed while compaction bands develop in a porous sandstone sample (Holcomb and Olsson 2003; Vajdova et al. 2004a).

4.1 Bifurcation analysis and discrete element modeling of compaction bands

Important insights into the mechanics of strain localization has been gained from the continuum analysis of its inception as a bifurcation in the constitutive response of a porous medium (Rudnicki and Rice 1975; Bésuelle and Rudnicki 2004; Paterson and Wong 2005). Critical conditions for the onset of localization and orientations of the high-angle shear and compaction bands can be derived as functions of the constitutive parameters (Olsson 1999; Issen and Rudnicki 2000). This is illustrated in Fig. 9 using Bentheim sandstone data with reference to the non-associative model

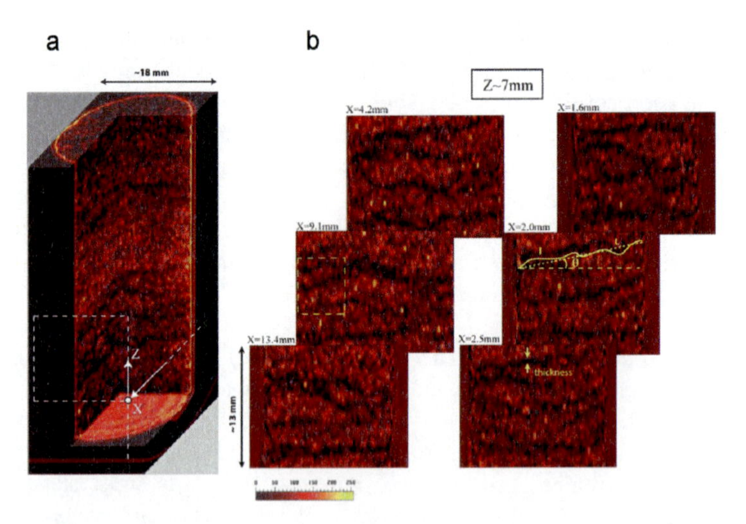

Fig. 8 **a** 3D block reconstruction of a shear-compacted sample of Diemelstadt sandstone using X-ray CT imaging (Louis et al. 2006). The dark tortuous bands represent discrete compaction bands that cut across the sample. **b** Six serial sections for the rectangular area marked in white in (a). These images show the complex geometry of compaction bands in the third dimension.

of Rudnicki and Rice (1975). In this isotropic hardening model three constitutive parameters (in addition to two elastic moduli) were used for characterizing the inelastic and failure behavior: an internal friction parameter μ, a dilatancy factor β, and a hardening modulus h. The flow law is associative only if $\beta = \mu$.

As elaborated by Wong et al. (1997) these constitutive parameters can be extracted from conventional triaxial compression data. In Fig. 9 parameter values for μ and β inferred by Baud et al. (2006) from initial yield data at the onset of shear-enhanced compaction are plotted with the failure mode indicated. In addition values inferred from three experiments for dilatant faulting at low confinement were also included. The data show that as Bentheim sandstone undergoes the brittle-ductile transition, the failure mode evolves from shear band to compaction band as the constitutive parameters β and μ decrease with increasing effective pressure. The dashed lines mark the different modes of strain localization as predicted by bifurcation analyses (Rudnicki and Rice 1975; Olsson 1999; Issen and Rudnicki 2000). Bifurcation analysis provides a useful framework for understanding compaction localization, in the sense that the data agree qualitatively with its prediction that the transition of failure mode from dilation band to shear band to compaction band is associated with decreasing β and μ.

However there is significant discrepancy between the quantitative predictions on the critical parameters at the onset of strain localization and experimental data. The values of β and μ inferred from laboratory data mostly fall in the domain for which the bifurcation analysis predicts shear band formation, and yet the failure mode in many of the Bentheim samples involved discrete compaction bands. The discrepancy is possibly due to the limitation of the Rudnicki-Rice model to comprehen-

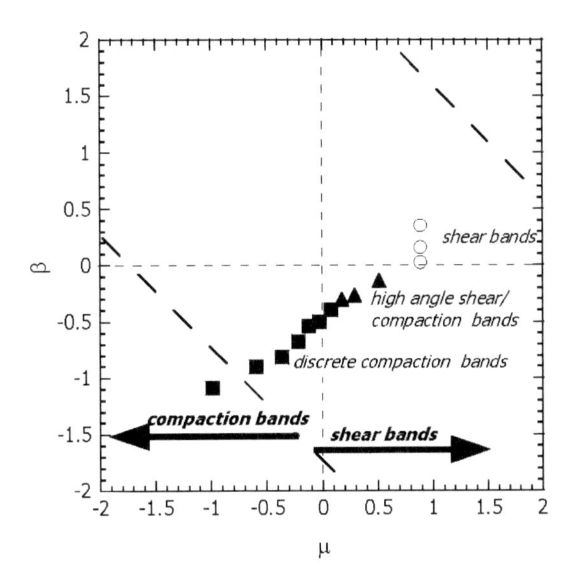

Fig. 9 Constitutive parameters for non-associated model of Rudnicki and Rice (1975): dilatancy factor β as a function of internal friction coefficient μ for Bentheim sandstone, with failure mode as indicated. Failure mode (compaction band, shear band and dilation band) predicted from bifurcation analysis are separated by the diagonal lines.

sively capture the partitioning of several operative damage mechanisms, including stress-induced microcracking, pore collapse and grain crushing. There is also an intrinsic limitation in that the continuum analysis addresses only the onset of constitutive instability in an initially homogeneous material, but not the subsequent propagation behavior of the localization structures or development of their geometric complexities.

A promising approach is the use of discrete element method for simulation of compaction band development (Li and Holt 2002; Marketos and Bolton 2007; Wang et al. 2008; Marketos and Bolton 2009). The simulations of Wang et al. (2008) indicate that the development of discrete compaction bands is promoted in a relatively homogeneous granular aggregate (Fig. 10), while diffuse band growth and distributed cataclastic flow are preferred modes of compaction in a more heterogeneous system. Their results agree with the suggestion of Klein et al. (2001) that discrete compaction band is predominant in the Bentheim sandstone possibly because of its relatively homogeneous mineralogy (with 95% quartz, 3% orthoclase and 3% kaolinite) and well-sorted grain sizes. However, the development of discrete compaction bands has also been documented in the Diemelstadt and Bleurswiller sandstones, at significantly lower stress levels $\sim$ 1/2 and 1/4, respectively of those in Bentheim sandstone (Fig. 4). While they all have comparable porosity and average grain size, unlike the Bentheim sandstone the Diemelstadt sandstone has 26% feldspars and Bleurswiller sandstone has 30% feldspars and 20% of oxide-micas. In spite of the variation in modal compositions, these sandstones seem to have pore spaces that are quite homogeneous, as characterized by Louis et al. (2007) using the coefficient of variation of their X-ray CT data.

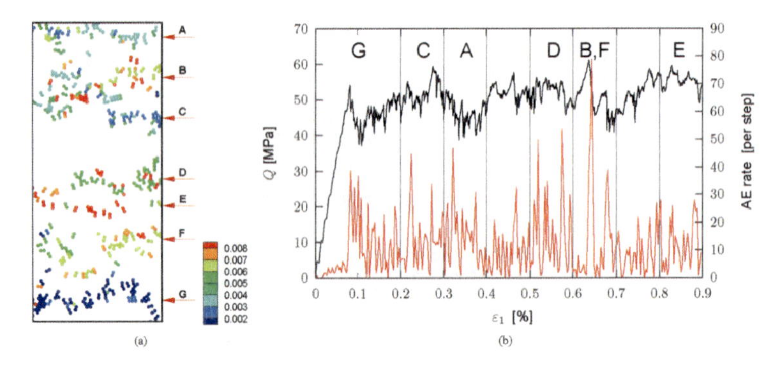

Fig. 10 Discrete element modeling of the sequential development of discrete compaction bands in a well sorted assembly of bonded disks (Wang et al. 2008). At least seven bands can be identified as distinct tabular clusters in **a** the spatial distribution of damage, which can be identified with stress drops and AE surges in **b**, showing the differential stress and AE rate as functions of axial strain. Color codes in figure are in units of axial strain.

4.2 Stress conditions for the propagation of compaction bands

In many respects, the geometric attributes of discrete compaction bands as observed in the laboratory are morphologically similar to compaction bands that have been documented in the field. Notwithstanding these similarities, Sternlof et al. (2005) emphasized that there are at least two apparent discrepancies between the failure modes and structures observed in the laboratory and in the field. First, the stress level for compaction band formation that they inferred in the Valley of Fire is lower than that required in the laboratory for the development of discrete compaction bands (Fig. 4). Second, the dimensions (thickness and length) as well as the damage intensity they measured in the field seem appreciably lower than those in the laboratory deformed samples.

Aspects of these discrepancies were explained in a consistent manner by the recent analysis of Tembe et al. (2008). Synthesizing the field and laboratory data, they observed that the thickness and length of compaction bands obey a quadratic relation, where the thickness scales approximately with the square root of the band length. The mechanical basis for such a scaling relation was provided by Rudnicki's (2007), who investigated the fracture mechanics of the propagation of a combined "anti-crack/anti-dislocation". Tembe et al. (2008) demonstrated that if Rudnick's (2007) model applies to the propagation of compaction bands, then their thickness and length would indeed obey the quadratic relation, and furthermore the stress level involved is predicted to be inversely proportional to compaction band thickness. As illustrated in Fig. 11, the laboratory and field data on stress and compaction band thickness are consistent with this prediction of an "anti-crack/dislocation" model, with values of critical strain energy release rate in the range of 2-80 kJ/m^2, comparable to laboratory estimates of the nominal compaction energy of 6-43 kJ/m^2 inferred from measurements on circumferentially notched samples (Vajdova and Wong 2003; Tembe et al. 2006). Schultz (2009) recently reported field measurements of length and thickness of compaction bands in Navajo Sandstone from Utah that are in basic agreement with this interpretation.

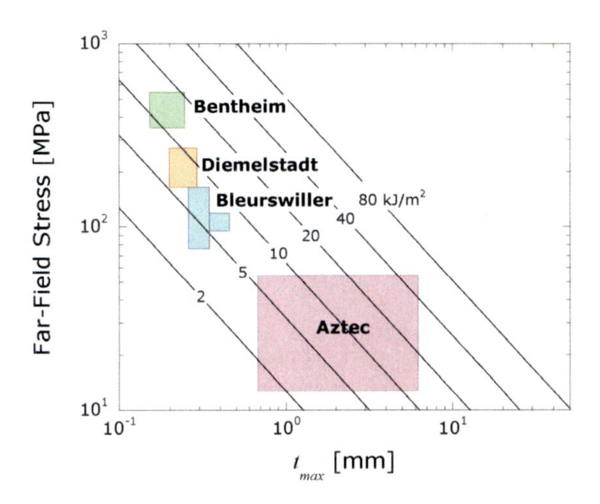

Fig. 11 Rudnicki's (2007) anti-crack/anti-dislocation model predicts the far-field stress would be proportional to the reciprocal of the compaction band half-thickness t_{max}. The solid black lines are the stresses predicted assuming a plastic strain of 0.14 for a range of critical strain energy release rates. The boxes are laboratory and field data.

5 Discussion

In this paper, we have reviewed recent experimental data on the phenomenology of mechanical compaction and strain localization, microstructural characterization of the spatial evolution of damage, as well as the micromechanical modeling of these phenomena in porous sandstone. A methodology integrating these different approaches have provided useful insights into the physics and chemistry of mechanical compaction. On one hand, conventional concepts from plasticity theory in conjunction with bifurcation analysis provide first-order macroscopic description of the inelastic and failure behavior and the intriguing phenomenon of compaction bands. On the other hand, damage mechanics and discrete element modeling provide useful tools for analyzing the micromechanics. The propagation and geometric attributes of compaction bands are further elucidated by the use of fracture mechanics.

We have focused our review on sandstone. However, it should be noted that significant advances have recently been made in the fundamental understanding of mechanical compaction of carbonate rocks, which represent more than half of the Earth's oil reserves. As noted by Vajdova et al. (2004b), inelastic compaction and cataclastic flow in a porous carbonate rock are qualitatively similar to a siliciclastic rock such as sandstone. Nevertheless, there are fundamental differences in the micromechanics. As elaborated here for sandstone, inelastic compaction derives primarily from grain crushing initiated by the stress concentrations at grain contacts. In contrast, inelastic compaction in limestone is typically associated with cataclastic pore collapse, that seems to initiate from stress concentrations at the periphery of the relatively equant pores. With the development of inelastic compaction, crystal plasticity processes (such as twinning and possible dislocation slip) can also be activated. Unfortunately it is beyond the scope of the present paper to review this important topic.

Acknowledgements Emmanuelle Klein, Laurent Louis, Sheryl Tembe, Veronika Vajdova, Baoshan Wang and Wei Zhu have contributed to the research reviewed here. We have also benefited from discussions with Yong Chen, Christian David, Joanne Fredrich, Yves Guéguen, Kathleen Issen, Philip Meredith, John Rudnicki and Wenlu Zhu. This research was partially supported by the Office of Basic Energy Sciences, Department of Energy under grant DE-FG02-99ER14996.

References

1. Baud P, Klein E, Wong T-f (2004) Compaction localization in porous sandstones: spatial evolution of damage and acoustic emission activity. J Struct Geol 26: 603-624
2. Baud P, Vajdova V, Wong T-f (2006) Shear-enhanced compaction and strain localization: Inelastic deformation and constitutive modeling of four porous sandstones. J Geophys Res 111: B12401, doi: 10.1029/2005JB004101
3. Bésuelle P, Rudnicki JW (2004) Localization: shear bands and compaction bands. In: Guéguen Y, Boutéca M (ed) Mechanics of Fluid-Saturated Rocks, Elsevier Academic Press, Amsterdam
4. Boutéca M, Sarda J-P, Schneider F (1996) Subsidence induced by the production of fluids. Rev Inst Franc Petr 51: 349-379
5. Carroll MM (1991) A critical state plasticity theory for porous reservoir rock. In: Massoudi M, Rajagopal KR (ed) Recent Advances in Mechanics of Structured Continua, ASME AMD Vol.117, Book No. G00617
6. Curran JH, Carroll MM (1979) Shear stress enhancement of void compaction. J Geophys Res 84: 1105-1112
7. DiMaggio FL, Sandler IS (1971) Material model for granular soils. J Eng Mech Div, ASCE 97: 935-950
8. Einav I (2007) Breakage mechanics - Part I: Theory. J Mech Phys Solids 55: 1274-1297
9. Fortin J, Stanchits S, Dresen G, Guéguen Y (2006) Acoustic emission and velocities associated with the formation of compaction bands in sandstone. J Geophys Res 111: B10203, doi:10.1029/2005JB003854

10. Fredrich JT, Deitrick GL, Arguello JG, deRouffignac EP (2000) Geomechanical modeling of reservoir compaction, surface subsidence, and casing damage at the Belridge diatomite field. SPE Res Eval & Eng 3: 348-359
11. Hamiel Y, Lyakhovsky V, Agnon A (2004) Coupled evolution of damage and porosity in poroelastic media: theory and applications to deformation of porous rocks. Geophys J Int 156: 701-713
12. Hill RE (1989) Analysis of deformation bands in the Valley of Fire State Park, Nevada. M.S. Thesis, University of Nevada, Las Vegas
13. Holcomb DJ, Olsson WA (2003) Compaction localization and fluid flow. J Geophys Res 108: 2290, doi:10.1029/2001JB000813
14. Issen KA, Rudnicki JW (2000) Conditions for compaction bands in porous rock. J Geophys Res 105: 21529-21536
15. Klein E, Baud P, Reuschle T, Wong T-f (2001) Mechanical behaviour and failure mode of Bentheim sandstone under triaxial compression. Phys Chem Earth (A) 26: 21-25
16. Li L, Holt RM (2002) Particle scale reservoir mechanics. Oil & Gas Science and Technology - Rev IFP 57: 525-538
17. Louis L, Baud P, Wong T-f (2007) Characterization of pore-space heterogeneity in sandstone by X-ray computed tomography. In: David C, Ravalec M (ed) Rock Physics and Geomechanics in the Study of Reservoirs and Repositories, Geological Society of London Special Publication 284
18. Louis L, Wong T-f, Baud P, Tembe S (2006) Imaging strain localization by X-ray computed tomography: discrete compaction bands in Diemelstadt sandstone. J Struct Geol 28: 762-775
19. Mair K, Elphick SC, Main IG (2002) Influence of confining pressure on the mechanical and structural evolution of laboratory deformation band. Geophys Res Lett 29: 10.1029/2001GL013964
20. Marketos G, Bolton MD (2007) Quantifying the extent of crushing in granular materials: a probability-based predictive method. J Mech Phys Solids 55: 2142-2156
21. Marketos G, Bolton MD (2009) Compaction bands simulated is discrete element models. J Struct Geol 31: 479-490
22. Menéndez B, Zhu W, Wong T-f (1996) Micromechanics of brittle faulting and cataclastic flow in Berea sandstone. J Struct Geol 18: 1-16
23. Mollema PN, Antonellini MA (1996) Compaction bands: A structural analog for anti-mode I cracks in aeolian sandstone. Tectonophysics 267: 209-228
24. Olsson WA (1999) Theoretical and experimental investigation of compaction bands in porous rock. J Geophys Res 104: 7219-7228
25. Olsson WA, Holcomb DJ (2000) Compaction localization in porous rock. Geophys Res Lett 27: 3537-3540
26. Paterson MS, Wong T-f (2005) Experimental Rock Deformation - The Brittle Field, 2nd Edition. Spinger-Verlag, New York
27. Ricard Y, Bercovici D (2003) Two-phase damage theory and crustal rock failure: the theoretical "void" limit, and the prediction of experimental data. Geophys J Int 155: 1057-1064
28. Rudnicki JW (2007) Models for compaction band propagation. In: David C, Ravalec M (ed) Rock Physics and Geomechanics in the Study of Reservoirs and Repositories, Geological Society of London Special Publication 284
29. Rudnicki JW, Rice JR (1975) Conditions for the localization of deformation in pressure sensitive dilatant materials. J Mech Phys Solids 23: 371-394
30. Segall P (1989) Earthquakes triggered by fluid extraction. Geology 17: 942-946
31. Schofield AN, Wroth CP (1968) Critical State Soil Mechanics. McGraw Hill, New York
32. Schultz RA (2009) Scaling and paleodepth of compaction bands, Nevada and Utah. J Geophys Res 114: B03407, doi:10.1029/2008JB005876
33. Sternlof KR, Rudnicki JW, Pollard DD (2005) Anticrack-inclusion model for compaction bands in sandstone. J Geophys Res 110: B11403, doi:10.1029/2005JB003764
34. Tembe S, Baud P, Wong T-f (2008) Stress conditions for the propagation of discrete compaction bands in porous sandstone. J Geophys Res 113: B09409, doi:10.1029/2007JB005439

35. Tembe S, Vajdova V, Wong T-f, Zhu W (2006) Initiation and propagation of strain localization in circumferentially notched samples of two porous sandstones. J Geophys Res 111: B02409, doi:10.1029/2005JB003611
36. Townend E et al. (2008) Imaging compaction band propagation in Diemelstadt sandstone using acoustic emission locations. Geophys Res Lett 35: L15301, doi:10.1029/2008GL034723
37. Vajdova V, Baud P, Wong T-f (2004a) Permeability evolution during localized deformation in Bentheim sandstone. J Geophys Res 109: B10406, doi:10.1029/2003JB002942
38. Vajdova V, Baud P, Wong T-f (2004b) Compaction, dilatancy and failure in porous carbonate rocks. J Geophys Res 109: B05204, doi:10.1029/2003JB002508
39. Vajdova V, Wong T-f (2003) Incremental propagation of discrete compaction bands: Acoustic emission and microstructural observations on circumferentially notched samples of Bentheim sandstone. Geophys Res Lett 30 (14): 1775, doi:10.1029/2003GL017750
40. Wang B, Chen Y, Wong T-f (2008) A discrete element model for the development of compaction localization in granular rock. J Geophys Res 113: B03202, doi:10.1029/2006JB004501
41. Wong T-f, Baud P (1999) Mechanical compaction of porous sandstone. Oil & Gas Science and Technology - Rev IFP 54: 715-727
42. Wong T-f, Baud P, Klein E (2001) Localized failure modes in a compactant porous rock. Geophys Res Lett 28: 2521-2524
43. Wong T-f, David C, Menéndez B (2004) Mechanical compaction. In: Guéguen Y, Boutéca M (ed) Mechanics of Fluid-Saturated Rocks, Elsevier Academic Press, Amsterdam
44. Wong T-f, David C, Zhu W (1997) The transition from brittle faulting to cataclastic flow in porous sandstones: Mechanical deformation. J Geophys Res 102: 3009-3025
45. Wu XY, Baud P, Wong T-f (2000) Micromechanics of compressive failure and spatial evolution of anisotropic damage in Darley Dale sandstone. Int J Rock Mech Min Sci 37: 143-160
46. Zhang J, Wong T-f, Davis DM (1990a) Micromechanics of pressure-induced grain crushing in porous rocks. J Geophys Res 95: 341-352
47. Zhang J, Wong T-f, Yanagidani T, Davis DM (1990b) Pressure-induced microcracking and grain crushing in Berea and Boise sandstones: acoustic emission and quantitative microscopy measurements. Mech Mater 9: 1-15
48. Zhu W, Wong T-f (1997) The transition from brittle faulting to cataclastic flow: Permeability evolution. J Geophys Res 102: 3027-3041

Long term behaviour and size effects of coarse granular media

E.E. Alonso, L. Oldecop, and N.M. Pinyol

Abstract Deformation mechanisms of rockfill are explained by particle breakage and subsequent re-arrangement of the granular structure. Breakage is the result of crack propagation within particles. The crack propagation velocity is controlled by crack geometry, stress intensity and Relative Humidity (RH). RH controlled compression and triaxial tests on compacted gravels led to the formulation of a plastic hardening model which incorporates basic features of the underlying basic mechanisms. Hydro-mechanical interactions in field applications are illustrated through the analysis of the response of a 40 m high rockfill embankment subjected to rainfall action. The paper concludes in an analysis of scale effects.

1 Introduction

In series of experimental investigations in the 60's and 70's the basic deformation mechanism of rockfill was identified as the breakage of the particles and subsequent rearrangement of the granular structure. Indices to quantify the amount of particles breakage were proposed by Marsal (1973) and Hardin (1985). Large scale triaxial tests reported by Fumagalli (1969), Marachi *et al.* (1972), Marsal (1973), Charles and Watts (1980) showed that the strength envelope was curved, particularly at low stress levels. Nonlinear strength envelopes were proposed. For instance, De Mello (1977) suggested,

$$\tau_f = A(\sigma)^b \tag{1}$$

E.E. Alonso
Departament of Geotechnical Engineering and Geosciences, UPC, Spain , e-mail: eduardo.alonso@upc.edu

L. Oldecop
Universidad Nacional de San Juan, Argentina

N.M. Pinyol
Departament of Geotechnical Engineering and Geosciences, UPC, Spain

where A and b are empirical coefficients. This nonlinearity was attributed to particle breakage during compression and shearing stages.

In parallel with field observations in rockfill dams it was also found that flooding rockfill specimens subjected to one-dimensional compression lead to a sudden settlement (collapse) attributed also to the breakage of particles due to rock weakening induced by wetting (Terzaghi, 1960). Well known experiments were published by Sowers *et al.* (1965), Nobari and Duncan (1973) and Marsal (1973). Wetting tests performed during triaxial testing on large diameter specimens (Veiga Pinto 1983; Naylor *et al.*, 1986) also identified the partial collapse during the combined effect of mean and deviatoric stress and the reduction of strength of flooded samples if compared with the "dry" ones. Figure 1 shows the strength envelope of slate gravel (Pancrudo slate) tested in a large diameter triaxial cell under relative humidity control. Increasing the relative humidity of the specimen led to a progressive reduction of the shear strength.

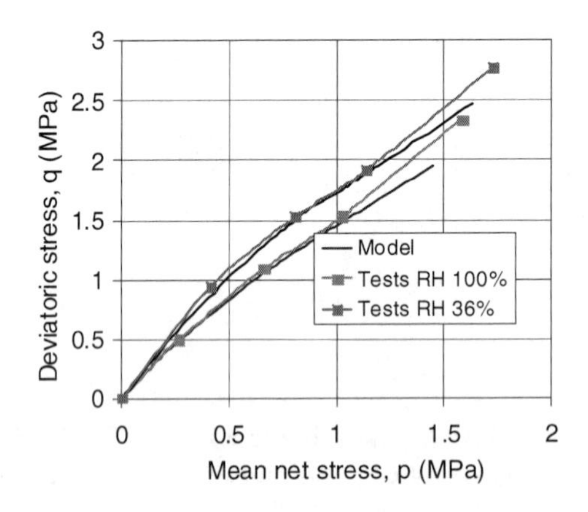

Fig. 1 Strength envelopes of Pancrudo slate for RH= 36% and RH = 100% (Chávez 2004).

The envelopes in Figure 1 were fitted with Equation (1). Grain size distributions at the end of the triaxial tests performed at different RH's (Fig. 2) indicated that particle breakage (measured by the Hardin index) increased with RH. In the figure the breakage index is plotted against the triaxial confining stress. The breakage index for $\sigma_3 = 0$ indicates the amount of breakage induced by compaction (an energy equivalent to Normal Proctor was used). Since particle breakage is the dominant issue in rockfill behaviour, this phenomenon will be examined now in more detail.

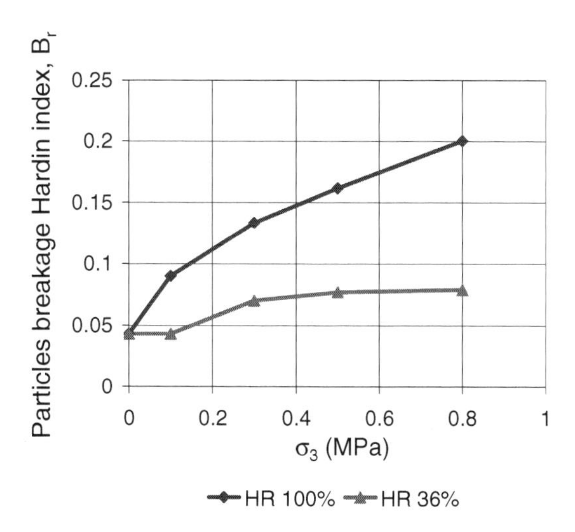

Fig. 2 Hardin particle breakage determined in RH controlled triaxial tests on Pancrudo slate. Chávez, 2004.

2 Particle Breakage

Figure 3a is a photograph of a specimen of hard crushed sandstone gravel ready to be tested in 30 cm diameter oedometer cell with RH control (Oldecop and Alonso 2000). Gravel particles were quite uniform in size and ranged between 20 and 30 mm. Specimens (Fig. 3a) were not compacted initially. They were subjected to a cycle of loading up to 2.3 MPa and then unloading. A detail of the specimen after testing is shown in Figure 3b. Breakage is either concentrated at the grain to grain contacts or it divides the gravel into pieces of significant size. Quite often the failure plane crosses diametrically the initial grain. In other cases (Fig. 3c) a more complex fracturing is found.

Granular assemblies have been intensely analyzed by means of the Distinct Element Method since the initial work of Cundall and Strack (1979). They found that the externally applied stress to a grain assembly is distributed among a number of heavily loaded chains of particles. The particles in those chains receive concentrated loads at some contacts. The remaining particles within the mass are only slightly loaded and they contribute to stabilize the main loading chains. It is therefore reasonable to start the discussion on particle breakage by considering a single particle, diametrically loaded, as shown in Figure 4. The observed grain fracturing in tests suggests that particles break because of the propagation of cracks. In fact, the classical Griffith theory to explain the observed strength of rock specimens relies on a fracture mechanics concept, which attaches a fundamental role to the size of an initial defect or discontinuity. Let's assume then that the particle, idealized as a disk of diameter D and width B has a central crack of size 2a aligned with the two opposite concentrated forces. The particle loading is similar to the arrangement of a classical Brazilian test. A uniform tensile stress, normal to the crack, acts on the vertical plane.

A "mode I" type of crack propagation will control the evolution of crack length. Crack propagation in linear elastic mechanics is associated with the concepts of material toughness and stress intensity factor. The latter is defined as:

$$K_i = \beta_i \sigma_i^* \sqrt{\pi a_i} \tag{2}$$

where a_i is the half length of crack, β_i is a dimensionless factor which depends on (a_i/B), σ_i^* is the stress which would act across the plane of the crack if the particle is not cracked. The stress σ_i^* is related to the overall stress acting on a "specimen" of granular material. Therefore σ_i^* increases with the applied stress to the granular medium.

When K_i, (i stands for a given crack) approaches a material constant, K_c (units: MPa m$^{1/2}$), known as toughness, the crack i characterized by $2a$, propagates at a certain (small) velocity for values $K_i < K_c$. This is the so-called subcritical crack propagation. Crack growth rates reported in several experi ments are collected in terms of (K/K_c) in Figure 5. The experimental data may be approximated by a simple exponential relationship:

$$V = V_0 (K/K_c)^n \tag{3}$$

which is a dimensionless version of the Charles (1958) proposal.

V_0 and n are model parameters. Note that a velocity V_0 = 01 m/s is suggested in Figure 5 for $K = K_c$.

The exponent n changes with the prevailing suction. The data for basalt, granite and marble given in Figure 5 is used to plot, in Figure 6, an approximate relationship between the exponent n and the RH. Date on synthetic quartz and glass suggests that n increases fast when the RH reduces to low values. In other words, under conditions of extreme dryness, crack propagation velocity tends to an extremely low value.

The crack propagation velocity V_i, can be expressed, in view of Equations (2) and (3):

$$V_i = \frac{da_i}{dt} = A_i a^{n/2} \tag{4}$$

where

$$A_i = V_0 \left(\frac{\beta_i \sigma_i^* \sqrt{\pi}}{K_c} \right) \quad . \tag{5}$$

Equation (4) can be integrated if A_i is assumed to be constant, an approximation which is justified in Oldecop and Alonso (2007). Since the values of n are large (20-200) the time for breakage of the particle is given by:

$$t_i^b = \frac{2}{n-2} \frac{a_{0i}}{V_{0i}} \tag{6}$$

where

Fig. 3 a) Specimen of hard sandstone crushed gravel in oedometer cell. b) Grain breakage after testing. c) Detail of broken particle.

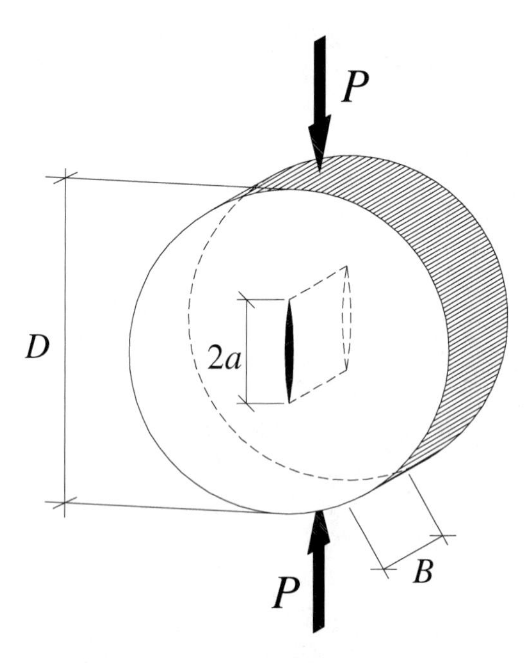

Fig. 4 A disk model for a cracked particle (Oldecop and Alonso 2007)

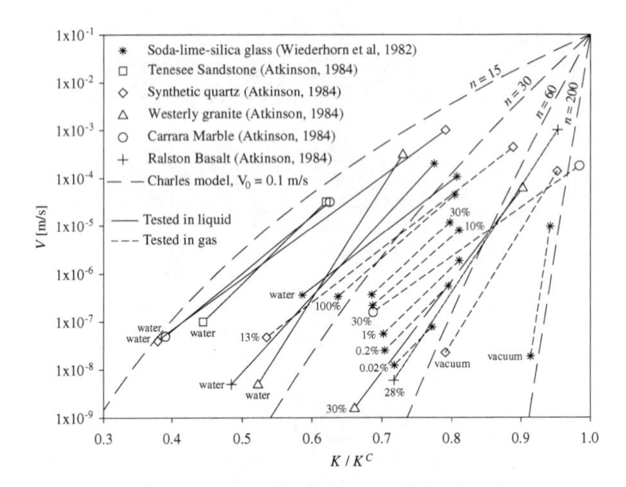

Fig. 5 Crack propagation rates for different rocks, quartz and glass. The testing condition immersed in water, imposed RH or vacuum) is shown next to each curve. Also indicated are curves of Charles model. (Oldecop and Alonso 2007)

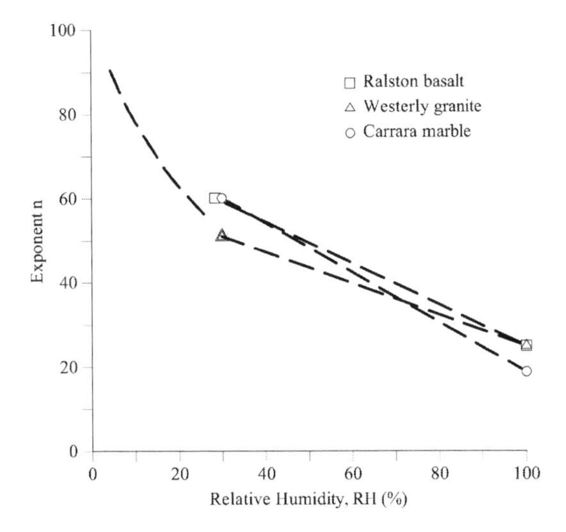

Fig. 6 Relationship between exponent n of Charles law and Relative Humidity.

$$V_{0i} = V_0 \left(\beta_i \sigma_i^* \sqrt{\pi a_{0i}} / K_c\right)^n \tag{7}$$

is the propagation velocity of a crack of length a_{0i}.

The relationship 6 has been plotted in Figure 7 for the following set of values: D = 500 mm (this is a relatively large boulder), V_0 = 0.1 m/s (taken from the plot in Fig. 5), σ = 7 MPa and K_c = 1 MPa·m$^{0.5}$. The plot shows the breakage time for different half lengths of initial defects and for varying relative humidity. The relationship between n and relative humidity has been taken from Figure 6. Note the "compressed" logarithmic scale of the time axis. The plot shows what is to be expected from a wetting episode (increasing RH): a very significant reduction (several orders of magnitude) in the breakage time. Breakage times in the order years may reduce to a few seconds. This is equivalent in practice to a sudden breakage and, if the entire granular body is considered, to a sudden deformation (typically a volumetric compression or collapse).

For a given rock type (characterized by K_c and to a certain extent by a_{0i}) and size of the particle, the breakage time is also very sensitive to the value of the applied stress (because of the high value of exponent n in Eq. 7). Therefore, an increase in σ will result in the immediate rupture of some of the grains and therefore in an immediate deformation even if the initial granular structure is kinematically "locked".

The final point is that stressed fissures are always increasing in size, even if the rate is very low. It implies, at the scale of a rockfill, that creep is always present.

Since RH plays such a significant role in particle breakage the next natural step is to perform experiments with suction control (total suction and RH are related through the psychrometric relationship).

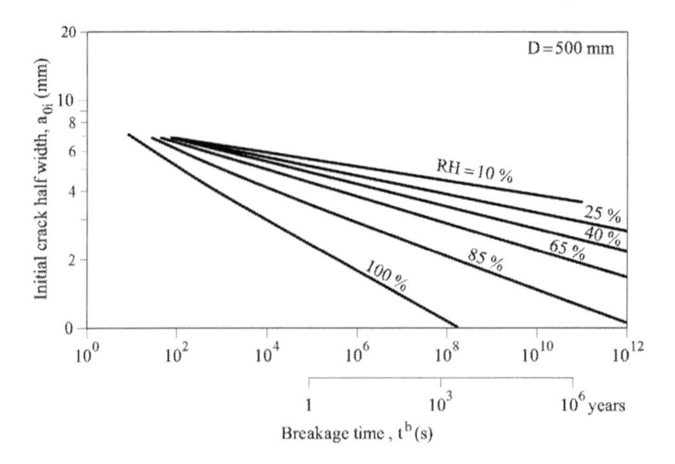

Fig. 7 Breakage times for varying initial crack lengths and prevailing suction.

3 Suction controlled experiments on rockfill

Oldecop and Alonso (2001, 2007) have reported the results of large diameter (30 m) oedometer tests in a specially built cell described in Oldecop and Alonso (2005). Some results are reproduced in Figure 8.

The tested material was a uniform Pancrudo slate gravel compacted at Normal Proctor energy.

The strain time records for two values of RH (50% and 100%) and varying vertical applied stress are consistent with the deformation framework offered by the propagation of cracks in particles. Every load increment leads to an acceleration of the deformation phenomena. After a few minutes the strain log-time curves exhibit steady state creep behaviour. The slope of these curves,

$$\lambda^t = \frac{d\varepsilon}{d(\ln t)} \tag{8}$$

increases with applied stress (and it eventually becomes constant). Reducing the suction (wetting) leads also to an increased creep rate. This is shown in Figure 9. For a limited range of stress (say 0-15 MPa) the data in Figure 9 show that the creep coefficient may be described by a relationship,

$$\lambda^t = \mu p \left(1 - \beta \ln(\frac{s + p_{atm}}{p_{atm}})\right) \tag{9}$$

which shows the dependence of λ^t on confining stress and suction s. p_{atm} is the atmospheric pressure and (μ, β) are model parameters.

If deformations are selected at given time (an arbitrary decision) stress deformation plots may be built for any stress suction path applied to the sample. This is the case of Figure 10 which shows the compressibility of compacted gravel of

Pancrudo slate for an extended range of applied suction and vertical stress. The plot shows the increase in compressibility as RH increases, the trend of samples wetted to end in the saturated compression line and the linear strain-stress (natural scale) behaviour in a relatively wide strain interval (0-8%). In addition, for stress smaller than a given threshold value (around 0.2 MPa), the compression behaviour seems independent of applied suction (i.e. independent of water content). These features were identified and interpreted in Oldecop and Alonso (2006). They proposed an elastoplastic model to describe this behaviour which was consistent with observations and with the underlying fracture propagation framework. Compression curves in Figure 10 are similar to the compression behaviour of a "regular" unsaturated soil. There are similarities in the behaviour of an unsaturated soil and a rockfill but also significant differences pointed out by Alonso (2006). One of the important differences is the existence, in the case of rockfill, of stress and suction states in which water content (or suction) changes do not induce any straining.

This is shown in Figure 11 which provides the Loading Collapse yield curves for the Pancrudo rockfill.

Note that (total) suction in the case of rockfill does not have any direct capillary or mechanical effect on the granular aggregate. The capability of the compression model to simulate loading, unloading, wetting and reloading behaviour is shown in Figure 12.

Chávez (2004) developed a large diameter, RH control triaxial apparatus and conducted a program of tests on Pancrudo slate. In this way the previous information, based in one dimensional compression tests was completed. Some significant results are given here.

The shape of the yield surface (for a saturated material) was determined by an "unloading-probing" procedure described by Poorooshasb *et al.* (1964) (Fig. 13) A distinct "cap" was found. Its shape and the plastics strain increment vectors suggest that rockfill hardening is due to volumetric as well as to deviatoric plastic components. The material is not associated, especially in the dilatant regime in the vicinity of the deviatoric limiting conditions.

Chávez and Alonso (2004) developed an elastoplastic model for rockfill which was largely based on the previous compressibility developments to describe the isotropic behaviour. The model was based on a number of experimental observations:

- Critical state conditions were accepted at the end of tests. Suction contributes to maintaining higher void ratios and stress ratios for a given confining stress.
- Limiting deviatoric states, also controlled by suction, were nonlinear.
- Two yield loci were proposed: a "deviatoric" one (q/p = constant) and an "isotropic" or "cap" one (Fig. 14).
- The deviatoric behaviour was described by a hyperbolic hardening rule which is able to model softening. Hardening was described by an effective plastic work and a plastic deviatoric strain. The effective plastic work was a procedure to take into account the rockfill degradation due to particle breakage.
- The plastic potential was based on a modified Rowe's dilatancy rule.

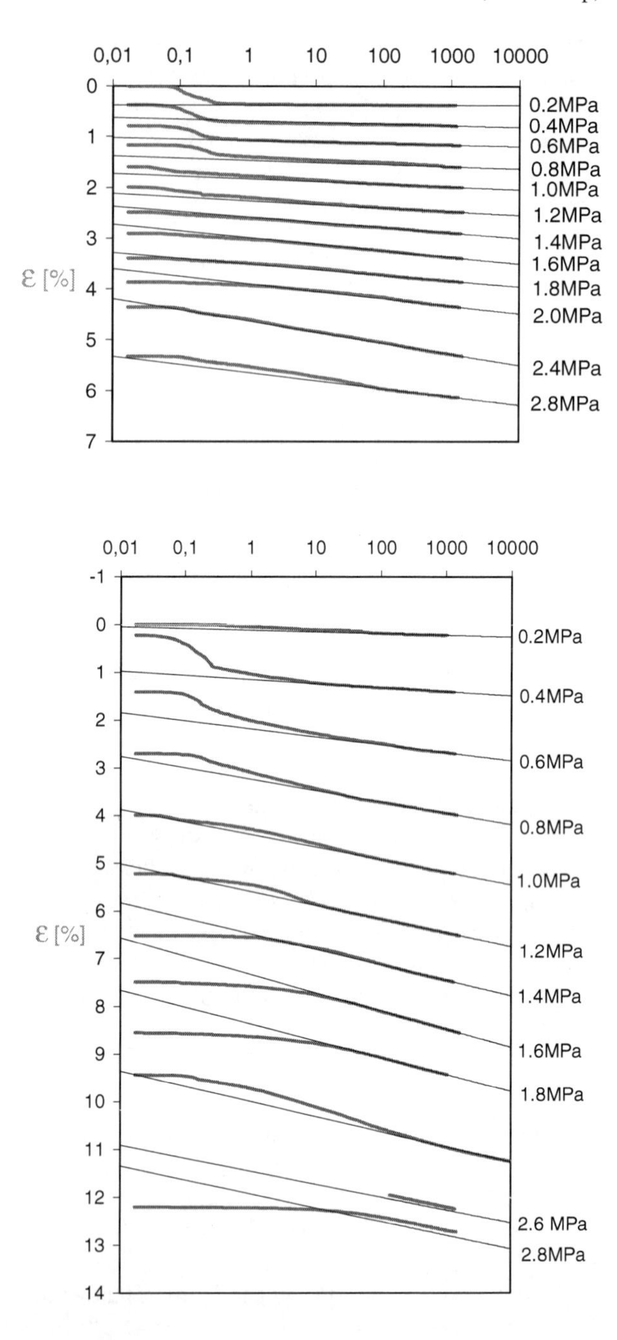

Fig. 8 Strain-time records of oedometers tests on compacted Pancrudo slate at two Relative Humidities (50%, upper figure and 100%, lower figure)

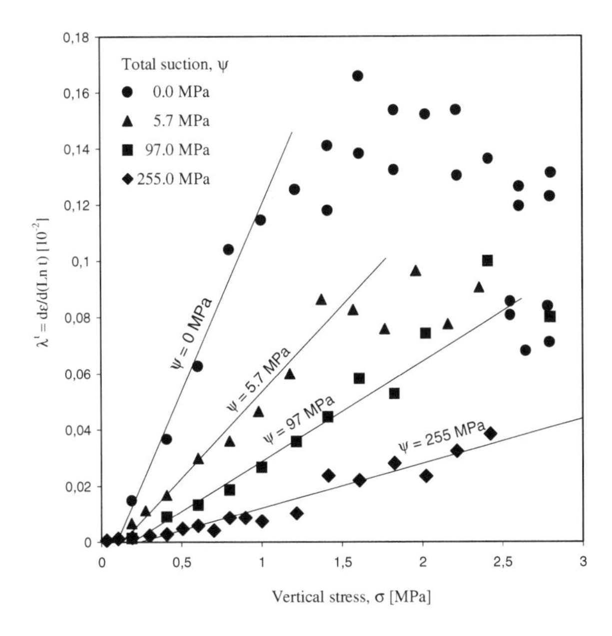

Fig. 9 Time dependent compressibility index of Pancrudo slate. Effect of applied stress and suction (Oldecop and Alonso 2007)

An example of a comparison between model performance and experimental results is shown in Figure 15.

4 Modelling the behaviour of a rockfill embankment subjected to rainfall

An interesting case is offered by the response of a 40 m-high rockfill embankment (Fig. 16), which is part of the high speed railway link between Madrid and Sevilla in Spain. The rockfill embankment is made of compacted schists. Settlements were recorded at the embankment upper platform. Figure 16 shows the strong correlation between rain intensity and rate of settlement. The correlation is good for the three periods of heavy rain in 1994, 1996 and 1997. However, the final heavy rain period, represented in Figure 16 did not induce any significant acceleration of settlements. This case will be analyzed in more detail here with the purpose of investigating the effect of some key parameters controlling the field performance of compacted rockfill structures. Figure 16 also points out that there is a "base" settlement rate not affected by rainfall. It may be properly identified as a pure creep rate. The implication is that rockfill deformation models should include a creep component if

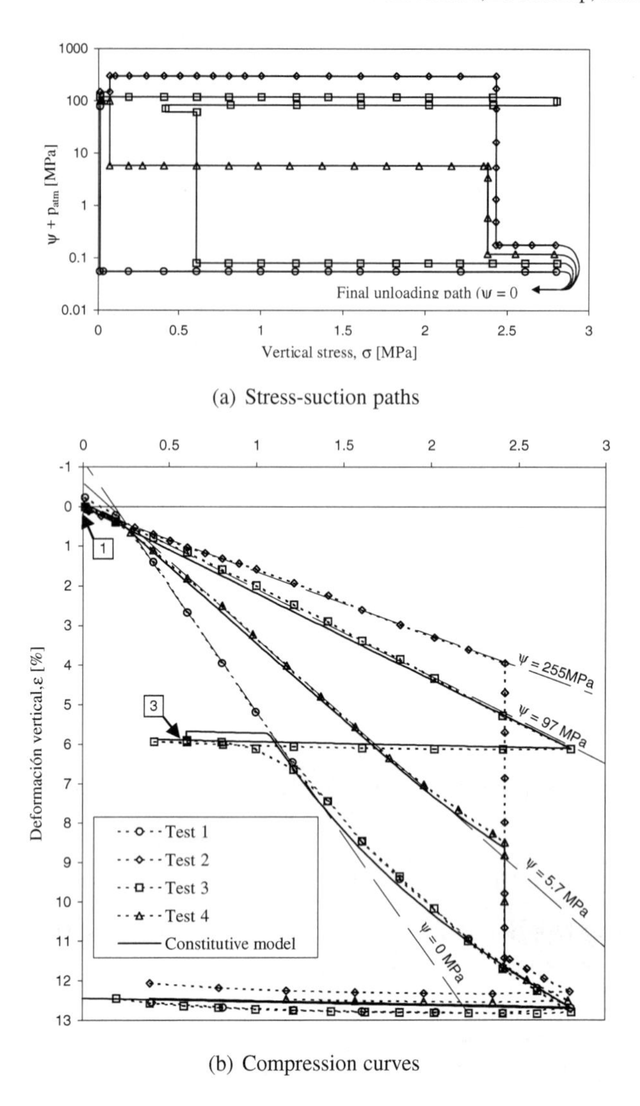

Fig. 10 Oedometer tests on Pancrudo slate. Square-enclosed numbers indicate the point of flooding. (Oldecop and Alonso 2003)

they are intended for long term simulations. The laboratory experiments mentioned before indicate that this basic creep behavior is also controlled by relative humidity.

A 40 m-high embankment having symmetrical slopes (1V:3H) and a crest width of 8 m was discretized and analyzed by means of CODE_BRIGHT (Olivella *et al.* 1994, 1996). The elastoplastic formulation of the model used in this paper has been described in Oldecop and Alonso (2001) and Alonso *et al.* (2005).

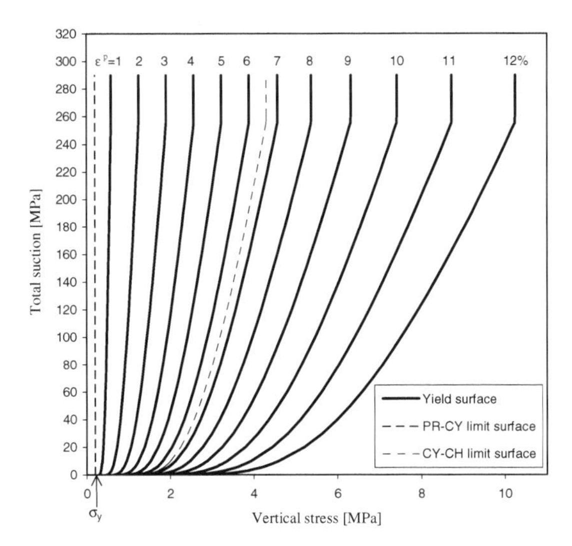

Fig. 11 Yield surfaces, corresponding to different plastic strain levels. Limit surfaces are shown between particle rearrangement (PR) and clastic yielding (CY) stages and between clastic yielding and clastic hardening (CH) stages (see Oldecop and Alonso 2006).

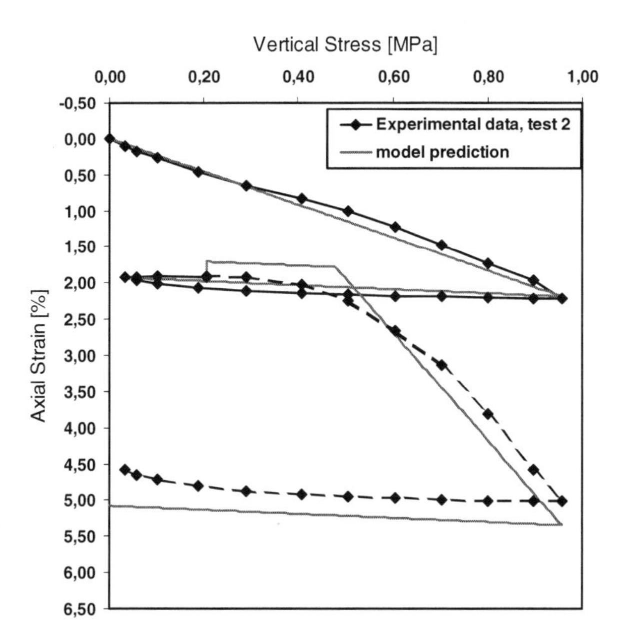

Fig. 12 Comparison of testing data on Pancrudo slate and model simulation (Alonso, 2006)

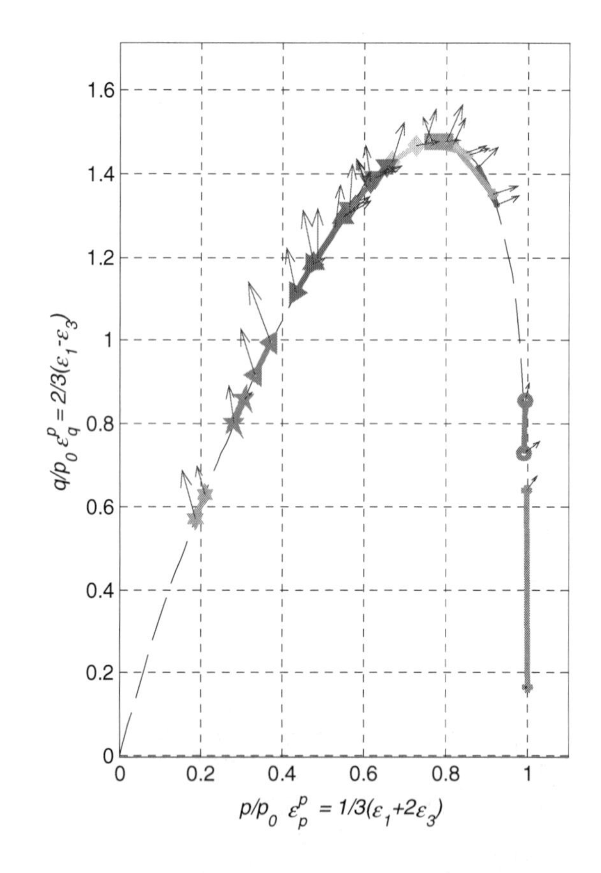

Fig. 13 Yield locus of compacted Pancrudo slate for RH = 100%. Triaxial tests reported by Chávez (2004)

The rockfill was described by the set of constitutive parameters reported in Table 1. Parameters correspond to the "inner" rockfill of Beliche dam (Alonso *et al.*, 2005). The outer rockfill was described as compacted fractured schist. The properties of the embankment material are not available (only a succinct description was given by Soriano & López, 1999). However, the comprehensive data set available for Beliche dam, for a material which is similar to the rockfill used in the railway embankment, provides a good starting point to perform the calculations in this case.

The analysis performed has centered only in three parameters: the air entry value of the water retention curve of the compacted schist, p_0, the creep parameter, μ, and the intrinsic permeability. Cases solved are grouped in Table 2. The initial state of the rockfill embankment is characterized by an initial mean yield stress, $p_0^* = 0.04$ MPa. A constant initial suction $s_0 = 1.5$ MPa was assigned to the entire structure. CODE_BRIGHT then performs an equilibrium calculation (impervious

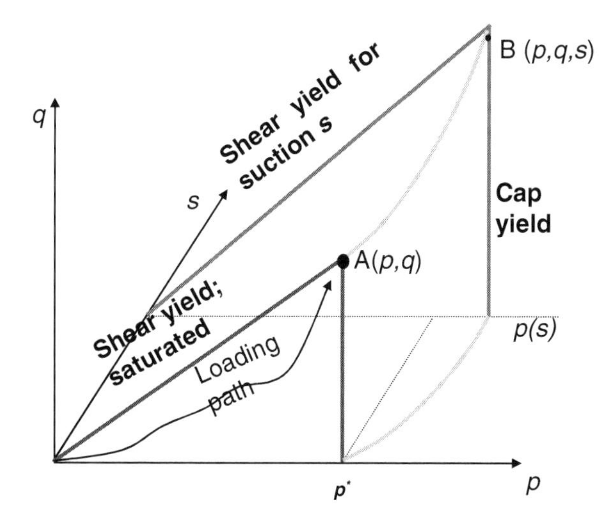

Fig. 14 Yield surfaces of a model reported by Chávez and Alonso (2003)

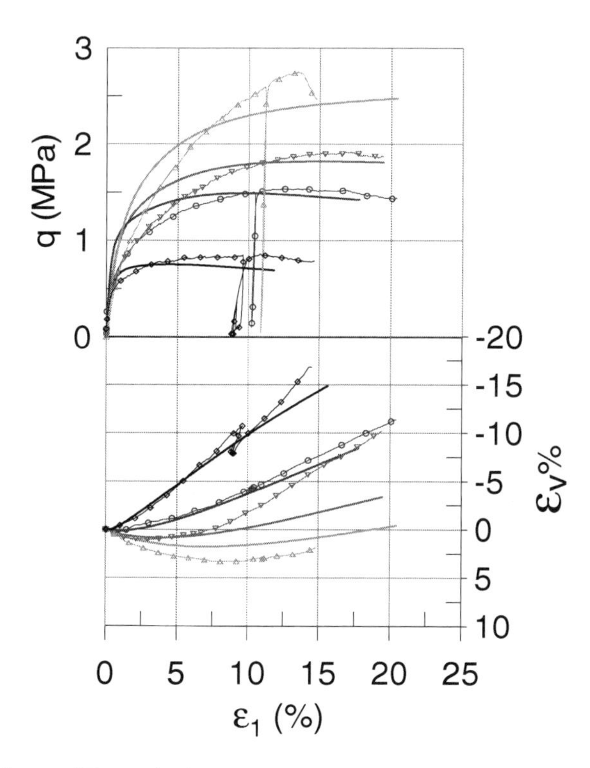

Fig. 15 Triaxial tests of Pancrudo slate gravel at RH = 36% and model predictions. (Chávez and Alonso 2003)

boundaries were assumed at this stage). The initial void ratio is $e_0 = 0.5$. The calculated response of the embankment is given in Figure 17 in terms of the evolution of suction of a representative point at the base of the embankment and the settlement rate. Also included in the figure are the measured settlement rates. This case fails to reproduce the actual behaviour. Most of the collapse takes place during the first 1.5 years, when relatively mild rainfall affected the structure. Later, suction maintains an essentially constant value and no further collapse is predicted by the model.

Table 1 Constitutive parameters of rockfill (mechanical and hydraulic)

I. ELASTIC BEHAVIOUR			
Elastic modulus	E	MPa	150
Poisson's ratio	ν	-	0.3
II. PLASTIC BEHAVIOUR			
Plastic virgin instantaneous compressibility	$\lambda^i - \kappa$	-	0.025
Virgin clastic compressibility for saturated conditions	λ_0^d	-	0.028
Parameter to describe the rate of change of clastic compressibility with total suction	α_s	-	0.010
Slope of critical state strength envelope for dry conditions	M^{dry}	-	1.75
Slope of critical state strength envelope for saturated conditions	M^{sat}	-	1.30
Parameter that controls the increase in cohesion with suction	k_s	-	0
Threshold yield mean stress for the onset of clastic phenomena	p_y	MPa	0.01
Parameter that defines the non-associativeness of plastic potential	α	-	0.3
III. HYDRAULIC PROPERTIES			
Intrinsic permeability	k	m/s	Analyzed parameter
Van Genuchten parameter describing air entry value	p_0	MPa	Analyzed parameter
Van Genuchten parameter describing mid slope of retention curve	λ	-	0.33

Table 2 Rockfill embankment. Cases analyzed.

Case	Air entry value, p_0 (MPa)	Creep coefficient, μ (MPa^{-1})	Intrinsic permeability (m^2)
A	0.01	0.0012	10^{-9}
B	0.10	0.0012	10^{-9}
C	0.10	0.0012	10^{-8}
D	0.10	0.0012	10^{-10}
E	0.10	0.0048	10^{-9}

Consider, however, Case B. The only change introduced is to increase the air entry value ($p_0 = 0.1$ MPa). The material is now able to store more water for the same applied suction. Figure 18 provides now the results of calculations. The agreement between model and measurements has vastly improved. Rains in 1993 and 1994 induce a moderate collapse as recorded. However, the extreme rain events at the beginning of 1996 and 1997 produce strong settlement accelerations, well matched by

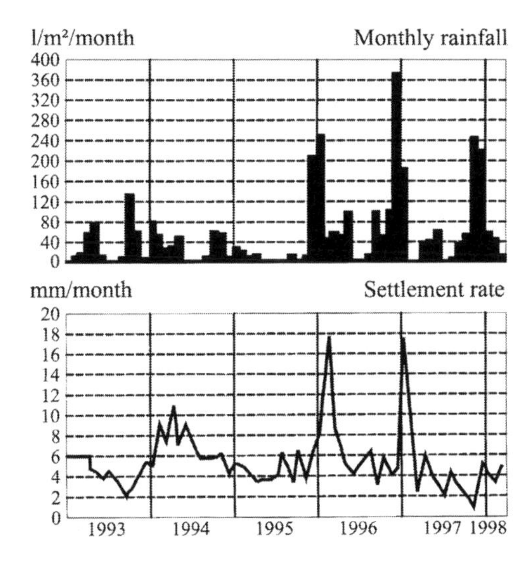

Fig. 16 Rainfall record and surface settlement rate of a 40 m high rockfill embankment. (Soriano and López 1999)

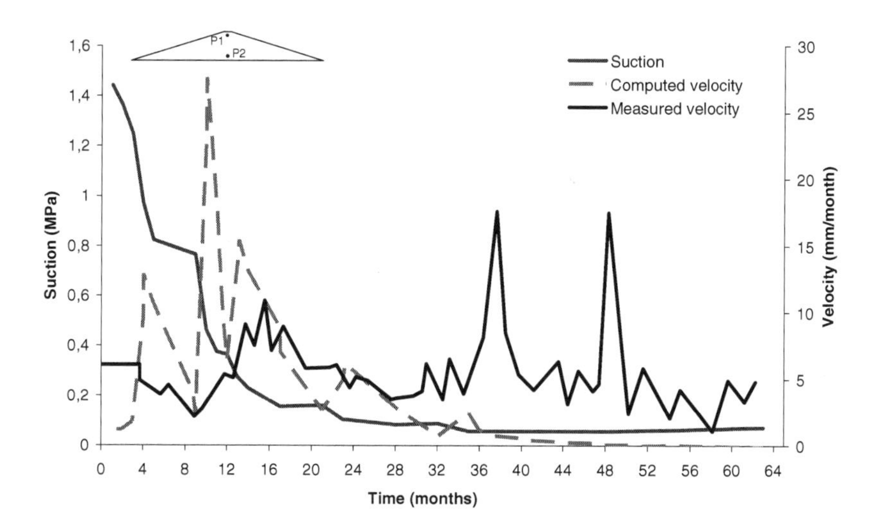

Fig. 17 Case A. Comparison of measured and calculated settlement rates (point P1). Also indicated is the suction change at point P2

the model. At the beginning of 1997, after the heavy rains of 1996, the suction in the embankment is already very low and the collapse potential of the embankment has disappeared. This explains the lack of reaction against the additional strong rains at the end of 1997. Now the rockfill is able to held water more effectively and suction is consistently higher than the values calculated for Case A. In all the remaining cases, the value of p_0 has been fixed at 0.1 MPa, because it provides a good fit to field measurements.

Calculated settlement rates in Figure 18 underpredict measurements in periods of moderate or no rain. Suction is essentially constant during these time intervals and the reason for the sustained settlement rate may be attributed to pure creep. Increasing creep would, therefore, help to reproduce better the field record. In Case E the creep coefficient μ has been increased four times, with respect to Case B. The embankment response is given in Figure 18.

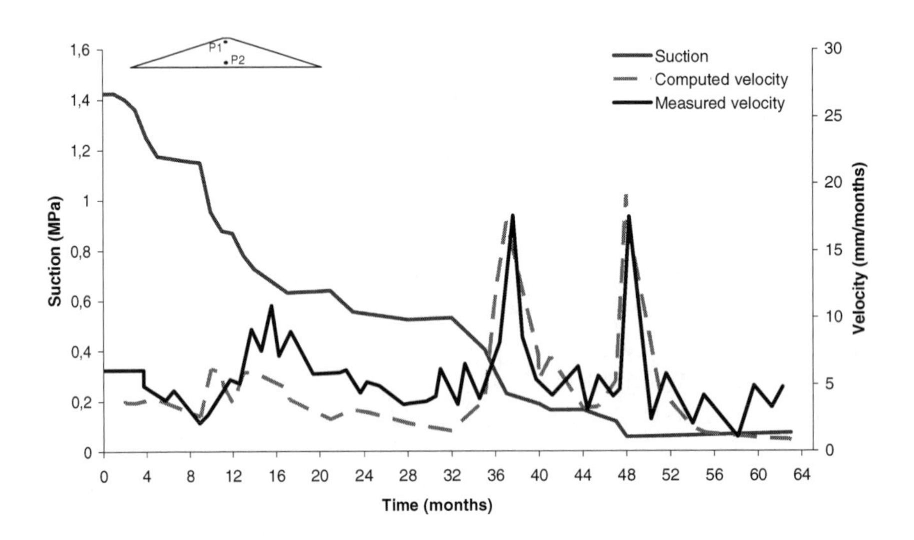

Fig. 18 Case E. Comparison of measured and calculated settlement rates (point P1). Also indicated is the suction change at point P2

The basic creep rate is now increased and the agreement with measurements increases. Interestingly, increasing the creep rate leads to a reduction of the collapse-induced effects even if the calculated suction response has not changed. The reason for this reduction of collapse-induced effects is due to the hardening experienced by the yield locus. Creep strains are considered plastic and, therefore, they harden the current yield locus. Since the stress state does not change during the creep-induced hardening, any reduction in suction results in smaller collapse. In this way the peak settlement rates measured at the beginning and the end of 1996 are now well reproduced. However, collapse rates measured in 1993-1994 are now underestimated for the same reason.

An additional insight into the collapse mechanisms associated with a given rainfall record and specific water retention properties of the rockfill is given in Figure 19. The figure shows the stress paths in a (p : net mean stress; s: suction) plane of the reference point inside the rockfill embankment for Cases A and B. The points highlighted in the plot identify the particular times shown in the figure. Also indicated is the position of the yield locus at some particular instances. The amount of collapse is directly related to the amount of the yield locus displacement along the p axis. The stress path for Case B is displaced towards the right because of the increased net vertical stress associated with the higher water content of the rockfill (controlled by the water retention curve). The strong displacement of the yield curve of Case A during the early wetting (rainfall in 1993-1994) implies large collapse deformations. The low suction reached afterwards is already an indication of very limited collapse potential. In contrast, the rockfill in Curve B looses suction more gradually and the displacement of the yield curve is smaller. At the end of the first rainy season the current suction still maintains a significant collapse potential. Note also that, due to the shape of the yield locus, changes in suction in the low range of values are particularly effective to induce significance collapse.

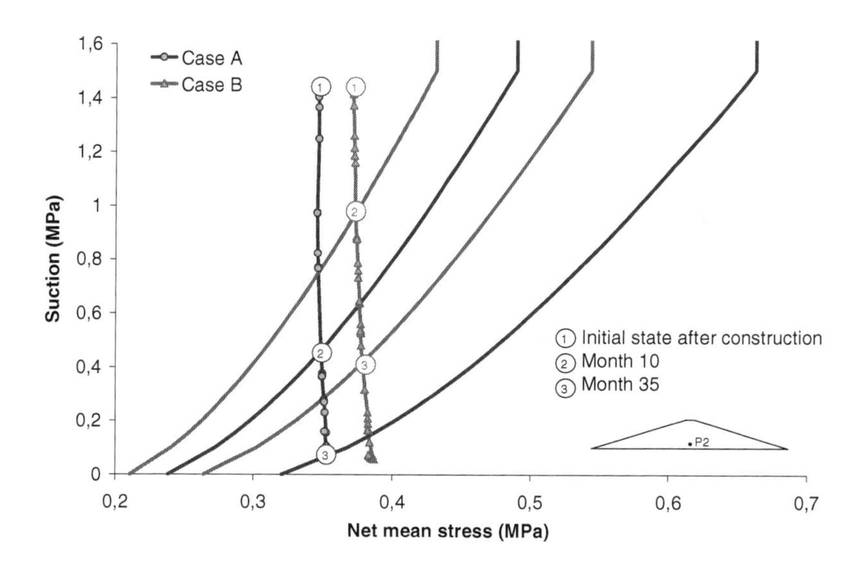

Fig. 19 Stress paths in point P2 for cases A and B. Also indicated is the position of the yield locus.

This case shows that the water retention curve of the material has a significant effect on the development of collapse settlements in time for a given rainfall regime. However, the intrinsic has a negligible effect. The time at which moisture change-induced settlements actually take place is controlled by some key rockfill properties (namely those characterizing the water retention properties) but also by the particular rainfall regime. Extreme events lead to additional collapse settlements only if they imply a further increase in relative humidity over the maximum value previously reached. Creep deformations are not weather-independent because of the effect of relative humidity on creep rates. Creep deformations affect water-induced collapse because they are truly irreversible volumetric strains that harden the rockfill. A hardened rockfill reacts with a reduced collapse potential against any further wetting. In practice, distinguishing between moisture-induced or pure creep long term deformations is not straightforward and probably requires the concourse of a model to analyze a given settlement record. However, water induced deformations tend to concentrate on singular rainfall events at least during a limited time period after the commissioning of the rockfill structure.

5 Scale effects

Experimental evidence indicates that the strength of rock specimens decreases when the size of the sample increases. A common explanation has a probabilistic nature: the probability of finding a longer crack at the worst orientation increase with the size of the specimen.

If tensile stress conditions are reproduced to force fracture propagation in type I, a similar result is observed. Figure 20 reproduces some results published by Lee (1992). The figure shows the test set-up (an irregularly shaped particle is diametrically loaded by two parallel polished plates) and the force-displacement records. Local failures are first detected at the plate-rock contacts before the final diametrical failures takes place. Tensile stresses along the diametrical surface are derived through the Brazilian expression and then plotted against the average particle size (Fig. 20c) Three rocks are represented. The relationship:

$$\sigma_f \approx d^{\alpha} \tag{10}$$

where σ is the tensile strength and α is a coefficient ranging between -0.34 and -0.42 fits experimental results.

Fracture mechanisms provide also a simple explanation for this experimental finding. Crack propagates when the stress intensity factor K reaches the material toughness K_c. Therefore, in view of Equation 2, the tensile stress which leads to fracture propagation is

$$\sigma = \frac{K_c}{\beta\sqrt{\pi a_i}} \sim a_i^{-0.5} \tag{11}$$

Now the idea is, again, that the larger the particle diameter the longer the initial defect, which may naturally occur in statistical terms. In the limit, the size of the specimen marks the maximum size of a defect or crack for that particular specimen. Therefore:

$$\sigma_f \sim d^{-0.5} \tag{12}$$

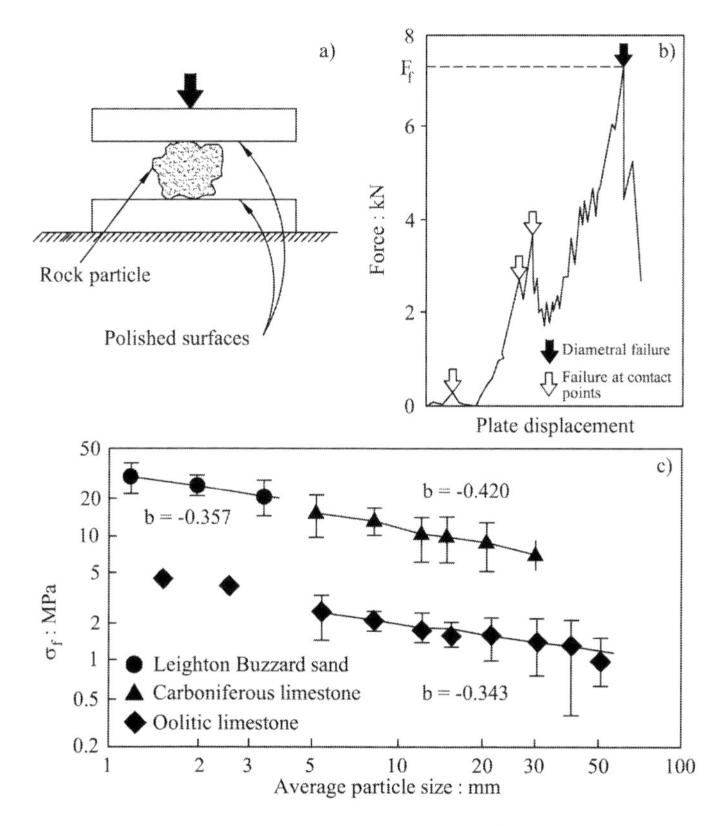

Fig. 20 Tests on rock particles (McDowell and Bolton 1998), quoting Lee 1992) a) Particle tensile strength test set-up; b) Typical load-deflection plot; c) Mean tensile strength as a function of particle size

Coefficients α determined experimentally by Lee (1998) are close to the theoretical (-0.5) coefficient in (12). Consider now the relationship between external stress σ_{ext} and the tensile stress σ_t in a diametrically loaded particle. In the Brazilian test, the concentrated load P is proportional to the tensile stress and to the square of the diameter:

$$P \sim \sigma_t d^2 \tag{13}$$

and in view of (12)

$$P \sim d^{-0.5} d^2 \sim d^{3/2} \tag{14}$$

will be the fracturing load.

In a simple arrangement of spheres (in plan view, Fig. 21) the equilibrium implies

$$\sigma_{ext}\ 1^2 = \sum P = P\frac{1}{d}\frac{1}{d} = \frac{P}{d^2} \tag{15}$$

Therefore:

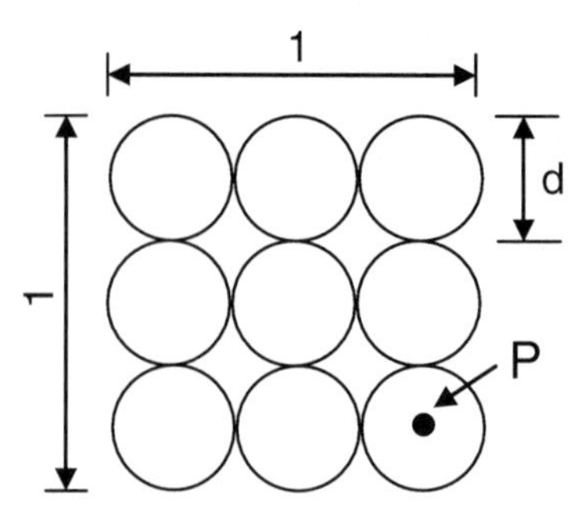

Fig. 21 Simple arrangement of spheres. Plan view

$$P = d^2 \sigma_{ext} \tag{16}$$

Equations (14) and (16) result in

$$d^2 \sigma_{ext} \sim d^{3/2} \tag{17}$$

and:

$$\sigma_{ext} \sim d^{-0.5} \tag{18}$$

In other words, the external stress capable of fracturing particles of diameter d is inversely proportional to the square root of the diameter. The bigger the particles, the easier for them to break, for a given overall stress field. If instead of Equation (11), the starting point is Equation (10), derived from some experimental evidence, then (17) transforms into:

$$\sigma_{ext} \sim d^{-\alpha} \tag{19}$$

where α is probably in the range 0.3-0.5. Then, it could be accepted that mechanical properties of rockfill which are controlled by particle breakage would be affected by a scale effect. Equations (18) and (19) may be used to find the scaling factor.

Take for instance the compressibility of rockfill. In view of the linearity of stress and strain in Figure 10, a compressibility coefficient (for any given suction) may be simply defined as:

$$\sigma_{ext} \sim d^{-\alpha} \tag{20}$$

and in view of (19):

$$\lambda \sim d^{\alpha} \tag{21}$$

Therefore if the compressibility of two granular aggregates of diameters d and d_0 are compared

$$\lambda^d = \lambda^{d_0} \left(\frac{d}{d_0} \right)^{\alpha} \tag{22}$$

Consider in Figure 22 some recent results of gravel compressibility. Samples of hard limestone gravel of uniform size (40-30 mm, 30-20 mm, 25-20 mm and 20-10 mm) were tested in a 30 cm diameter oedometer. Figure 22a shows the linear compressibility coefficient measured under saturated conditions (Ortega 2008). Two particle packings were tested: a loose one ($e_0 = 0.947$) and a dense one ($e_0 = 0.502$). Figure 22b shows the scaled compressibility, following Equation (22) taking in each case d_0 as the minimum of the (maximum) particle size tested. The plot shows that the exponent $\alpha = 0.5$ fits almost perfectly the experiments when $e_0 = 0.947$. For the dense packing ($e_0 = 0.502$) an exponent $\alpha = 0.3$ is better suited. It seems therefore that the degree of compaction has an effect on the scale rule.

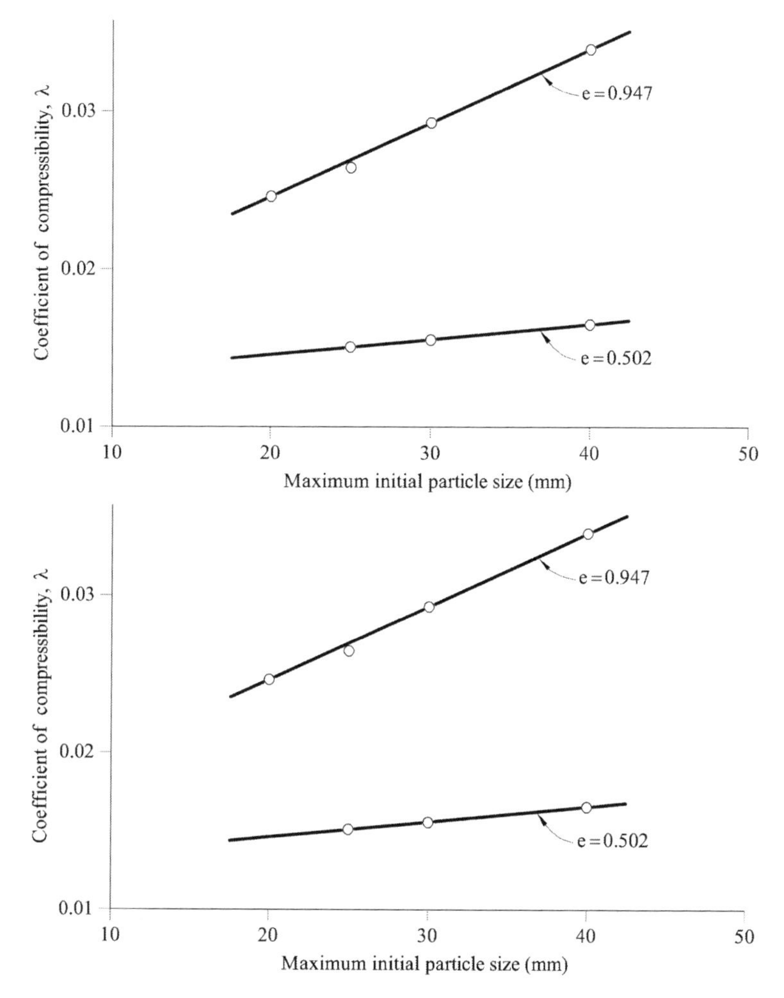

Fig. 22 a) Compressibility coefficient of uniform limestone gravel in terms of maximum particle size (Ortega 2008). b) Scaled plot.

Frossard (2009) has discussed the scale effects in a similar manner. When dealing with strength (Eq. 1) if the scale rule is applied to both the shear stress τ_f and the confining normal stress σ it can be written:

$$\frac{\tau_f}{(d/d_0)^{-\alpha}} = A_0 \left(\frac{\sigma}{(d/d_0)^{-\alpha}} \right)^b \tag{23}$$

and therefore:

$$\tau_f = A_0 \sigma^b \left(\frac{d}{d_0} \right)^{-\alpha(1-b)} \tag{24}$$

This provides the shear strength of a rockfill of particle size d if the strength envelope for a rockfill having particle size d_0 is known ($\tau_{f0} = A_0 \sigma^b$).

The discussion on scale effects is more complex when non-uniform grain size distributions are considered. Then, some of the assumptions leading to the derivation presented above are probably far from being fulfilled, even in an approximate manner. On the other hand a full characterization of the rockfill requires a number of constitutive parameters and not just strength and compressibility.

However, if a constitutive model is available, tests could be performed on different grain size distributions, model parameters could be derived and trends regarding the variation of constitutive parameters with some indices describing the grain size distribution could be derived. This approach was followed by Ramon *et al.* (2008), who tested specimens of Pancrudo slate gravel having the grain sizes given in Figure 23. Three of them are characterized by common D_{50} value and varying "fines" content. The reference constitutive model was described in Oldecop and Alonso (2001).

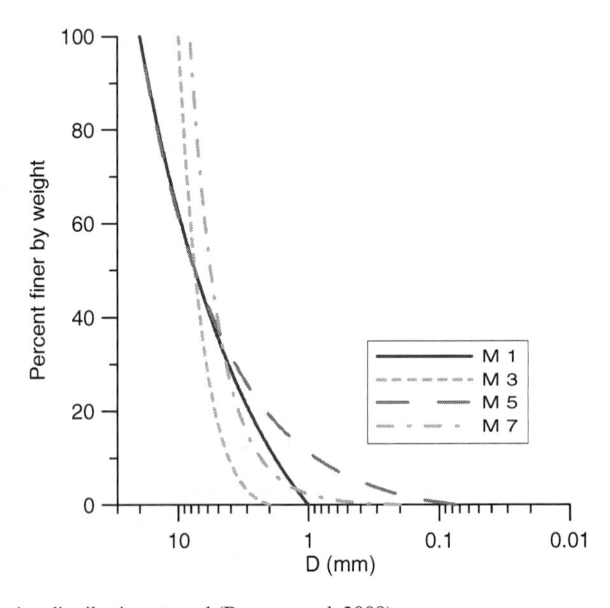

Fig. 23 Grain size distributions tested (Ramon *et al.* 2008)

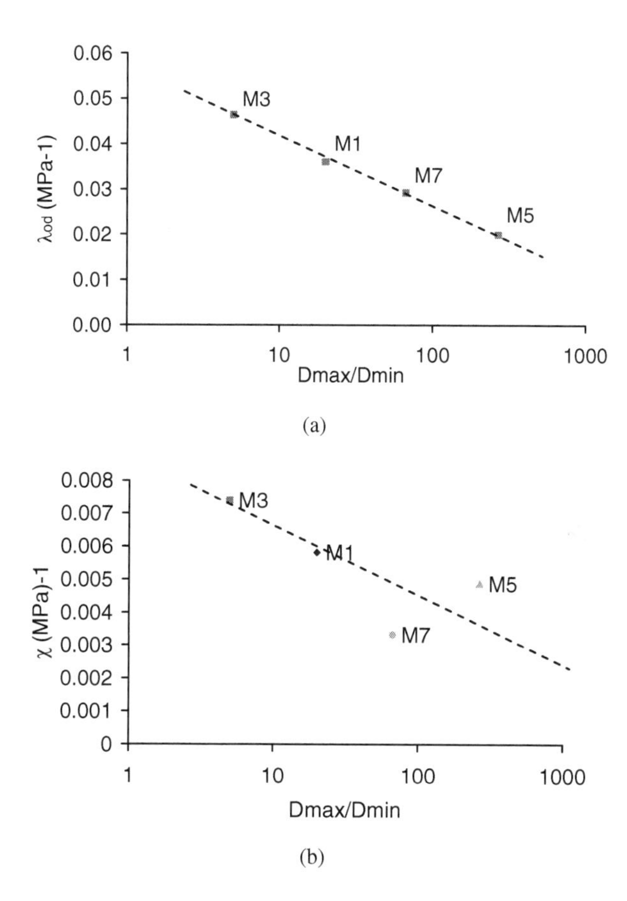

Fig. 24 a) Compressibility parameter λ_0^d from the non saturated oedometer tests. b) Parameter χ plotted against ratio D_{max}/D_{min} for the analysed grain size distributions. (Ramon *et al.* 2008)

It was found that the best parameter to identify the grain size distribution in this case was the ratio D_{max}/D_{min}. The compressibility parameter χ decreased continually with increasing D_{max}/D_{min} (Fig. 24a). Parameter χ (Fig. 24b) describes the intensity of collapse. It seems to reduce also with D_{max}/D_{min} (Fig. 24b). However, no definite trends were found for the elastic compressibility parameter κ. The creep index λ^t (Eq. 8) maintained a constant ratio with the relationship with the standard compressibility and, therefore, it follows a similar relationship with grain size distribution.

Scale effects are an open question and an interesting research subject of very significant relevance in practice because of the difficulty and limitations of laboratory testing.

6 Conclusions

The paper presents some background information on the model developed by the authors to reproduce the effect of water content changes and creep effects on rockfill compressibility. Both phenomena contribute to the long term deformation of rockfill structures. Relative humidity changes control the rate of particle breakage and therefore the development of collapse strains when the rockfill is wetted.

True creep strains depend on confining stress and suction. The model described in the paper is consistent with creep tests under suction control performed on shale gravels, under oedometric conditions. The elasto-visco-plastic constitutive model developed was implemented into a general purpose finite element program for coupled flow-deformation analysis under saturated/unsaturated conditions.

A six year record of settlement rate of a 40 m-high compacted schist rockfill embankment provided a good case history to check the capabilities of the model developed and to investigate the effect of some relevant variables.

It was found that the water retention curve of the material has a significant effect on the development of collapse settlements in time for a given rainfall regime. Creep deformations are not weather-independent because of the effect of relative humidity on creep rates. Creep deformations affect water-induced collapse because they are truly irreversible volumetric strains that harden the rockfill. A hardened rockfill reacts with a reduced collapse potential against any further wetting.

Scale effects are analyzed through a simple model inspired in fracture mechanics principles and some conceptual representation of the granular media. Expressions for the correction of constitutive parameters to take into account particle size have been derived. In particular, expressions for the coefficient of compressibility and the rockfill strength are given. The results of large diameter tests on samples characterized by different grain size distributions support the derived relationships.

References

1. Alonso, E.E.(2003). Exploring the limits of unsaturated soil mechanics: The behavior of coarse granular soil and rockfill. The 11th Buchanan Lecture. University of Texas A&M
2. Alonso, E. E., S. Olivella and N. Pinyol (2005) A review of Beliche Dam. Géotechnique, Vol. 55, No. 4, 267-283
3. Charles, R. J. (1958). Static fatigue of glass. Journal of Applied Physics, 29, 1549-1560.
4. Chávez, C. (2004), Estudio del comportamiento triaxial de materiales granulares de tamaño medio; con énfasis en la influencia de la succión. Tesis doctoral Universidad Politécnica de Cataluña, España.
5. Chávez, C. and Alonso, E.E. (2003) A constitutive model for crushed granular aggregates which include suction effects. Soils and Foundations, Vol. 43, No.4, 215-227.
6. Cundall, P. A. & Strack, O. D. L. (1979). A discrete numerical model for granular assemblies. Géotechnique, Vol. 29, No. 1, 47-65.
7. De Mello, V.F.B. (1977) Seventh Rankine Lecture: Reflections on design decisions of practical significance to embankment dams. Géotechnique, Vol. 27, No. 3, 279-356.
8. Frossard, E. (2009) On the structural safety of large rockfill dams. Proc. 23rd Conf. on Large Dams. Brasilia. In press.

9. Fumagalli, E. (1969). Tests on cohesionless materials for rockfill dams. J. Soil Mech. Found. Engng Div., ASCE 95, No. SM1, 313±330.
10. Hardin, B. O. (1985) Crushing of soil particles. Jnl. Of Geotech. Engng., ASCE, Vol. 111, No. 10, 1177-1192.
11. Marachi, N. D., Chan, C. K., Seed, H. B. & Duncan, J. M. (1969).Strength and deformation characteristics of rockfill materials,Department of Civil Engineering, Report No. TE-69-5.University of California.
12. Marsal, R. J. (1973). Mechanical properties of rockfill. In Embankment dam engineering. Casagrande volume (eds R. C. Hirschfeld & S. J.Poulos), John Wiley & Sons, New York, 109±200.
13. McDowell, G. R. and M. D. Bolton (1998) On the micromechanics of crushable aggregates. Géotechnique, Vol. 48, No. 5, 667-679
14. Naylor, D. J., Maranha das Neves, E., Mattar, Jr., D. Veiga Pinto, A. A. (1986). "Prediction of construction performance of Beliche Dam", Géotechnique, Vol. 36, No. 3, 359-376.
15. Nobari, E. S. & Duncan, J. M. (1972). Effect of reservoir filling on stresses and movements in earth and rockfill dams, Department of Civil Engineering, Report No. TE-72-1. University of California.
16. Oldecop, L.A. and E.E. Alonso (2001) A model for rockfill compressibility. Géotechnique, Vol. 51, No. 2, 127-139.
17. Oldecop, L.A. and E.E. Alonso (2003) Suction effects on rockfill compressibility. Géotechnique, Vol. 53, No.2, 289-292
18. Oldecop, L.A. and E. E. Alonso (2004) Testing rockfill under relative humidity control. Geotechnical Testing Journal, Vol. 27, No. 3, 269-278.
19. Oldecop, L. y E.E. Alonso (2007). Theoretical investigation of the time-dependent behaviour of rockfill. Géotechnique, Vol. 57, 289-301.
20. Olivella, S., Carrera, J. Gens, A. and Alonso, E. E. (1994) "Nonisothermal multiphase flow of brine and gas through saline media". Transport in Porous Media, 15, 271-293.
21. Olivella, S., Gens, A., Carrera, J. Alonso, E. E. (1996). "Numerical formulation for simulator (CODE_BRIGHT) for coupled analysis of saline media". Engineering Computations, Vol. 13, No. 7, 87-112.
22. Pinyol, N.M., E. E. Alonso and L. A. Oldecop (2008) Hydric and creep effects on long term deformations of rockfill embankments. Zhu, Liu & Qiang (eds). Proceedings of the 1st International Conference on Long Time Effects and Seepage Behavior of Dams (LTESBD08). Hohai University, Nanjing, China, May 30-June 2, 2008.
23. Poorooshasb, H., I. Holubec and A. Sherbourne (1966) Yielding and flow of sand in triaxial compression. Part I. Canadian Geotech. Jnl., Vol. 3, No. 4, 179-190
24. Ramon, A., E. E. Alonso and E. E. Romero (2008) Grain size effects on rockfill constitutive behaviour. Proc. of the 1st Europ. Conf. on Unsaturated Soils. Taylor and Francis. 341-347.
25. Soriano, A. & Sánchez, F. J. (1999) Settlements of railroad high embankments. Proc. XII European Conf. on Soil Mech. and Geotech. Eng., Netherlands.
26. Sowers, G. F., Williams, R. C. & Wallace, T. S. (1965). Compressibility of broken rock and settlement of rockfills. Proc. 6th ICSMFE, 2, Montreal: 561-565.
27. Terzaghi, K. (1960). Discussion on Salt Springs and Lower Bear Riverdams. Trans. ASCE 125, pt 2, 139±148.
28. Veiga Pinto, A. A. (1983). Previsao do comportamento estrutural de barragens de enrocamento. PhD thesis, Laboratorio Nacional de Engenharia Civil, Lisbon.

Nature – A Very Clever Experimentalist

Gary D. Couples

Abstract Human experimentalists who investigate geo-materials are quite clever. They have developed a range of lab apparatus that allows them to determine material parameters that define robust constitutive laws to describe how geo-materials deform. They can also design models of processes that provide important insights into the operation of geo-material systems. Unfortunately, we human experimentalists must operate on finite specimens, and this constraint has guided us into adopting a perspective that is too focused on boundary conditions, along with developing a false view that we can impose arbitrary states that are valid. Nature never creates artificial boundaries, and the natural conditions for any location are always evolved and compatible with the deformation state of the surrounding geo-materials. This makes Nature a more clever experimentalist that we humans, but we can improve our approach by recognising the intellectual trap associated with the notion that boundary conditions can be arbitrarily fixed.

1 Introduction

In this paper, I consider some of the lessons we might learn from thinking about the way that Nature conducts experiments on geo-materials. In contrast with human-controlled experiments, Nature does not create artificial material boundaries, and Nature does not impose any sort of boundary conditions. Instead, natural deformations occur in a fashion that is, well, natural, with conditions that evolve as part of the deformation process.

I take the opportunity of writing this paper to summarise a personal perspective on that endeavour which consists of lab-based experimental studies concerning the deformation processes that characterise geo-materials. My prototype geo-material

Gary D. Couples
Institute of Petroleum Engineering, Heriot-Watt University, Edinburgh EH45 9HZ Scotland, e-mail: gary.couples@pet.hw.ac.uk

is a porous sedimentary rock, but my comments apply equally to crystalline rocks and to soils. The lab-based experimental activities which are the subject of this discussion have been underway in a fashion that we could call "modern" for nearly one hundred years. They have provided us with quantitative information and, importantly, with critical insights. We use the derived understanding to explain what has happened in the past, and to predict what will happen in the future under specified circumstances. Using this knowledge, we can sometimes avoid unwanted geo-material responses by controlling the ways that we perturb the natural state, or by altering the natural state via direct action. Such practical outcomes of the acquisition of knowledge about geo-materials justify the support of a research activity that is intellectually very satisfying for many of us.

Can we do better? I certainly think so. Important elements in making progress, in my view, will be to better appreciate the artificiality of certain aspects of our experimental protocols, and to recognise that our standard description of the way that we control our experimental conditions is misleading – it is almost certainly the case that non-experimentalists interpret our stated protocols in a fashion that is not correct.

In contrast with the usual research paper, which is normally based on hard information and analysis, this contribution is written more in the style of an essay. Many of the topics included in this paper are treated in considerable detail by other contributors to this volume, and I urge readers to explore these subjects via the extensive summaries and reference lists contained in these companion papers. I do list references in the case of specific issues, and to provide a starting point for a study of the past experimental work that uses physical models, but my intent is to keep this text as simple and clear as possible. In this paper I aim to highlight a crucial difference between natural deformations and those we create in the laboratory, and to suggest a slight shift of perspective that will allow us to better interpret our lab-based activities. I begin by considering the valuable outcomes of experimental studies of geo-materials, attempting to highlight the broad conceptual areas that have emerged from the many decades of work. I then consider the critical contrasts between what we do in the lab and what Nature does in the real world, making the point that we have adopted a form of language to describe our lab-based achievements that obscures these dissimilarities.

2 What Do Experimentalists Provide?

In addition to continued developments of lab technology, experimentalists provide us with observations concerning material behaviours, and with models of deformation processes. Based largely upon lab information, we have progressively created the idea that geo-materials respond to loading in different modes that are governed by key parameters such as the material composition, its solidity, the mean stress, the pressure of the pore fluids (or their saturations), the rate of loading, etc. Observations of the physical outcomes of such experiments reveal a range of macroscopic and microscopic spatially-ordered changes, such as microfractures, collapse of pores, grain crushing, etc. – often arranged in localised bands of greater strain (Fig. 1).

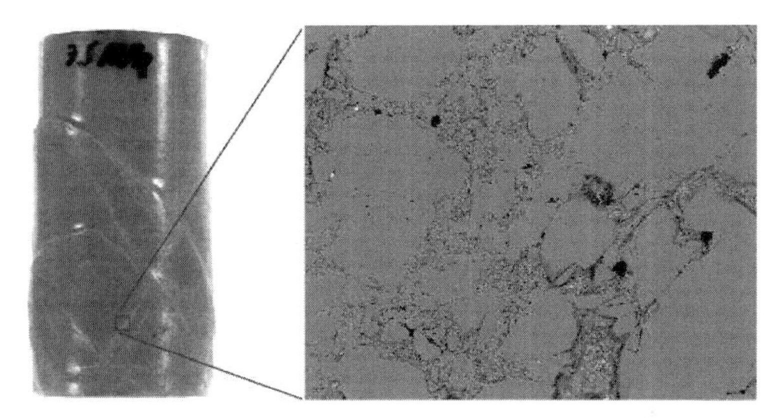

Fig. 1 Example of localised deformation. Left image illustrates the macroscopic pattern of localised shear bands that have developed during the loading of a porous sandstone sample. This sample was subjected to a radial confining pressure of 75 MPa, separated from the specimen via a plastic membrane (its shiny surface is visible in the photograph). Note the development of multiple bands arranged in a pseudo-radial fashion, each inclined to the axis of the specimen by a similar angle. Experiment performed by Helen Lewis. Right image is a back-scattered SEM scan of a shear band from a comparable sample taken from a location like the one indicated. Field of view is about 0.3 mm. Note the occurrence of grain crushing, reducing the grain size from about 0.1 mm to sizes about two orders of magnitude smaller. Such cataclastic deformations locally reduce the flow characteristics within a band, and, by their distribution over the sample, the bulk properties of the whole sample.

In combination with theorists, the experimental observations can be synthesised into relationships that are known as constitutive laws. These relationships describe how macroscopic parameters such as stress and strain are functionally related during a deformation. The measurement of the parameters (sometimes called material properties) that define such constitutive laws is one of the major contributions of experimentalists. In fact, the formulation of constitutive laws is primarily governed by the types of parameters that can be measured in the lab.

A simple model – and it needs to be remembered that it is only a model – is the case where the material is assumed to act like an elastic solid (which is again only a model) until some critical state is reached, when the material fails. The Mohr-Coulomb expression of this idea has been, and continues to be, a useful means of making predictions (Handin 1969). In some circumstances, those predictions (which may amount to an OK/not OK? choice) are all that is needed, since the desire is to identify and then avoid the failure state arising (one might think of the use of this method in wellbore stability analysis).

The current state-of-the-art in material description goes considerably beyond Mohr-Coulomb. This modern model seeks to specify the conditions (mean effective stress, and stress differences) that define yielding of the material – which we can usefully take to be a change of state (the definition of yielding can take several forms: cessation of pseudo-elasticity, or beginning of significant permanent strain) – along with how the material behaves after yielding (including changes in mechanical properties). The porosity state, or void ratio, is inherent in the material description. Other physical attributes of the material, such as its flow properties, its thermal conductivity, its acoustic characteristics, its electrical and magnetic properties (which are affected by the porosity changes) can be functionally associated with the geomechanical responses. The constitutive law can be formulated in several ways, depending on how it is to be implemented within a simulation/calculation. Because this material concept involves all of the elastic, viscous (rate-dependent in general terms), and plastic aspects of porous geo-materials, it seems sensible to call this class the poro-elasto-visco-plastic material description (Fig. 2). Experimentalists are actively working to expand our understanding of the complex interactions that can be expressed within these realistic geo-material formulations, and we will almost certainly continue to be surprised at the interesting and anomalous responses that will be reported from their observations in the years to come. Presumably, we will also be pleased to be able to use this expanding knowledge to make more and better predictions about the responses of geo-materials.

The lab-based work noted above, which is focused on material properties, has been incredibly valuable to us, but there is another major (and sometimes-forgotten) contribution from experimentalists. This work concerns the demonstration of deformation processes affecting systems of geo-materials. Although some might not agree with my classification, I include within this activity those lab-based efforts that seek to understand the processes that lead to the appearance of macro-scale fractures and shear bands. My reason for treating experiments of this type differently, compared to those that seek simply to measure material properties, is that the specimen evolves during the experiment, with local regions of the material undergoing significant alterations of their state. Wibberley et al (2007) document a surprising evolution of material properties in a naturally-deformed high-porosity sandstone. They show that shear bands undergo a compactional (volume loss) deformation followed by a dilational one, and present a cogent geomechanical argument to explain this process in terms of the material "needing" to compact to gain sufficient strength and stiffness to be able to support the dilational and softening responses that occur later. Experimentally-created shear bands in comparable materials exhibit similar deformation textures, but I am not aware of real-time evidence from the lab to prove that interpretation. If I provisionally accept the hypothesis that geo-materials can evolve in this fashion, this sequential and spatially-ordered set of events defines a system – one that is self-organised, and which crucially is dependent on the strong non-linearities that are a primary aspect of realistic geo-materials.

Other examples of this type of experimental work are clearly associated with systems – where a system is taken to mean a spatial arrangement of multiple parts (the multiple parts might emerge from an original continuum), subjected to loading. A non-comprehensive listing includes the shear-zone experimental studies that were aimed at discovering the conditions and material controls that govern stable versus episodic sliding responses (Teufel and Logan 1978), with the initial studies being

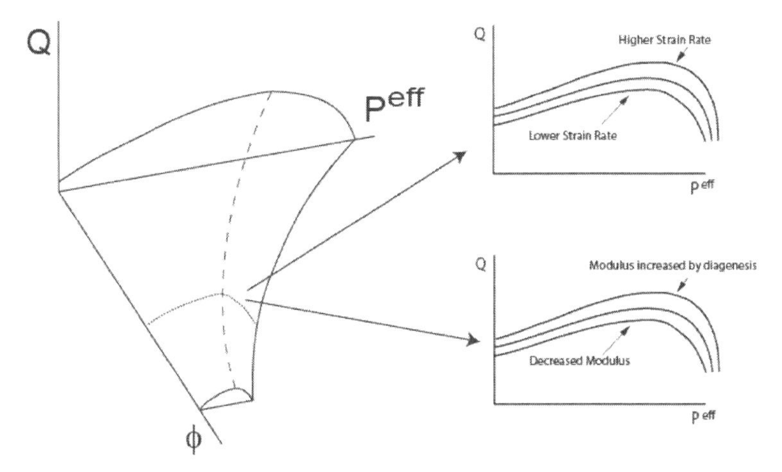

Fig. 2 Graphical depiction of a poro-plastic material model. The surface in the left image defines a set of conditions (a "state") under which the material undergoes yielding. States that plot under the surface (or "inside" of it) are not at yield, and the material is typically assumed to react in an elastic fashion under those conditions. The yield surface is plotted in a *P*-*Q*-porosity space. The *P*-axis is the mean stress, here plotted as P^{eff}, which is the value obtained by subtracting the pore-fluid pressure from the nominal mean stress. The Q-axis can be defined slightly differently, depending on the exact formulation of the constitutive law selected. In general, it is simple to think of Q as being an indicator of the magnitude of the stress differences, which one might associate with the material seeking to avoid distortions of its shape. Geologists think of the porosity (as indicated here) as a useful measure of the proportions of void space in the material, but a similar image could be shown using the void ration instead. The dashed line highlights the local crest of the yield surface in a *P*-*Q* plane, and this boundary often delimits the strain-softening response region (towards lower *P*-values) from the strain-hardening region. These regions tend to be associated with dilational and compactional volumetric strains, respectively. The simple plots on the right side indicate how the yield surface might vary as a function of strain rate (upper image) or in response to cementation of the material during its burial, which amounts to an alteration of the material type (lower image). The poro-plastic concept is able to include almost every factor that we understand in terms of being important to the mechanical characteristics of geo-materials.

prompted by the desire to better predict earthquake processes. That work led to the design of further process models (Fig. 3) in which a synthetic gouge material was emplaced between pieces of rock that were then forced to slide under a loading arrangement, providing an opportunity to examine the spatial patterns of deformation fabrics that developed during simple shear displacements as a function of gouge composition and stress state (Logan 2007). I strongly believe that it is appropriate to say that these experiments are models of a process – with the limitations associated with any model, but a model nonetheless. The initial work of Logan and colleagues has led to the design of further experimental models of shear-related processes (e.g. Takahashi 2003).

Continuing with my list of landmark experiments using rock materials, I want to emphasise the buckle fold models of Handin et al (1972; 1976). These experiments need to be considered as process models that capture some aspects of the proto-

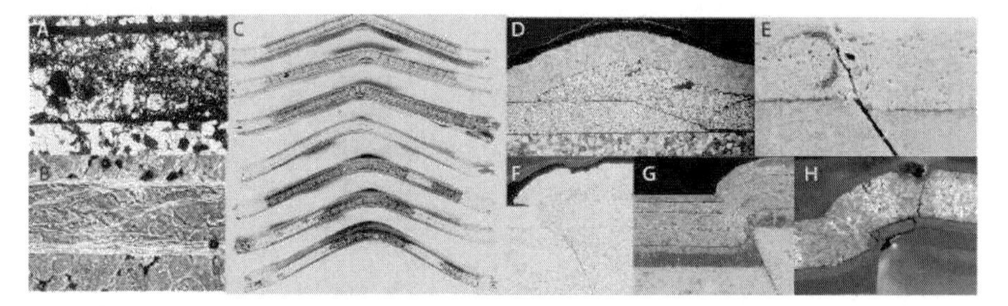

Fig. 3 Illustration of a range of process models involving rock materials deformed under confining pressure as part of a system. (A-B): shear-zone studies investigating the textures ("fabrics") that emerge during the shearing of artificial gouge material caused by movement of rock pieces that were shaped to have a smooth surface inclined to the axis of the cylindrical sample configuration. These images are rotated into the orientation of the sliding surface, and both have a clockwise sense of shear. The shear zone is about 1 mm thick. (A): indicates a compaction+shear response that occurs under higher pressure and temperature conditions. (B): indicates a dilation+shear response that occurs under milder conditions. Drawn after Logan (2007). (C): suite of thin rock layers subjected to end-loading that caused them to distort in a buckling-like mode. Drawn after Handin et al (1972). Layers were about 20 cm long. (D): thrust ramp simulation model, using the approach described by Serra (1977). The lower portion of this image shows an intact block of rock that acts as "basement" in the experiment. The lower of the upper layers (each about 4 mm thick) was pre-cut to have a gently-inclined surface, while the uppermost layer was initially intact. By shortening the upper layers relative to the lowest block, a thrust fault forms, along with a ramp anticline equivalent. Note the occurrence of arrays of shears (curving zones) that affect the middle layer as it translates through the ramp. (E): normal fault model, using a technique similar to that described by Patton et al (1998). The lower part of the image indicates an assembly of two pre-cut pieces of sandstone, overlain by an intact limestone layer (about 7 mm thick). Using an extensional loading scheme (axial stress reduced below confining pressure), the pre-cut surfaces slide and induce elongation of the overlying layer. (F): upthrust model, after Friedman et al (1976). In this experiment, the lowest unit is composed of two pre-cut pieces, which, when shortened, cause loading of the overlying layers. The reverse fault propagates into the overburden by a combination of fault+fold processes. Layers about 1 cm thick. (G): fault-fold model. This configuration is similar to that shown in (F), except that there is a layer of lead interposed between the pre-cut pieces and the intact layers. Note how the ductility of the lead allows a fold-dominated response. (H): forced fold model (after Couples et al 1994), using a different loading arrangement of steel blocks designed by Weinberg (1979) that translate and rotate. Note that faulting is contained within the folded layers, and is part of that imposed deformation, rather than being directly linked to the loading blocks.

type deformation concept, and they give us important insights. That experimental team examined other folding and faulting processes using rock materials and inventive loading schemes to create process models that proved to be very significant for generating new understanding of important types of natural deformations. They looked at forced folds (Friedman et al 1976; Weinberg 1979; Chester et al 1988; Couples et al 1994), thrust-fault ramps (Serra 1977; Chester et al 1991), strike-slip fault systems (Bartlett et al 1981), normal faults (Patton et al 1998), and aspects of the flexural-slip process in multi-layer folds (Friedman et al 1980; Couples and Lewis 1998). With the advances that have been made in terms of numerical simulation capabilities, some of this work now can be partially duplicated or extended

(Crook et al 2006; Couples et al 2007), but the experimental underpinning was and remains a crucial element of the investigations.

We must also remember the wet-clay models of Cloos (1955); the sandbox models of Hubbert (1951), Sanford (1959), McClay (1990) and others since; and the investigations of density-driven tectonic processes studied by Ramberg (1981) using models loaded in a centrifuge. Substantial modern work in analogue modelling uses a variety of experimental materials including glass beads and gels that allow scaling; such models can reveal insights into the complex deformation and linked depositional processes operating within large-scale tectonic systems (e.g. Adam et al 2005). The work of a large group of model-based experimentalists must be recognised within any summary of the value of experimental contributions.

3 Why Makes Nature So Clever?

My consideration of this question is undertaken in two situations – namely the cases of material properties and of models. In both cases, the cleverness of Nature is related to the fact that Nature is not constrained by finite experiments with their artificial boundaries. Nature also has a profound disregard for uniformity. Although both cases (properties and models) refer to similar aspects of Nature, it seems useful to examine the main question within each circumstance separately.

3.1 Material Properties

Let us consider the typical lab-based experimental protocol in which a finite specimen is placed into an apparatus that has been designed to replicate some portion of the assumed states that we think govern the material responses. We usually assume that our sample preparation has not induced any boundary-related artefacts, and we want to assume that the sample is homogeneous across the length-scale of the specimen. My experience is that homogeneity of geo-materials is often not the case in lab-sized samples – variations in natural deposition and subsequent processes typically result in textural differences that an experienced geoscientist immediately recognises on visual inspection. Even aggregates of loose grains cannot be simply assembled into a homogeneous arrangement – when they are packed into a tube to make the "specimen", there is a self-organised response in which the grain arrangements adjust to carry the imposed loads via force-chains, with parts of the assembly being less compacted (the same process of self-organised load distribution affects most aggregates, and is why the construction of road-bases or the back-filling of excavations is undertaken in small "lifts"). For some experiments involving rocks or soils, the spatial variations of initial material state may occur at a scale that is much smaller than the sample size, and we can posit that the variability is homogenised by a bulk measurement – but this may not be a suitable view for all deformation modes.

Our apparatus design is intended to allow us to create conditions within the sample that represent a set of control values within a parameter space. Establishing a given temperature ought to be easy, but is often not so, and maintaining a particular temperature for a long-term experiment is decidedly challenging. Pore fluids represent a further challenge, especially if we let them flow through the sample (to permit geochemical processes to be studied, or to determine the permeability). The engineering issues associated with high-pressure apparatus demand that pore fluids enter/leave the machine via small tubes, but we want them to flow across the whole section of the specimen. So we introduce a porous disc at the specimen ends and hope that this distributes fluid across the whole end – but it often does not work that way.

It seems that the experimental community is fairly content that it is a straightforward matter to generate a given initial state of stress in a specimen. Given the noted heterogeneity of materials, this may not be true, but even if we assume that variations lead to some average value, there is still an issue. Common experimental arrangements identify "sides" or passive boundaries and "ends" or active boundaries. On the sides we often impose a loading provided by a fluid under pressure, separated from the specimen by a membrane, with an alternate arrangement being a fixed-displacement condition associated with a strong platen. The first of these loadings supposedly leads to a fixed (radial) stress and the second supposedly leads to a plane strain condition. Although the radial pressure boundary is imposed onto the specimen, the radial stress component inside may not everywhere be equal to that boundary traction (see comments below). And observed deformations on one face of a plane strain test do not necessarily appear as exact projections on the opposite face. There may well be differences between the states we think are generated and those that actually exist.

The axial or active loading is more problematic. In most machines, the active loading is mechanically related to the movement of a piston-like part that is actuated by a motor and gears, or by a hydraulic ram. The important point is that the part moves and imposes a displacement onto the end of the specimen (or onto the face of various discs that are part of the assembly). Adopting an elastic model for the components of the apparatus, we can measure the slight distortion of the (very stiff) piston and calculate the stress state in it. The piston is attached to the specimen (directly or indirectly), and we then assume that the axial stress in the piston is imposed onto the end of the specimen. That may be true at the specimen boundary, but it may not be correct elsewhere in the specimen (see comments below). We are usually able to control the advance of the piston part by means of some rule, such as assuming that we will impose a constant strain rate or a constant stress.

The experiment then operates, and we make measurements of various things. Some measurements are made in real-time as the experiment proceeds, and others are determined afterwards (these are often called post-mortem measurements). As a minimum, we would expect to record the stress-strain response. The axial stress is determined externally to the specimen, as noted above, and the radial stress is assumed to equal the pressure of the confining fluid (in that particular configuration). The axial strain is often calculated based on displacements of the apparatus (corrected for differences that occur between the measurement point and the specimen). We can apply strain gages to locations on the specimen surface, which give additional readings at a few points. If we assume a distribution of the specimen strains, we can extrapolate the point strains into volumetric strains. If we measure the flux of pore fluid, we can calculate an effective permeability, but this may be a composite number that represents flow through regions of the specimen with variations of strain (especially if strongly localised deformation occurs). We can listen for acoustic emissions that occur during loading, but even if these are located by acoustic arrays, we must adopt a model to interpret their actual cause (as we must do in assigning a process model in earthquake first-motion studies). We can measure the velocity of high-frequency acoustic waves that cross the sample, actuated by piezoelectric transducers that we integrate into our sample+machine system.. Real-time imaging of deformation processes by X-ray computed tomography methods is just becoming possible, but these studies remain limited by the resolution/size trade-offs associated with this technique (along with extreme limitations in terms of construction of the apparatus).

Post-mortem studies provide essential information about the spatial locations of deformation in a specimen, and permit its post-deformation property distributions to be considered. Optical examination of thin sections, or SEM imaging, provide crucial ground truth concerning the locations and types of deformation processes that operated during the experiment (but we usually have to make a judgement to exclude some observed deformation, attributing it to changes during unloading and handling). We can subject the deformed sample to new experiments that allow us to determine fluid flow, acoustic, electrical and magnetic changes. Tomographic investigations can highlight the locations of deformation-induced changes, and characterise them. As an example, Desrues et al (2007) show that the internal response of a cylindrical specimen is quite surprising, with the evolution of multiple strain (and presumably stress) states that are far from uniformly distributed through the specimen (Fig. 4).

Using these methods, we can reconstruct a deformation process by assembling the correct data and relating this information to the measured mechanical evolution. Such integrative studies underpin the modern perspective on geo-material behaviour. However, all is not perfect. The reproducibility of experimentally-determined properties is not very robust. This variability could be due to initial heterogeneities of the materials, or to the preparation of samples, or to the operation of the apparatus, or to measurement errors. Technical improvements in equipment and procedures will address some of the issues, but we do not have clear approach to design a method to quantify the role of natural variability (if we can agree that this is an issue!).

So, experimentalists are quite clever. But Nature can do some things that we cannot manage to do. Nature has experimented with a much wider range of geomaterials than we will ever study, and has examined a comprehensive suite of conditions that far exceeds what we can accomplish in the lab. The outcomes of Nature's experiments are located everywhere, and we can use these to construct a compre-

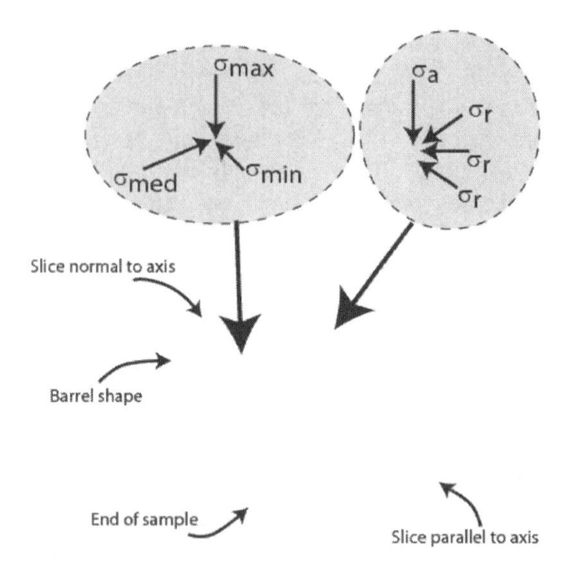

Fig. 4 Interpreted distribution of shear-compacted regions within part of a cylindrical sample, imaged via post-mortem X-ray tomography (after Desrues et al 2007). The ends of the sample (only one end depicted here) exhibit a cone of (relatively?) un-deformed material, with an assembly of conjugate shears developing between the cone and the sample exterior. These shears intersect the sample surface and we see them much like the pattern depicted in Figure 1. Internally, they indicate self-organised complexity. If we use a naive idea to relate the extant state of stress to the observed deformation fabric, we might guess that the stress state in the part of the sample that is deforming via the shears bands is quite different to the assumed radially-symmetric state that might exist simultaneously in the cone region. After deformation localises into a system of features, there may be considerable spatial variability of the stress states (see discussion in Couples et al 2007).

hensive picture of how geo-materials behave. Perhaps the most important thing that Nature has done is to avoid the constraints of finite specimens. Nature never uses artificial boundary conditions, and Nature never imposes an arbitrary state onto a material. Instead, all materials are located within a system of deformation and the states are evolved states that emerge from interactions of the components of the entire system. In my opinion, this makes Nature a much more clever experimentalist than I could ever hope to become. Unfortunately, Nature has not provided us with a simple way to extract numerical values for the parameters we wish to use in our constitutive laws – but there may be a strategy for overcoming this disappointment (see below).

3.2 Deformation Systems

It is probably obvious that my admiration of Nature as the great experimentalist is derived from Nature's absolute stricture only to experiment with *systems* of geo-materials – never with isolated and artificial pieces. No artificial boundaries are ever created, and there is no notion in Nature of boundary conditions or initial conditions. Those are artefacts of our human approach in which we (must) define finite regions to examine. In Nature, there is simply an ongoing process that leaves a record of its operation. Nature always includes multiple length scales – in fact, every scale (Fig. 5). Nature works with materials that are spatially heterogeneous, and which exhibit temporal variations that indicate the specifics of their individual histories. Perhaps of primary importance is that Nature *never* operates with an inappropriate constitutive law – simple responses that we observe in Nature can be interpreted as being the result of actions within a specific set of conditions and circumstances, with the same material able to exhibit a fuller range of responses when involved in a more complex setting.

Fig. 5 Illustration of the multi-scale aspects of natural deformations, here depicting fracture-dominated responses. (A): photo-micrograph of mm-scale quartz grains that have been comprehensively altered by means of microfractures that emanate from grain contact points. (B): bedding surface of a carbonate rock illustrating a well-connected array of hairline fractures (picked out by lichen using the extra water-carrying capacity of the locally-damaged rock to support their growth). Note end of pen for scale. (C): layer-normal fracture systems affecting a folded sequence of limestones and dolostones in northwestern Montana, with almost no shale in this rock column. The fractures occur in local assemblages that are bounded by stratigraphic limits that indicate the locations of bedding-plane slip that operated during the flexure of this sequence. Image in (B) was observed on one of the bedding surfaces shown here. (D): large-scale fracture corridors developed in carbonate rocks in the Italian Dolomites. Note the strong localisation into highly-fractured zones that traverse as much as 100 m vertical sections of rock. These zones can extend for km, and contain all scales of deformation features within them.

Although Nature has not yet revealed to us a simple way to extract the numerical values we seek to obtain for our constitutive laws and property predictions, I think that there is a way to make more use of what Nature has provided for us. Once we appreciate that every deformation in Nature is part of a system of geo-materials, we can characterise our observations from Nature into relative positions within some continuum of a geo-material response space. My personal capability to do this is, of course, limited, but I have found that I can create a global semi-quantitative geo-material understanding that allows me to make good predictions about relative responses within systems. When I need to translate that understanding into parameters for use in a simulation or quantitative prediction, I can typically achieve satisfactory results by bounding the problem between anchor points where lab data may be available for selected "type" materials. I strongly believe that the responses of geo-materials are governed by their textures – which is to say that the arrangements of grains of various composition, along with cements and other components, and the shape and distribution of the pore spaces, determine the geomechanical properties.

Although I say that I have some degree of success in predicting the characteristics and responses of geo-materials, a strong disclaimer is needed. Predictions do not always work out in the way that I expect. These failings provide opportunities to update the information base. Perhaps a more significant point is that clever experimentalists continue to report facts that cause me to be surprised about what deformation is, and what its effects can be. But the most salutary lessons come from Nature, from whence new observations can cause me to undertake major alterations of my conceptual understanding. Nature is the most clever experimentalist of all.

I have written these comments as a personal story that describes how I have been able to construct my own perception of the panoply of geo-material responses within the paradigm of the poro-elasto-visco-plastic material description. I suspect that others may well have done the same. Is it possible to formalise these methods to create a spectrum of numerical parameters for the constitutive properties of all geo-materials? This parameter distribution must be constrained by the always-limited lab data and by simple observations about the textures and compositions of "type" materials, but this approach offers a potential solution to the need to make predictions about materials that we have not, and possibly cannot, test in the lab. Construction of this database would be a worthy research goal.

4 Summary Remarks

Nature does not impose conditions. The conditions under which a deformation occurs in Nature are governed by the whole system. Although we as human experimentalists must be able to operate with finite specimens, with the attendant need to deal with boundary conditions, we can choose to avoid an intellectual trap. We ought to expunge from our vocabulary the phrase: "impose specified boundary conditions onto the sample". Instead, we should phrase our work as: "seek to generate a specific state within the specimen". This change in phrasing will help us to remember that the specimen and apparatus form a system, and that part of the system (the apparatus) may prevent the specimen from doing what it would do in a natural situation. It also helps us to remember that the specimen may well react in such a way as to generate a heterogeneous internal state that is not the same as the one we assume based on our external controls. It may even be useful to recall that stress is the dependent variable, describing an arrangement of tractions that are *generated by the material* as it attempts to avoid being distorted. Being able to re-arrange an equation for our convenience (as we learn to do in linear elasticity) does not mean that we can simply ignore this dependency. The advice to avoid the word "impose" applies to the initialisation of experiments and to the interpretation of what happens during the experiment. We may never be quite as clever as Nature, but we can improve what we say that we do.

We also ought to increase our efforts to read the Great Storybook composed by Nature. This book (or maybe it is better to think of it a series of linked volumes) contains a number of important story-lines, along with many sub-plots and local character studies. We need to be cautious in how we read and use this book. We can extract a single paragraph and interpret it. But just as in a real book, that can be misleading. The context of the story needs to be considered before we can derive value from each paragraph. And, like real books, some paragraphs or sentences may have escaped the watchful eye of the editor, and simply be irrelevant.

Our efforts be great experimentalists will hopefully continue. There are so many things to investigate. Some researchers will continue to provide us with critical numbers, and some will probably persist in surprising all of us with new observations and ideas, and this is great news. If these efforts can be pursued within a holistic view of geomechanics, the resulting outcomes will be more valuable. We can all be inspired to be ever more clever in how we undertake our studies. Nature will always be there to inspire us.

Acknowledgements Over the course of a long (and hopefully continuing) career, I have been influenced by many individuals who have inspired me to think or think again about aspects of geomechanics. Some have been decidedly positive, whilst others have influenced my thinking even in disagreement. Amongst the former group, I especially acknowledge my mentors (John Handin, Mel Friedman, John Logan and Dave Stearns) from the Center for Tectonophysics at Texas A&M University, along with fellow students who were always ready to gently or otherwise uncover intellectual shortcomings. A special thanks to my long-term collaborator Helen Lewis for providing support and criticism in appropriate measures.

References

1. Adam, J., Urai, J.L., Wieneke, B., Oncken, O., Pfeifferm K., Kukowski, N., Lohrmann, J., Hoth, S., van der Zee, W., Schmatz, J., 2005, Shear localisation and strain distribution during tectonic faulting–new insights from granular-flow experiments and high-resolution optical image correlation techniques, Journal of Structural Geology, **27**, 283–301.
2. Bartlett, W. L., Friedman, M., and Logan, J. M., 1981, Experimental folding and faulting of rocks under confining pressure, Part IX. Wrench faults in limestone layers, Tectonophysics, **79**, 255–277.

3. Chester, J. S., Logan, J. M., and Spang, J. H., 1991, Influence of layering and boundary conditions on fault-bend and fault-propagation folding, Geological Society of America Bulletin, **103**, 1059–1072.
4. Chester, J. S., Spang, J. H., and Logan, J. M., 1988, Comparison of thrust fault rock models to basement-cored folds in the Rocky Mountain Foreland, in: C. J. Schmidt and W. J. Perry, eds., Interaction of the Rocky Mountain Foreland and the Cordilleran Thrust Belt. Geological Society of America Memoir 171, 65–74.
5. Cloos, E., 1955, Experimental analysis of fracture patterns. *Geological Society of America Bulletin*, **66**, 241–256.
6. Couples, G.D., Stearns, D.W., and Handin, J.W., 1994. Kinematics of experimental forced folds and their relevance to cross-section balancing. *Tectonophysics* **233**, 193–213.
7. Couples, G.D., Lewis, H., Olden, P., Workman, G.H. and Higgs, N.G., 2007. Insights into the faulting process from numerical simulations of rock-layer bending. In: Lewis, H. and Couples, G.D. (eds) *Relationships Between Damage and Localization*, Geological Society of London, Special Publications **289**, 161–186.
8. Crook, A.J.L., Willson, S.M., Yu, J.G., and Owen, D.R.J., 2006. Predictive modelling of structure evolution in sandbox experiments, Journal of Structural Geology, **28**, 729–744.
9. Desrues, J., Bésuelle, P., and Lewis, H., 2007, Strain localization in geomaterials. In: Lewis, H. and Couples, G.D. (eds) *Relationships Between Damage and Localization*, Geological Society of London, Special Publications 289, 47–73.
10. Friedman, M., Handin, J. M., Logan, J. M., Min, K. D., and Stearns, D. W., 1976, Experimental folding of rocks under confining pressure: Part II- Faulted drape folds in multilithologic layered specimens, Geol. Soc. Amer. Bull., **87**, 1049–1066.
11. Friedman, M., Hugman, R.H.H., and Handin, J. 1980. Experimental folding of rocks under confining pressure, Part VIII – Forced folding of unconsolidated sand and of lubricated layers of limestone and sandstone. *Geological Society of America Bulletin*, **91**, 307–312.
12. Handin, J., 1969, On the Coulomb-Mohr failure criterion. *Journal of Geophysical Research*, **74**, 5343–5348.
13. Handin, J., Friedman, M., Logan, J. M., Pattison, L., and Swolfs, H. S., 1972, Experimental folding of rocks under confining pressure: I. Buckling of single-layer rock beams , in *Flow and Fracture of Rocks*, Amer. Geophys. Union Monograph – *The Griggs Volume*, 1–28.
14. Handin, J.W., Friedman, M., Min, K.D., and Pattison, L.J., 1976, Experimental folding of rocks under confining pressure: Part II. Buckling of multilayered rock beams. *Geological Society of America Bulletin*, **87**, 1035–1048.
15. Hubbert, M.K., 1951, Mechanical basis for certain familiar geologic structures, *Geological Society of America Bulletin*, **62**, 355–372.
16. McClay, K.R., 1990, Extensional fault systems in sedimentary basins: a review of analogue model studies, *Marine and Petroleum Geology*, **7**, 206–233.
17. Patton, T.L, Logan, J.M., and Friedman, M., 1998, Experimentally generated normal faults in single- and multilayer limestone beams at confining pressure, *Tectonophysics*, **295**, 53–77.
18. Ramberg, H., 1981. *Gravity, Deformation and the Earth's Crust* (2nd Edition). Academic Press, London, 452p.
19. Sanford, A.R., 1959, Analytical and experimental study of simple geologic structures. *Geological Society of America Bulletin*, **70**, 19–52.
20. Serra, S., 1977, Styles of deformation in the ramp regions of overthrust faults. In: *Twenty-Ninth Annual Field Conference–1977–Wyoming Geological Association Guidebook*, 487–498.
21. Takahashi, M., 2003, Permeability change during experimental fault smearing, *J. Geophys. Res.*, **108**(B5), 2234, doi:10.1029/2002JB001984.
22. Weinberg, D.M., 1979, Experimental folding of rocks under confining pressure, Part VII – Partially scaled models of drape folds. *Tectonophysics*, **54**, 1–24.
23. Wibberley, C. A. J., Petit, J.-P., Rives, T., 2007, The mechanics of fault distribution and localization in high-porosity sands, Provence, France. In: Lewis, H. and Couples, G.D. (eds) *Relationships Between Damage and Localization*, Geological Society of London, Special Publications **289**, 19–46.

Author Index